NON-EQUILIBRIUM
DYNAMICS *of*
SEMICONDUCTORS
and
NANOSTRUCTURES

NON-EQUILIBRIUM DYNAMICS of SEMICONDUCTORS and NANOSTRUCTURES

Edited by
Kong-Thon Tsen

Taylor & Francis
Taylor & Francis Group

Boca Raton   London   New York

A CRC title, part of the Taylor & Francis imprint, a member of the
Taylor & Francis Group, the academic division of T&F Informa plc.

Published in 2006 by
CRC Press
Taylor & Francis Group
6000 Broken Sound Parkway NW, Suite 300
Boca Raton, FL 33487-2742

International Standard Book Number-10: 1-57444-696-7 (Hardcover)
International Standard Book Number-13: 978-1-57444-696-8 (Hardcover)
Library of Congress Card Number 2005050558

This book contains information obtained from authentic and highly regarded sources. Reprinted material is quoted with permission, and sources are indicated. A wide variety of references are listed. Reasonable efforts have been made to publish reliable data and information, but the author and the publisher cannot assume responsibility for the validity of all materials or for the consequences of their use.

Library of Congress Cataloging-in-Publication Data

Non-equilibrium dynamics of semiconductors and nanostructures / edited by Kong Thon Tsen.
    p. cm.
Includes bibliographical references and index.
ISBN 1-57444-696-7 (alk. paper)
    1. Semiconductors. 2. Nanostructures. 3. Picosecond pulses. 4. Quantum wells. I. Tsen, Kong Thon.

QC610.9.N64 2005
621.3815'2--dc22
                             2005050558

Taylor & Francis Group
is the Academic Division of T&F Informa plc.

Visit the Taylor & Francis Web site at
http://www.taylorandfrancis.com

and the CRC Press Web site at
http://www.crcpress.com

Preface

This book consists of recent new developments in the field of ultrafast dynamics in semiconductors and nanostructures. It consists of eight chapters. Chapter 1 reviews spin dynamics in a high-mobility, two-dimensional electron gas. Chapter 2 deals with generation, propagation, and nonlinear properties of high-amplitude, ultrashort strain solitons in solids. Chapter 3 is devoted to nonlinear optical properties of nano-scaled artificial dielectrics. Chapter 4 discusses optical properties of hexagonal and cubic GaN self-assembled quantum dots. Chapter 5 reviews ultrafast, non-equilibrium electron dynamics in metal nanoparticles. Chapter 6 discusses the generation of monochromatic acoustic phonons in GaAs. Chapter 7 presents optical studies of carrier dynamics and non-equilibrium optical phonons in nitride-based semiconductors. Chapter 8 demonstrates electromagnetically induced transparency in semiconductor quantum wells.

There are many books on the market devoted to the review of certain fields. This book is different from those in that the authors not only provide reviews of the fields, but also present their own important contributions to the fields in a tutorial way. As a result, researchers who are already in the field of ultrafast dynamics in semiconductors and its device applications, as well as researchers and graduate students just entering the field, will benefit from it.

Editing a book with eight different chapters involving authors in several countries is not an easy task. I would like to thank all authors for their patience and cooperation. I would also wish to express appreciation to my wife and children for their encouragement, understanding, and support.

K.T. Tsen, Editor
Tempe, Arizona
February 2005

Contributors

Yong-Hoon Cho
Department of Physics
Chungbuk National University
Cheongju, Korea

Le Si Dang
Laboratoire de Spectrometrie Physique
Universite Joseph Fourier
Grenoble, France

Jaap I. Dijkhuis
Atom Optics and Ultrafast Dynamics
Department of Physics and Astronomy
Debye Institute
University of Utrecht
Utrecht, Netherlands

H. Grebel
Electronic Imaging Center
New Jersey Institute of Technology
Newark, New Jersey

Richard T. Harley
School of Physics and Astronomy
University of Southhampton
Southampton, U.K.

Anthony J. Kent
School of Physics and Astronomy
University of Nottingham
Nottingham, U.K.

Otto L. Muskens
Atom Optics and Ultrafast Dynamics
Department of Physics and Astronomy
Debye Institute
University of Utrecht
Utrecht, Netherlands

Mark C. Phillips
Sandia National Laboratories
Albuquerque, New Mexico

Nicola M. Stanton
School of Physics and Astronomy
University of Nottingham
Nottingham, U.K.

K.T. Tsen
Department of Physics and
  Astronomy
Arizona State University
Tempe, Arizona

Fabrice Vallée
Universite Bordeaux
Talence, France

Hailin Wang
Department of Physics and
  Oregon Center for Optics
University of Oregon
Eugene, Oregon

Contents

Chapter 1   Spin Evolution in a High-Mobility, Two-Dimensional
Electron Gas..1

Richard T. Harley

Chapter 2   High-Amplitude, Ultrashort Strain Solitons in Solids 15

Otto L. Muskens and Jaap I. Dijkhuis

Chapter 3   Nonlinear Optical Properties of Artificial Dielectrics
in the Nano-Scale..49

H. Grebel

Chapter 4   Optical Properties of Hexagonal and Cubic GaN Self-Assembled
Quantum Dots 69

Yong-Hoon Cho and Le Si Dang

Chapter 5   Ultrafast Non-Equilibrium Electron Dynamics in Metal
Nanoparticles...101

Fabrice Vallée

Chapter 6   Generation and Propagation of Monochromatic Acoustic
Phonons in Gallium Arsenide...143

Anthony J. Kent and Nicola M. Stanton

Chapter 7   Optical Studies of Carrier Dynamics and Non-Equilibrium
Optical Phonons in Nitride-Based Semiconductors.......................179

K.T. Tsen

Chapter 8   Electromagnetically Induced Transparency in Semiconductor
Quantum Wells..215

Mark C. Phillips and Hailin Wang

Index...251

1 Spin Evolution in a High-Mobility, Two-Dimensional Electron Gas

Richard T. Harley

CONTENTS

1.1 Introduction 1
1.2 Review of Theoretical Ideas on Spin-Dynamics in a 2DEG 2
    1.2.1 Mechanisms in III–V Semiconductors 2
    1.2.2 Electron Spin Dynamics in the Strong Scattering Regime 4
    1.2.3 Collision-Free Regime 6
1.3 Experimental Investigation of Electron Spin-Dynamics in a 2DEG 6
    1.3.1 Samples and Experimental Techniques 6
    1.3.2 Experimental Results 10
1.4 Conclusions 12
References 13

1.1 INTRODUCTION

The possibilities for application of the spin degree of freedom, in addition to (or instead of) the charge of electrons in semiconductor devices, are attracting increasing attention.[1] Ideas exist for purely semiconductor as well as for hybrid metal–semiconductor devices that may allow functions not provided by existing metallic spintronics[2,3]; in particular, incorporation of direct-gap III–V materials can give optoelectronic or "spinoptronic" functions. Semiconductor quantum wells and two-dimensional electron gases (2DEGs) are likely to play an important role in these developments. To exploit the behavior of electronic spins, it is necessary to understand the fundamental physics of spin dynamics in such quantum structures.

This chapter describes recent[4] and continuing optical experiments aimed at elucidating the mechanisms of electron spin relaxation in very high-mobility 2DEGs in gallium arsenide (GaAs). The topic is important because it was long expected that the electron spin relaxation rate should be proportional to the electron mobility,[5–7] suggesting that high mobility and long spin memory, each potentially desirable

1

properties for spintronic applications, are not compatible. The experiments described here[4] have led to a modification of this view and to renewed theoretical predictions[8,9] that are providing new insights into the mechanisms of spin dynamics of 2DEGs. However, as we will see, there is tantalizingly little experimental data and much remains to be done.

1.2 REVIEW OF THEORETICAL IDEAS ON SPIN-DYNAMICS IN A 2DEG

1.2.1 MECHANISMS IN III–V SEMICONDUCTORS

In direct-gap semiconductors optical measurements,[1,10] both cw and time-resolved, are extremely effective probes of the electronic spin because absorption of circularly polarized photons allows injection of populations of spin-polarized carriers whose evolution may be probed using a variety of polarization-sensitive optical methods. In the experiments we describe, a population of electrons is prepared in the conduction band of a GaAs quantum structure with a preferential alignment of spins along a particular axis (z), namely the growth axis in the sample; and we examine the evolution with time of the z-component of the total spin $\langle S_z \rangle$ in zero applied magnetic field. In the case of exponential decay, the decay time is termed τ_s and is the longitudinal relaxation time with respect to the growth axis.[7]

Three basic mechanisms have been identified for spin relaxation of conduction electrons in zinc-blende semiconductors[7,10]: (1) spin-flips associated with electron scattering resulting from spin-orbit coupling, known as the Elliott-Yaffet (EY) mechanism; (2) spin-flips induced by exchange interaction with an unpolarized population of holes, known as the Bir, Aronov, and Pikus (BAP) mechanism; (3) spin precession in flight between scattering events in the effective magnetic field represented by conduction band spin-splitting which we[4] term the D'yakonov, Perel' and Kachorovskii (DPK) mechanism.[5,6] The spin-splitting results from the combined effects of spin-orbit coupling and inversion asymmetry of the crystallographic and/or device structure.[7]

For n-type samples and particularly 2DEGs, neither the BAP nor the EY mechanism is important. The former requires relatively high hole concentrations,[10,11] which are not present in a 2DEG. The latter is, in general, not only intrinsically weaker than the DPK mechanism, but also gives spin-relaxation rate $(\tau_s)^{-1}$ proportional[10] to the electron momentum scattering rate, $(\tau_p{}^*)^{-1}$, so that it becomes unimportant in high-mobility material where scattering is minimized. On the other hand, spin reorientation via precession, as envisaged in the DPK mechanism, becomes more efficient for weak momentum scattering and thus is dominant for a high-mobility 2DEG sample.

The in-flight spin precession, central to the DPK mechanism, arises because, in general, the z-component of angular momentum is not an eigenvalue of the Hamiltonian. The individual electron spin vectors can therefore be considered to precess with an effective Larmor vector $\mathbf{\Omega}_k$ whose magnitude corresponds to the conduction band spin-splitting and which varies in direction according to the electron's wavevector $\mathbf{k}$. There are three theoretical contributions to $\mathbf{\Omega}_k$ in zinc-blende quantum structures[7]: (1) the Dresselhaus or Bulk Inversion Asymmetry (BIA) term due to inversion asymmetry of the zinc-blende structure; (2) the Rashba or Structural Inversion Asymmetry (SIA) term due to inversion asymmetry of the device structure; and (3) the Natural Interface Asymmetry (NIA) term due to asymmetry associated with the chemical bonding within interfaces. In a GaAs/AlGaAs quantum well structure,[7,12,13] only BIA and SIA terms are important and the former is usually the largest, the latter being less important because it vanishes if the device structure has inversion symmetry. For a (001)-oriented well, the precession vector lies in the plane of the well and, in a perturbation approximation, has the form[14]:

$$\Omega_k = \frac{2}{\hbar}\left\{\left[a_{42}k_x\left(\langle k_z^2\rangle - k_y^2\right) - a_{46}E_z^{eff}k_y\right]\vec{x} + \left[a_{42}k_y\left(k_x^2 - \langle k_z^2\rangle\right) + a_{46}E_z^{eff}k_x\right]\vec{y}\right\} \quad (1.1)$$

where $\langle k_z^2\rangle$ is the mean squared electron wavevector along the growth direction that is associated with quantum confinement; k_x and k_y are components of the in-plane electron wavevector; and $\vec{x}$ and $\vec{y}$ are unit vectors along the (100) and (010) axes, respectively. The coefficients a_{42} and a_{46} define the strengths of BIA and SIA terms and have values of 1.6×10^{-29} eV m^3 and 9.0×10^{-39} C m^2, respectively.[14] E_z^{eff} is an effective electric field in the growth direction that vanishes for a symmetrical structure and depends on band bending and band edge offsets.[14]

Usually, a collision-dominated or "motional-narrowing" regime can be assumed,[5–7,10] wherein $\langle|\Omega|\rangle\tau_p^* \ll 1$ with $\langle|\Omega|\rangle$ an average precession frequency for the electron population. Thus, the spin precession is interrupted frequently by momentum scattering, and the spin reorientation proceeds as a sequence of many, randomly oriented, small spin rotations. In this regime, spin relaxation is much slower than the average precession period and is actually inhibited by momentum scattering because the angle through which a spin will tip, on average, between collisions is inversely proportional to the scattering rate. There is exponential spin relaxation along a particular axis with rate (see Ref. 10)

$$\tau_s^{-1} = \langle\Omega^2\rangle\tau_p^* \quad (1.2)$$

where $\langle\Omega^2\rangle$ is the average square component of Ω_k perpendicular to the axis. This formula contains the well-known and counterintuitive feature of the DPK mechanism, namely that increasing electron scattering produces slower spin relaxation.

There are two important questions concerning the DPK mechanism for 2DEGs in (001)-oriented quantum wells considered here. First, we examine whether, as has been assumed,[5–7,10] the momentum scattering time, τ_p^*, appropriate for spin relaxation that appears in Equation 1.2 can be equated to the drift momentum relaxation time, τ_p, from the electron mobility. Second, we consider the possibility of an experimentally observable collision-free regime; that is where scattering is sufficiently weak that the electron spins precess more or less freely.[15] Such a regime can be revealed experimentally by choosing the conditions so that quasi-free precession of individual electron spins will give oscillatory evolution of $\langle S_z\rangle$. One reason why these questions are particularly interesting in 2DEGs is that the electron mobility and hence τ_p can assume remarkably high values. For example, in GaAs, a mobility of 10^6 cm^{2}V^{-1}s^{-1}, a readily achievable value in good samples at low temperatures, indicates $\tau_p = 38$ ps, which turns out to be long compared to $(\langle|\Omega|\rangle)^{-1}$ in many situations.

1.2.2 Electron Spin Dynamics in the Strong Scattering Regime

The usual assumption of DPK theory[5–7,10] has been that the scattering time, τ_p^*, for randomization of spin precession is closely related to the transport scattering time, which determines the mobility, τ_p. It has only been realized recently[4,8,9] that this is, in principle, incorrect and breaks down completely in 2DEGs. The point is that precession of an electron spin can be as effectively randomized by scattering from another electron via the Coulomb interaction, as by scattering from thermal vibrations or defects, and yet electron-electron scattering does not directly relax an electron drift current and can only affect the mobility very weakly, for example via U-processes or by changing the distribution of momenta within the electron population.[16] Therefore, in general, $\tau_p^* \leq \tau_p$.

The practical significance of this difference clearly depends on the relative strengths of the electron-electron scattering and the remaining scattering processes that do affect the mobility. In undoped structures, it should be a good approximation to equate τ_p^* with τ_p. However, experimental tests of this expectation have been relatively few (see Ref. 7) because measurements of electron mobility are difficult when the sample is insulating. In n-type structures, electron-electron scattering will give a completely different dependence of the spin-relaxation time on both temperature and electron concentration than that which would be predicted by only considering other scattering mechanisms. An early indication of the likely importance of electron-electron scattering was given by Glazov and Ivchenko,[8] who showed that, theoretically, in nondegenerate n-doped bulk GaAs, the electron-electron scattering is considerably more effective at randomizing the spin precession than scattering of the electrons by the ionized donors giving rise to the free electrons in the first place. In degenerate electron gases, electron-electron scattering tends to be inhibited by the Pauli exclusion principle and vanishes altogether at T = 0K. Thus, in a bulk sample there should be a crossover, as the temperature is reduced, to a regime where electron-electron scattering plays only a minor role in spin dynamics compared to ionized impurity scattering. In a high-quality, modulation-doped 2DEG, unlike in a bulk sample, impurity scattering is not correlated with the electron density. It can be reduced to a very low level and, as demonstrated by the experiments described below,[4] electron-electron scattering remains dominant in the spin-dynamics down to very low temperatures.

A qualitative picture of the expected temperature dependence of τ_s for a 2DEG can be obtained by considering the two factors in Equation 1.2, as illustrated schematically in Figure 1.1; as indicated, in a 2DEG of sufficient quality, electron-electron scattering will cause a striking peak in the spin-relaxation time at a temperature on the order of the degeneracy temperature $T_F \sim E_F/k_B$. First consider $\langle\Omega^2\rangle$.

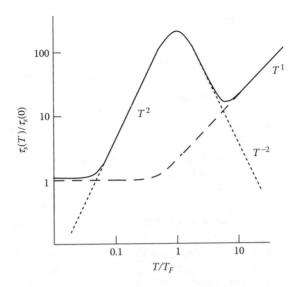

FIGURE 1.1 The solid curve is a schematic prediction of the temperature dependence of electron spin-relaxation time, τ_s, in a 2DEG based on the DPK mechanism (Equation 1.2). The dashed curve indicates the contribution of scattering processes that contribute to the electron mobility, and the dotted curve indicates the additional term in the spin relaxation time due to electron-electron scattering.

In a quantum well at relatively low temperatures, electron confinement energy will exceed the thermal energy; in Equation 1.1, $\langle k_z^2 \rangle$ is large compared to k_x^2 or k_y^2. Thus, to a first approximation, Ω_k is linear in the in-plane electron wavevector so that $\langle \Omega^2 \rangle$ is proportional to the average in-plane kinetic energy. For a degenerate electron gas, this is independent of T. For a nondegenerate 2DEG, the average in-plane kinetic energy is linear in T. Now consider the scattering rate $(\tau_p^*)^{-1}$. If we ignore electron-electron scattering, $(\tau_p^*)^{-1}$ will follow the inverse of the mobility, constant at low temperatures and increasing roughly as T^2 at high temperatures where phonon scattering takes over.[17] Combining this with the variation of $(\langle \Omega^2 \rangle)^{-1}$ gives a contribution to τ_s that is constant at low temperature and roughly proportional to T at high temperatures. This is indicated by the dashed curve in Figure 1.1. For a degenerate electron gas, the electron-electron scattering rate $(\tau_{e-e})^{-1}$ has the form $(T^2/E_F) \log_e(k_B T/E_F)$[18–20]; whereas for a nondegenerate electron gas, $(\tau_{e-e})^{-1} \sim T^{-1}$ (see Ref. 8). Combining these with $(\langle \Omega^2 \rangle)^{-1}$ and ignoring the weak logarithmic term gives a term in τ_s proportional to T^2 at low temperatures and proportional to T^{-2} at high temperatures. This contribution is shown as the dotted curve in Figure 1.1, and the combined effect of the different scattering processes is given by the solid curve. The details of the variation will also depend on the electron concentration N_S and the electron confinement energy E_{1e}. Exact calculations have been carried out by Glazov and Ivchenko,[8,9] which confirm these qualitative expectations. As we will see, only a very limited part of this expected behavior of τ_s has been explored experimentally to date.

1.2.3 COLLISION-FREE REGIME

For sufficiently weak scattering, DPK spin-dynamics will enter a collision-free regime in which the individual electron spins rotate many times between scatterings. A clear experimental signature of this regime would be oscillatory rather than exponential evolution of the average spin $\langle S_z \rangle$; but because Ω_k is a strong function of wavevector, one would not necessarily expect $\langle S_z \rangle$ to show oscillations because the precession of the individual electrons could rapidly get out of phase. Nonetheless, because Ω_k can be relatively independent of the direction of the electron's wavevector in the x-y plane, in a degenerate 2DEG the value of $|\Omega_k|$ at the Fermi level can be approximately constant and a population of spin-polarized electrons injected at the Fermi level can then give rise to an oscillatory signal.

Figure 1.2a shows $|\Omega_k|$ calculated, using Equation 1.1, at the Fermi wavevector of a 2DEG in a 10-nm GaAs quantum well with electron concentration $N_S \sim 1.86 \times 10^{11} \text{cm}^{-3}$, both for $E_z^{eff} = 0$ and for $E_z^{eff} = 15 \text{ kV cm}^{-1}$, a realistic value in a modulation doped structure. These conditions correspond with those for the experiments described below.[4] Figures 1.2b and 1.2c show the predicted values of $\langle S_z \rangle$ as a function of delay following injection of a small population of spin-polarized electrons at E_F and with a uniform distribution of wavevectors. The calculation does not include scattering of the electrons and thus reveals only the damping and beating that would be associated with residual anisotropy of Ω_k. It is clear from this that experimental conditions might exist in a 2DEG for which oscillations would, in principle, occur.

The effect of scattering will be to interrupt precession of the individual spins and cause additional damping of the oscillations of $\langle S_z \rangle$. As discussed above, at very low temperatures we expect to be able to estimate τ_p^* from the mobility because, in this regime, electron-electron scattering vanishes so that $\tau_p^* = \tau_p$. In Figures 1.2b and 1.2c, the predicted oscillation frequency is ~0.24 rad ps^{-1} so that in a sample with electron mobility of $10^6 \text{ cm}^2 \text{ V}^{-1}\text{s}^{-1}$ ($\tau_p \sim 38$ ps), we would have $\langle |\Omega| \rangle \tau_p \sim 9$. The spin evolution of electrons at the Fermi energy would thus be comfortably within a collision-free regime; on average, each electron spin would precess through 9 radians before being scattered; and experimentally, the evolution of $\langle S_z \rangle$ would show oscillations with a decay time of one to two cycles. Such behavior was originally predicted by Gridnev[15] and has now been observed,[4] as described below, although the sample used had roughly four times lower mobility and the oscillations showed a decay time of less than one cycle.

1.3 EXPERIMENTAL INVESTIGATION OF ELECTRON SPIN-DYNAMICS IN A 2DEG

1.3.1 SAMPLES AND EXPERIMENTAL TECHNIQUES

Measurements to investigate these qualitative expectations for spin-dynamics of a 2DEG, as yet, give an incomplete picture. They have been carried out on a sample[21] in which the 2DEG is confined in a (001)-oriented, 10-nm, one-side, n-modulation-doped GaAs/AlGaAs single quantum well structure grown by MBE. (The same

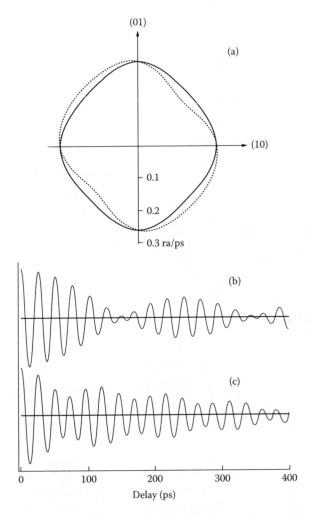

FIGURE 1.2 (a) Polar plot of the precession frequency, $|\Omega_\mathbf{k}|$, given by Equation 1.1 at the Fermi surface of a 2DEG in a 10-nm GaAs quantum well with electron concentration $N_S \sim$ 1.86 x 10^{11}cm^{-3} for $E_z^{e\!f\!f} = 0$ (solid curve) and for $E_z^{e\!f\!f} = 15$ kV cm^{-1} (dotted curve). (b and c) Predicted time evolution, in the absence of electron scattering, of $\langle S_z \rangle$ for a packet of spin-polarized electron injected at the Fermi energy for $E_z^{e\!f\!f} = 0$ and $E_z^{e\!f\!f} = 15$ kV cm^{-1}, respectively. The oscillation frequency is ~0.24 rad ps^{-1} in each case, and the decay and beating is due entirely to the anisotropy of $|\Omega_\mathbf{k}|$.

qualitative features of spin evolution were also observed for three other samples with nominally comparable mobilities and electron concentrations.) The structure was processed into a Hall bar FET with transparent Schottky gate for optical measurements and *in situ* Hall measurements of the experimental parameters; bias was set for maximum electron concentration, N_S, and mobility in the well. Figure 1.3 shows mobility, photoluminescence, photoluminescence excitation, and electroreflectance data at the lowest available temperature (1.8K). In Figure 1.3b, the photoluminescence (*PL*)

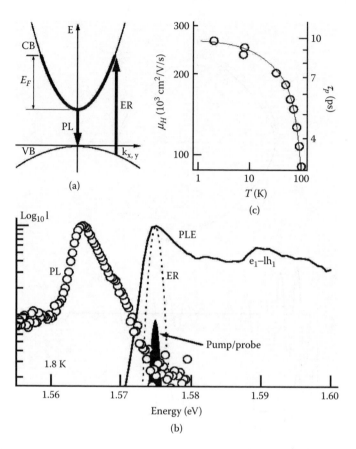

FIGURE 1.3 Properties of the 2DEG sample used for experimental investigation of spin dynamics. (a) Schematic band structure at T = 0K indicating the electron Fermi sea in the conduction band (CB) and optical inter-band transitions contributing to photoluminescence (PL) and electroreflectance (ER) spectra. (b) PL, ER, and photoluminescence excitation (PLE) spectra at 1.8K. Pump/probe indicates the energy and bandwidth of the time-resolved optical measurement of spin-dynamics. The energy separation of the PL and PLE/ER peaks is used to determine the depth of the Fermi sea, E_F. (c) Temperature dependence of the Hall mobility μ_H and corresponding transport scattering time τ_p.

peak (1.5643 eV) corresponds to the band-edge recombination (see Figure 1.3a), while the coincident peaks of luminescence excitation (*PLE*) and electroreflectance (*ER*) at 1.5749 eV indicate absorption at the Fermi edge; the Stokes shift of 10.6 meV gives $E_F = 6.64$ meV and $N_S = 1.86 \times 10^{11}$ cm^{-2}, giving the Fermi temperature $T_F = E_F/k_B \approx 70$K. The Hall mobility (Figure 1.3c) reaches 264×10^3 cm^2 V^{-1}s^{-1} at 1.8K equivalent to $\tau_p \approx 10$ ps. The Hall measurements also showed N_S as approximately independent of temperature below 100K. The photon energy used for time-resolved investigation of spin-dynamics is indicated by *ER* in Figure 1.3a; excitation at this energy results in a small photoexcited population of spin-polarized electrons in a narrow band of states near the Fermi energy (E_F). The pump beam intensity

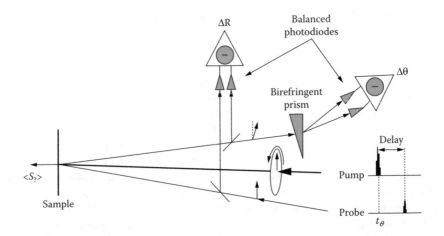

FIGURE 1.4 Schematic of pump-probe apparatus for optical investigation of spin-dynamics. Pulses are from a mode-locked Ti-sapphire laser (75-MHz repetition frequency), tuned to the interband transition (ER in Figure 1.3), time resolution ~2 ps. A circularly polarized pump injects electrons at t_0, spin-polarized along z. Delayed linearly polarized probe monitors pump-induced changes using two pairs of balanced photodiodes; change of population via reflection (ΔR) and spin component $\langle S_z \rangle$ via polarization rotation ($\Delta\theta$).

was 0.1 mW focused to a 60-micron diameter spot, giving an estimated photoexcited spin-polarized electron density of 2×10^9 cm^{-2}, less than 2% of the unpolarized electron concentration in the 2DEG; the probe power density was 25% of the pump.

The spin evolution was observed by monitoring the rotation of the plane of polarization of an optical probe pulse reflected from the surface of the sample at a variable delay following excitation by a circularly polarized pump pulse. This has been referred to in the past as the specular inverse Faraday effect (SIFE).[22] The experimental arrangement is illustrated in Figure 1.4. The pulses, pump, and probe are derived from a mode-locked Ti–sapphire laser and have the same wavelength tuned to the Fermi level in the sample (Figure 1.3a). The repetition frequency of the laser pulses is 75 MHz so that time evolutions on a scale of up to a few nanoseconds can be recorded and averaged without interference between "laser shots." The pulse lengths (~2 ps) were selected to provide a compromise between adequate time resolution and the corresponding spectral resolution (~0.3 meV). The pump-induced changes of probe intensity (ΔR) and polarization rotation angle ($\Delta\theta$) are recorded using pairs of identical photodiodes whose currents are subtracted at the inputs of two transimpedance amplifiers. For the determination of ΔR, beam splitters are used to direct small fractions (~4%) of the incident and reflected probe to the diodes; the intensities of the two beams incident on the diodes are adjusted to be exactly equal in the absence of the pump so that small pump-induced changes in reflected power can be amplified and recorded. Photodiodes have a very high dynamic range and frequency response up to many megahertz so that this balanced detector method has the advantage of very effective common-mode rejection of laser intensity fluctuations with bandwidth limited by the frequency response of the bare diodes. The amplifier characteristics affect only the amplified current difference and the transimpedance mode has the advantage that, for the gains typically required, output noise is dominated by Johnson noise in the amplifier feedback network so that the signal-to-noise ratio is proportional to $\sqrt{\text{gain}}$.[23] The overall sensitivity is limited, in principle, by shot noise, which can be minimized by lock-in detection. In practice, scattered pump light reaching the detectors is a serious source of extra noise but can also be minimized by lock-in techniques. Thus, the incident pump and probe beams are chopped at different audio frequencies and the lock-in detection is set at the sum of the two. For $\Delta\theta$, a similar balanced detector arrangement is adopted, but in this case the two beams are obtained using a birefringent, calcite, prism that is oriented in such a way as to spatially separate the components polarized at $\pm 45°$ to the polarization axis of the reflected probe in the absence of the pump. Thus, very small pump-induced rotations can be detected and amplified as intensity differences in the two detectors, again with broadband common-mode rejection of laser noise. The same sum-frequency lock-in detection is used to discriminate against the scattered pump light that reaches the detectors. For the detected probe beam intensities (typically 10 microwatts), the sensitivity of the measurements was $\sim 1{:}10^6$.

Care is required in interpretation of the ΔR and $\Delta\theta$ signals in terms of spin dynamics since they depend on the nonlinear mechanism by which the pump excitation changes the reflectivity. A variety of studies[24–27] have established that the main mechanism in the case of band-edge excitation is "phase–space filling" in which photo-excited electrons and/or holes occupy conduction and valence band states leading to a change of oscillator strength of the transition due to the Pauli exclusion principle. For measurements on 2DEGs, it is found[4,25–27] that ΔR can be used as a measure of the population of photoexcited carriers and $\Delta\theta$ as a measure of the net z-component of electron spin. Consequently, their ratio gives the pure spin dynamics of the electrons.

1.3.2 EXPERIMENTAL RESULTS

Figure 1.5 shows the experimentally measured probe polarization rotation $\Delta\theta$ as a function of pump-probe delay at several temperatures.[4] On the timescales of Figure 1.5, the pump-induced reflection signal ΔR was constant within experimental error, so that $\Delta\theta$ has the time dependence of $\langle S_z \rangle$ normalized for effects of electron–hole recombination. At the lowest temperature (1.8K), the spin polarization evolves as a heavily damped oscillation of frequency 0.19 ± 0.02 rad ps^{-1}. This behavior strongly suggests a quasi-collision-free regime in which the injected electron spins precess through a few radians before the spin polarization is lost. The frequency is in reasonable agreement with the calculated average precession frequency, 0.24 rad ps^{-1} (see Section 1.2.3). The fact that the oscillations are much more heavily damped than for the calculations of Figures 1.2b and 1.2c, which did not include electron scattering, indicates that, experimentally, scattering is the most important damping mechanism and that anisotropy of $|\Omega_\mathbf{k}|$ has essentially no role in the observed decay. At 10K, the evolution becomes exponential; and as the temperature is further increased, the decay slows dramatically. This is the qualitative trend expected for the transition to a collision-dominated regime as described by Equation 1.2 with steadily increasing momentum scattering rate, $(\tau_p{}^*)^{-1}$.

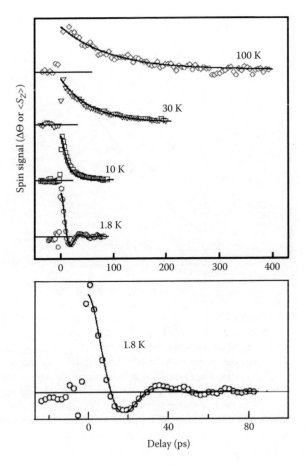

FIGURE 1.5 Spin-relaxation signals in zero magnetic field for ~10^9 cm^{-2} electrons injected optically at the Fermi level in a 2DEG in a 10-nm modulation-doped GaAs/AlGaAs quantum well containing 1.86×10^{11} cm^{-2} electrons ($E_F/k_B \sim 70$K). The evolution is clearly oscillatory at 1.8K and shows motionally slowed relaxation at higher temperatures. Solid curves are Monte Carlo simulations to extract spin-relaxation times τ_s as described in the text.

The solid curves in Figure 1.5 are simulations of these time evolutions using a simple Monte Carlo procedure[4] in which we arbitrarily assume an initial population of 10^5 spin polarized electrons injected at the Fermi energy with an isotropic distribution of in-plane wavevectors. The electrons undergo random elastic momentum scattering with rate $(\tau_p{}^*)^{-1}$ that can be adjusted in the simulation. For simplicity and because, as explained above, anisotropy of the spin-splitting plays essentially no role in the evolution of $\langle S_z \rangle$, each injected electron is assigned the same precession frequency, which is also an adjustable, but its axis of precession is changed at random, in accordance with the theoretical vector $\boldsymbol{\Omega}_k$ as its wavevector is changed by the scattering events.[6,14] The resultant z-component of spin $\langle S_z \rangle$ is computed at each time step. The 1.8K data is best fit using a frequency of $0.19 \pm 0\ 02$ rad ps^{-1} (as stated above) and $\tau_p{}^* = 6 \pm 1$ ps. This corresponds to $\langle |\boldsymbol{\Omega}| \rangle \tau_p{}^* \approx 1.2$, confirming

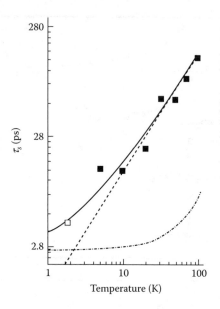

FIGURE 1.6 Spin-relaxation times τ_s (points) extracted from the data of Figure 1.5 by Monte Carlo simulation. Dot-dashed curve is $(|\Omega|^2\tau_p)^{-1}$, where $|\Omega|$ has the value, 0.19 rad ps^{-1}, used in the simulations and τ_p is the transport relaxation time obtained from Hall mobility measurements. The data converge on this curve at low temperatures, consistent with the expected equality of τ_p and τ_p^* at $T = 0$K. The dashed line is purely empirical, representing an additional contribution to τ_s linear in T. The solid curve is the sum of the dotted and dot-dashed curves.

breakdown of the collision-dominated regime. At higher temperatures, where the evolution is exponential, we assume the same value of frequency, $|\Omega|$, because we do not expect the frequency to change significantly with temperature[4] and adjust the scattering rate $(\tau_p^*)^{-1}$ to simulate the experimental decays. In Figure 1.6, the plotted points are the corresponding values of $\tau_s = (|\Omega|^2\tau_p^*)^{-1}$ given by the simulation. The dot-dashed curve shows $(|\Omega|^2\tau_p)^{-1}$, where τ_p is the transport relaxation time from Hall mobility measurements. For most of the temperature range, this curve is about an order of magnitude below the measured spin relaxation times; they approach one another only at the lowest temperature. At higher temperatures, τ_s increases rapidly whereas $(|\Omega|^2\tau_p)^{-1}$ remains relatively constant. These experimental results are strongly reminiscent of the qualitative predictions for effects of electron-electron scattering depicted in Figure 1.1, except that experimentally τ_s increases approximately as T^1 rather than as T^2. This data extends only to 100K, which is not sufficiently above $T_F = 70$K, to verify the maximum in τ_s predicted theoretically.

1.4 CONCLUSIONS

The spin-dynamics of electrons in high-mobility, two-dimensional electron gases present a series of interesting fundamental questions that are also relevant to spintronic applications. Experiments have thus far revealed a rather small fraction of what is predicted; there is a hint of a crossover from the normal high-temperature, collision-dominated spin-relaxation regime toward a collision-free regime of oscillatory spin evolution at low temperatures and also an indication of the expected strong influence of electron-electron coulomb scattering. The picture is far from complete, and it still remains to properly investigate the influence of electron concentration, temperature, and electron confinement energy.

References

1. Awschalom, D.D., Loss, D., and Samarth, N., Eds., *Semiconductor Spintronics and Quantum Computation*, Springer, Berlin, 2002.
2. Parkin, S.S.P., *IBM J. Res. Dev.*, 42, 3, 1998.
3. Moodera, J.S. and Mathon, G., *J. Magn. Magn. Mater.*, 200, 248, 1999.
4. Brand, M.A., Malinowski, A., Karimov, O.Z., Marsden, P.A., Harley, R.T., Shields, A.J., Sanvitto, D., Ritchie, D.A., and Simmons, M.Y., *Phys. Rev. Lett.*, 89, 236601, 2002.
5. D'yakonov, M.I. and Perel', V.I., *Sov. Phys. JETP*, 33, 1053, 1971.
6. D'yakonov, M.I. and Kachorovskii, V.Yu., *Sov. Phys. Semicond.*, 20, 110, 1986.
7. Flatté, M.E., Byers, J.M., and Lau, W.H., *Semiconductor Spintronics and Quantum Computation, Chapter 4: Spin Dynamics in Semiconductors*, D.D. Awschalom et al., Eds., Springer, Berlin, 2002.
8. Glazov, M.M. and Ivchenko, E.L., *JETP Lett.*, 75, 403, 2002; Glazov, M.M. and Ivchenko, E.L., *Journal of Superconductivity: Incorporating Novel Magnetism*, 16, 735, 2003; Glazov, M.M., *Physics of the Solid State*, 45, 1162, 2003.
9. Glazov, M.M., Ivchenko, E.L., Brand, M.A, Karimov, O.Z., and Harley, R.T., *Proc. Int. Symp. Nanostructures: Physics and Technology*, St Petersburg, June 2003.
10. Meier, F. and Zakharchenya, B.P., in *Optical Orientation*, Vol. 8, *Modern Problems in Condensed Matter Sciences*, Agranovich, K.M. and Maradudin, A.A., Series Eds., North-Holland, Amsterdam, 1984.
11. Fishman, G. and Lampel, G., *Phys. Rev. B*, 16, 820, 1977.
12. Krebs, O. and Voisin, P., *Phys. Rev. Lett.*, 77, 1829, 1996.
13. Olesberg, J.T., Lau, W.H., Flatté, M.E., Yu, C., Altunkaya, E., Shaw, E.M., Hasenberg, T.C., and Bogess, T.F., *Phys. Rev. B*, 64, 201301(R), 2001.
14. Jusserand, B., Richards, D., Allan, G., Priester, C., and Etienne, B., *Phys. Rev. B*, 51, 707, 1995.
15. Gridnev, V.N., *JETP Letters*, 74, 380, 2001.
16. Ziman, J.M., *Electrons and Phonons*, Clarendon Press, 1960.
17. See, for example, Weisbuch, C., and Vinter, B., *Quantum Semiconductor Structures*, Academic Press, 1991.
18. Giuliani, G.F. and Quinn, J.J., *Phys. Rev B*, 26, 4421, 1982; Zheng, L. and Das Sarma, S., *Phys. Rev. B*, 53, 9964, 1996.
19. Kim, D.S., Shah, J., Cunningham, J.E., Damen, T.C., Schmitt-Rink, S., and Schäfer, W., *Phys. Rev. Lett.*, 68, 2838, 1992.
20. Shah, J., *Ultrafast Spectroscopy of Semiconductors and Nanostructures*, Springer, New York, 1988.
21. For details of the structure, see Kaur, R., Shields, A.J., Osborne, J.L., Simmons, M.Y., Ritchie, D.A., and Pepper, M., *Phys. Stat. Sol.*, 178, 465, 2000.
22. Zheludev, N.I., Brummel, M.A., Malinowski, A., Popov, S.V., Harley, R.T., Ashenford, D.E., and Lunn, B., *Solid State Commun.*, 89, 823, 1994.

23. Horowitz, P. and Hill, W., *The Art of Electronics*, second edition, Cambridge University Press, 1989.
24. Schmitt-Rink, S., Chemla, D.S., and Miller, D.A., *Phys. Rev.* B, 32, 6601, 1985.
25. Snelling, M.J., Perozzo, P., Hutchings, D.C., Galbraith, I., and Miller, A., *Phys. Rev.* B, 49, 17160, 1994.
26. Worsley, R.E., Time-resolved Relaxation Processes in Quantum Wells Ph.D. thesis, Southampton University, unpublished, 1996.
27. Britton, R.S., Grevatt, T., Malinowski, A., Harley, R.T., Perozzo, P., Cameron, A.R., and Miller, A., *Appl. Phys. Lett.*, 73, 2140, 1980.

2 High-Amplitude, Ultrashort Strain Solitons in Solids

Otto L. Muskens and Jaap I. Dijkhuis

CONTENTS

2.1 Introduction 15
    2.1.1 Historical Perspective 16
    2.1.2 Nano-Ultrasonics 17
    2.1.3 Strain Solitons and Shock Waves 18
2.2 Theory of Strain Solitons 20
    2.2.1 Nonlinear Elasticity 20
    2.2.2 One-Dimensional Propagation 23
    2.2.3 Soliton Trains 24
    2.2.4 Shock Wave Development 26
    2.2.5 Discrete and Multidimensional Models 28
2.3 Simulations 29
    2.3.1 Soliton Trains in Sapphire 29
    2.3.2 Diffraction and Soliton Train Formation 31
    2.3.3 Stability of Individual Solitons 32
2.4 Brillouin-Scattering Experiments 34
    2.4.1 Introduction 34
    2.4.2 Setup 35
    2.4.3 Transition from Shock Waves to Soliton Trains 36
    2.4.4 Diffraction and the Formation of Solitons 39
2.5 Conclusions and Prospects 43
Acknowledgments 45
References 45

2.1 INTRODUCTION

This chapter reviews recent experiments on the development of ultrashort acoustic solitons from intense picosecond strain wavepackets. High-amplitude, bipolar wavepackets are injected by the impact of optical pulses from an amplified Ti-sapphire laser on a thin chromium film evaporated on the surface of the sapphire crystal.

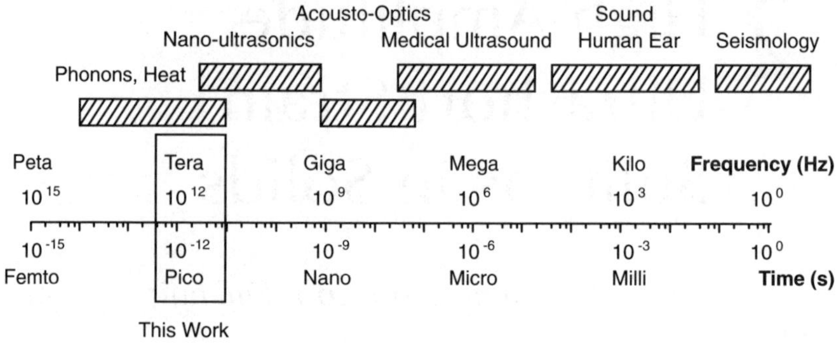

	Acousto-Optics		Sound	
Nano-ultrasonics	Medical Ultrasound	Human Ear	Seismology

Phonons, Heat ▨▨▨▨ ▨▨▨▨▨ ▨▨▨▨▨ ▨▨▨▨
▨▨▨▨▨▨ ▨▨▨▨

Peta	Tera	Giga	Mega	Kilo	**Frequency (Hz)**
10^{15}	10^{12}	10^{9}	10^{6}	10^{3}	10^{0}

10^{-15}	10^{-12}	10^{-9}	10^{-6}	10^{-3}	10^{0}
Femto	Pico	Nano	Micro	Milli	**Time (s)**

This Work

FIGURE 2.1 Acoustic-wave spectrum between 10^0 and 10^{15} Hz, with corresponding time scale in seconds. Shaded blocks denote the typical application windows of acoustic waves in technology and everyday life. Rectangle denotes the frequency range of the picosecond acoustic pulses that form the topic of this chapter.

Under these conditions, the combined action of nonlinearity of the lattice and phonon dispersion leads to the formation of trains of stable acoustic solitons after propagation distances of hundreds of micrometers. This was demonstrated by numerical simulations of the Korteweg-de-Vries equation and extensive Brillouin scattering experiments.[1,2] The ultrafast character of these soliton pulses was proved by experimentally demonstrating impulsive excitation of the 1-THz crystal-field transition in optically excited ruby.[3] From this we can conclude for sapphire that we generate millimeter-wide, nanometer-thick, supersonic disks with pressures as high as tens of kilobars. We will briefly speculate on the possible use of these extraordinary nanoultrasonic pulses for constructing a phonon laser, for generating terahertz surface plasmons in piezoelectric epitaxial layers, and perhaps for exciting nano-ultrasonic surface waves.

2.1.1 HISTORICAL PERSPECTIVE

The spectrum of acoustic waves that can be excited in condensed matter spans over 15 orders of magnitude. Figure 2.1 shows this acoustic frequency spectrum, together with some important applications in technology and everyday life. As the acoustic frequency scales up, the corresponding wavelength shifts down to regimes that are of interest for modern solid-state technological applications.

The generation of coherent, longitudinal acoustic waves, and more specifically the formation of short, stable acoustic pulses, has been an active topic for decades. Extension toward high frequencies of operation, although challenging, has for a long time merely been a technological issue. The development and optimization of thin-film piezoelectric resonators has enabled ultrasonic applications up to gigahertz frequencies.[4] However, conventional electro-acoustic methods reach their practical limit at this point, and several other schemes have been attempted to approach the regime of terahertz coherent phonons. Grill and Weis[5] reported surface generation of coherent phonons by direct conversion of infrared laser radiation in a piezoelectric crystal. The efficiency of this process, however, turned out to be greatly influenced

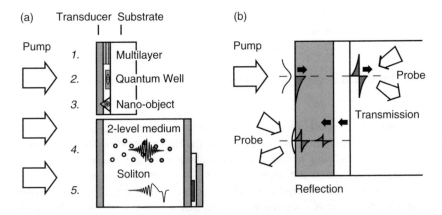

FIGURE 2.2 (a) Directions of research using nano-ultrasonic pulses, numbers refer to topics in text. (b) Principle of picosecond ultrasonics, using ultrafast thermoelastic excitation and subsequent detection of either the reflected echoes or of the transmitted wavepacket.

by the quality of the crystal surface, preventing its use for practical applications. Another line of research consists of the generation of a macroscopic acoustic field using electronic two-level systems, such as found in paramagnetic impurity ions. The interaction of gigahertz ultrasonic waves with these centers was developed experimentally by Shiren[6] and described theoretically by Jacobsen and Stevens.[7] Tucker[8] showed that, after population inversion of the electronic system, it may release its energy in the form of stimulated emission of phonons. An important step forward was made by the application of terahertz two-level systems found in transition metal ions, which have excited-state multiplets that can be simply inverted using optical radiation and that decay via a single-phonon transition. Many studies have been conducted on stimulated emission in these systems,[9–13] but most of them could be explained by incoherent rate equations, leaving controversy as to the degree of coherence reached in the acoustic field.

2.1.2 NANO-ULTRASONICS

A completely new field of opto-acoustics has emerged with the development of picosecond pulsed lasers.[14] Ultrafast optical excitation of metal films allows for relatively direct generation of pressure pulses, which can subsequently be used to probe nanometer-sized structures in an otherwise opaque substrate. This method of picosecond ultrasonics has developed since the mid-1980s into a well-established technique for studying multilayers, both in fundamental research and in the semiconductor industry. Only recently have new applications emerged that explore the use of nano-ultrasonic pulses in other directions. We categorize some of these novel research perspectives into the following main categories [c.f. Figure 2.2a]:

1. Analysis of thin-films, multilayers, and nanostructured media[14–16]
2. Excitation of semiconductor embedded structures, quantum wells, and piezoelectric materials[17–19]

3. Imaging of single nano-objects and of local elastic properties, using a combination of nano-ultrasonics and microscopy[20,21]
4. Coherent phonons in two-level media and the construction of a phonon laser[3]
5. Long-distance propagation, dispersion, diffraction, and soliton formation[1,2,22–26]

The principle of operation of the picosecond acoustic method is shown in Figure 2.2b. Ultrafast optical pulses are used to excite the electron gas via intra- or interband absorption in the first few nanometers of the absorbing layer. The excess energy in the promoted electrons is redistributed over the other carriers via electron–electron interactions within tens of femtoseconds and subsequently transferred to the lattice through electron–phonon coupling within less than a picosecond. Depending on the exact mechanism involved, either thermo-elasticity for metals or athermal bond-switching processes for semiconductors, this energy is converted into a coherent compression or expansion of the lattice. When the area of excitation is much larger than the optical skin depth, a plane-wave is launched in both directions perpendicular to the surface. The part traveling to the stress-free surface is reflected with a phase change, resulting in a release wave following the compressive wave, and vice versa. At interfaces or embedded objects inside the material, part of the bipolar wavepacket is reflected back toward the surface. The arrival of these acoustic echoes at the surface can be detected using time-resolved reflectometry or interferometry, yielding information on the internal structure.

2.1.3 STRAIN SOLITONS AND SHOCK WAVES

Whereas picosecond ultrasonics has focused for many years on the propagation of strain pulses over submicrometer distances, long-distance transport has been explored only recently. In transport experiments, information can be obtained on fundamental material properties that exert their influence only after long distances of propagation. Some established methods, well equipped at detecting strain over large distances, have recently been adapted to study terahertz coherent phonon wavepackets. Examples of these are the superconducting phonon bolometer (this volume, Chapter 5), the Brillouin scattering method described in this chapter,[1,2] and the ruby phonon spectrometer of Ref. 3. Additionally, picosecond time-resolved methods have recently demonstrated their value in long-distance transport experiments as well. Duquesne and Perrin[22] determined ultrasonic attenuation in a quasi-crystal using picosecond interferometry. Daly et al.[23] studied the influence of diffraction on small-area picosecond acoustic pulses and recovered two-dimensional information using a sophisticated imaging algorithm. In a series of experiments, Hao and Maris[24,25] demonstrated the influence of phonon dispersion on the propagation of picosecond strain wavepackets. They showed that after several hundreds of micrometers, the wavepackets are severely distorted by the dispersion of the longitudinal acoustic phonons. Thus, for the first time it seemed that the stability of short acoustic pulses is limited by fundamental physical restraints rather than by technological issues.

However, as we show later on, at larger strain amplitudes, nonlinear contributions emerge that can partially cancel the effect of the dispersive destruction of the pulse shape. In their experiments, Hao and Maris demonstrated the development of an archetypical *solitary* wave.[26] Solitons have been found in many other physical systems combining dispersion and nonlinearity but were never before demonstrated for picosecond strain pulses in crystalline solids. The solitons, which are of the compressive kind, travel with a supersonic velocity through the crystal. This property allowed their separation from the remaining wavepacket in time-resolved experiments.

These first investigations of the phonon solitons used nanojoule optical pulses to excite and to probe the picosecond strain packets. The optical power of such a system is, however, insufficient to enter far into the soliton regime. Investigations were limited to intermediate strain values of the order of 10^{-4} and to narrow strain pencils, due to the necessity of tightly focusing the pump laser. Furthermore, to improve the effect of acoustic nonlinearity, a thin-film transducer had to be constructed that delivered very high frequency strain pulses (of up to 300 GHz), requiring a very high-quality contact interface between the substrate and the transducer.

To extend the regime of strain amplitudes significantly, we developed new experiments on acoustic solitons, using high-power optical pulses from an amplified ultrafast laser system as the excitation source.[1-3] In this way we can routinely generate high aspect-ratio (up to $1:10^5$), plane-wave picosecond strain packets with very high amplitudes (0.4% strain in the metal film), which can be very suitable for technological applications. Further, we use a relatively thick, 100-nm chromium transducer that produces strain pulses with initial acoustic frequency components below 100 GHz. As we show below, this does not prevent the formation of terahertz acoustic solitons, as the large strains ensure a fast nonlinear evolution of the pulse into a shock wave structure.

Under high-power excitation, the formation of a shock front occurs in a distance of less than 100 micrometers, the distance over which the strain pulse steepens by the velocity increase of the high-amplitude peak with respect to the low-amplitude front. As the slope of the shock front increases, higher-frequency components are generated in the acoustic wave spectrum. At elevated temperatures, thermal scattering will result in strong dissipation of these high-frequency components, flattening out the shock. At low temperatures, however, thermal scattering is small and the high-frequency components of the pulse survive. As a result, phonon dispersion comes into play as terahertz frequencies are being generated at the steep front. The combined action of nonlinearity and dispersion leads to a breakup of the shock wave into a train of stable, ultrashort solitons. As the ratio between nonlinear and dispersive behavior grows with increasing strain amplitudes, the solitons shorten and the train grows up to many solitons, depending on the crystal, transducer materials, and excitation powers used.

The development of high-amplitude, ultrashort strain solitons opens up a variety of technological perspectives. A lot has been learned about the behavior of solids under high hydrostatic pressure under steady-state conditions.[29] High-pressure shock wave experiments with nanosecond time resolution have further revealed the equation of state (EOS) of many materials.[30] The novel perspective to generate, nondestructively, a high-amplitude strain pulse of *ultrashort* time duration allows for a novel domain of dynamical high-pressure experiments. The strains reached in our experimental work (0.1 to 1%) are close to the plasticity threshold for many solids, and irreversible behavior can be expected under these extreme conditions. High-pressure pulses of ultrashort time duration might therefore be applied to modify material properties with nanometer precision. The influence of high-amplitude, ultrashort strain pulses on important material properties, such as the nature of the vibrational modes in amorphous solids or the local structure around impurity or dopant ions in a semiconductor, has yet to be investigated. Semiconductor physics would certainly benefit from such a source of high-intensity acoustic strain. The acoustic wavelength of several nanometers in principle allows for a very high spatial resolution for imaging or manipulation of materials. The sensitivity for the local elasticity of the material provides a novel contrast mechanism for imaging nano-structures.[23]

2.2 THEORY OF STRAIN SOLITONS

This section begins with a brief overview of the classical theory of nonlinear elasticity for finite deformations. In contrast to the theory of infinitesimal elasticity, already well-developed in the beginning of the 20th century,[20] extension to interme-diate strains was taken up seriously in the early 1950s.[30] Closely related to the theory of fluid mechanics, solid-state models have been developed in the continuum limit and in the language of the bulk elasticity parameters. This approach results in a dispersionless theory, and lattice dispersion can only be included, *a posteriori,* in an artificial way. Therefore, we also mention the existence of paths leading to the nonlinear dispersive wave equation that starts from a microscopic picture. Then, after transformation to traveling coordinates, we readily arrive at the well-known Korteweg-de Vries (KdV) equation, which has stable solutions in the form of *solitary waves.*[31,32] It is shown that an initial compressive strain pulse breaks up into a train of such solitons. We consider the effect of temperature on the wavepacket develop-ment and briefly present the theory for shock waves using the Burgers equation. The section concludes with an extension of the nonlinear dispersive wave equation, to include diffraction.

2.2.1 NONLINEAR ELASTICITY

In this overview of nonlinear elasticity theory, we follow the derivation and nomen-clature as presented in the review of Wallace.[33] First, a set of curvilinear coordinates is introduced and the local deformation and strain variables are derived in this basis. Throughout the text we will use the Einstein summation convention.

Let us consider an undeformed solid in which a point is denoted by the vector $\vec{r} = r_j \vec{a}_j$ with respect to the basis $\vec{a}_j$ of the crystal. In an arbitrary deformed state, the position of the vector changes to $\vec{r}' = \vec{r} + \vec{u}(\vec{r}, t)$, with $\vec{u}(\vec{r}, t)$ the local displace-ment vector corresponding to the deformation. The above definitions also permit transformations that do not distort the internal material structure, such as uniform translation and rotation. An adequate definition of local deformation that filters out these simple transformations is the length change, or deformation Δ, of a local vector $\vec{r} = \vec{r}_2 - \vec{r}_1$ that can be written in the form:

$$\Delta = |\Delta \vec{r}'|^2 - |\Delta \vec{r}|^2 = 2\eta_{ij}\Delta r_i \Delta r_j. \tag{2.1}$$

The right-hand side of Equation 2.1 was obtained by expanding the squared terms using $\vec{u}(\vec{r}_2, t) - \vec{u}(\vec{r}_1, t) = \nabla \vec{u}(\vec{r}, t)\Delta \vec{r}$. The matrix connecting the initial configuration $\vec{r}$ to the deformation Δ is called the strain matrix η and follows from Equation 2.1 as:

$$\eta_{ij} = \frac{1}{2}\left(u_{ij} + u_{ji} + u_{ik}u_{kj}\right), \tag{2.2}$$

where $u_{ij} = \partial u_i / \partial r_j$ describes the displacement gradient matrix elements. The above description holds as long as variations in the displacement gradients take place on a length scale much larger than the differential elements $\Delta \vec{r}$. In practice, this means that the strain must be smooth on the scale of the interatomic distances in the solid.

The most important consequence of Equation 2.1 is that related quantities like the free energy at finite strain can be expressed in terms of the undeformed coordinates $\vec{r}$ and the strain matrix η. For example, one can expand the internal energy Φ per unit of mass in terms of strain according to:

$$\rho \Phi_S(\vec{r}, h) = \rho \Phi_0 + C_{ij}\eta_{ij} + \frac{1}{2}C_{ijkl}\eta_{ij}\eta_{kl} + \frac{1}{6}C_{ijklmn}\eta_{ij}\eta_{kl}\eta_{mn} + \dots \tag{2.3}$$

Here, ρ is the mass density and the subscript S denotes that the deformation takes place at constant entropy (i.e., under adiabatic conditions). The constants C are the elasticity constants of first, second, and third order, respectively, defined by the first- and higher-order partial derivatives of the internal energy to the strain elements $C_{ij} = \partial \Phi_S / \partial \eta_{ij}$, $C_{ijkl} = \partial^2 \Phi_S / \partial \eta_{ij}\partial \eta_{kl}$, etc.

For a simple, one-dimensional spring, one can calculate the force resulting from a compression by taking the derivative of the internal energy to the displacement (Hooke's law). In a similar vein, one can calculate the induced stresses for finite deformation using simple thermodynamic considerations.[33] For a stress-free initial state, the stress-strain relation reads

$$T_{ij} = \rho \frac{\partial \Phi_S(\vec{r}, \eta)}{\partial \eta_{ij}}. \tag{2.4}$$

The equation of motion for finite strains can now be formulated in terms of this stress tensor and the mass density ρ. The Euler-Lagrange equations of motion for the generalized coordinates $\vec{r}'(\vec{r}, t)$ in the independent variables $\vec{r}$ and t read

$$\frac{\partial}{\partial t}\frac{\partial L}{\partial \dot{r}_i} + \frac{\partial}{\partial r_k}\frac{\partial L}{\partial \alpha_{ik}} = 0, \tag{2.5}$$

with the abbreviation $\alpha_{ik} = \partial r_i'/\partial r_k$. The Lagrangian density can be written as the difference between kinetic and potential energy, which for adiabatic deformation gives

$$L = \frac{1}{2}\rho \dot{r}_i' \, \dot{r}_i' - \rho \Phi_S(\vec{r},t). \tag{2.6}$$

Combining Equation 2.5 and Equation 2.6, one arrives at the equation of motion:

$$\rho \ddot{r}_i' = \rho \frac{\partial}{\partial r_k}\left(\frac{\partial \Phi_S}{\partial \eta_{lm}}\right)\frac{\partial \eta_{lm}}{\partial \alpha_{ik}}. \tag{2.7}$$

Using the relation $\partial \eta_{lm}/\partial \alpha_{ik} = \delta_{il}\delta_{km}$ for a stress-free initial state, one can simplify the equation of motion, Equation 2.7, to a generalized form of Newton's law:

$$\rho \ddot{u}_i = \frac{\partial}{\partial r_k}T_{ik}. \tag{2.8}$$

Here we have replaced the higher-order derivative of $\vec{r}'$ by that of the displacement vector $\vec{u}$, which is allowed by its definition in Equation 2.1. With the combination of Equation 2.8, the stress-strain relation Equation 2.4 and the free energy expansion Equation 2.3, we now have all the ingredients to compose a nonlinear acoustic wave equation. Keeping only terms up to second order in derivatives of u, this equation of motion can be rearranged into the form

$$r\ddot{u}_i = \frac{\partial u_{jk}}{\partial r_l}\left(C_{ijkl} + u_{pq}A_{ijklpq}\right). \tag{2.9}$$

It should be noted that the term A_{ijklmn} contains a combination of third-order *and* second-order elastic constants:

$$A_{ijklpq} = C_{jlpq}\delta_{ik} + C_{ijql}\delta_{kp} + C_{jkql}\delta_{ip} + C_{ijklpq}. \tag{2.10}$$

The second-order coefficients in this expression are a consequence of the presence of the quadratic term in the definition of strain (Equation 2.2). Therefore, these contributions to the nonlinear term in the propagation equation (Equation 2.9) are sometimes referred to as *geometrical* nonlinearity, whereas the third-order constants are called the *physical* nonlinearity. One can convert the elastic constants in Equation 2.9 to tabulated values in Voigt notation by the reduction of pairs of indices.[34]

2.2.2 ONE-DIMENSIONAL PROPAGATION

In the case of an acoustic plane-wave along an axis of high symmetry z, the equation of motion, Equation 2.9, reduces to the simple form:

$$\rho u_{tt} = \gamma u_{zz} + \alpha u_z u_{zz}, \tag{2.11}$$

where the subscripts denote differentiation. The last term on the right-hand side is the quadratic nonlinearity, owing to the geometric nonlinearity and the cubic terms in the inter-atomic potential. The nonlinearity coefficient α depends only on the propagation direction in the crystal. For the [0001]-direction in a trigonal crystal like sapphire, for example, the two constants of Equation 2.11 can be found from Equation 2.10 as

$$\gamma = C_{33}, \alpha = \left(3C_{33} + C_{333}\right), \tag{2.12}$$

where the coefficients C_{33} and C_{333} are the second- and third-order elastic constants in the [0001]-direction, respectively. For most solids, the contribution of the third-order modulus is larger than the geometric term and has a negative sign, yielding a negative nonlinearity α.

Up to this point we have not taken into account any dispersion in the equation of motion. In the case of longitudinal acoustic lattice vibrations (LA phonons) in a trigonal crystalline solid in the direction of the c axis ([0001]), the dispersion due to discreteness of the lattice can be written up to third order as[35]

$$\omega = c_0 k - \beta k^3$$

$$\beta = \frac{c_0 a^2}{24} = \frac{c_0^3}{6\omega_{max}^2} \tag{2.13}$$

where $\omega_{max} = (4C_{33}/M)^{1/2}$ is the LA angular frequency at the edge of the Brillouin zone, and $c_0 = (C_{33}a^2/M)^{1/2}$ denotes the sound velocity. This dispersive correction can be put into Equation 2.11, leading to a fourth-order spatial derivative. At this point it is convenient to switch from the displacement coordinate u to a uniaxial component of the acoustic strain, further denoted as s. This is done by differentiation of Equation 2.11 with respect to the z-coordinate. Given the initial wavepacket at $t = 0$ of amplitude s_0 and shape $\phi(z)$, the resulting boundary value problem including dispersion reads

$$s_{tt} - c_0^2 s_{zz} - \frac{\alpha}{\rho}\frac{\partial}{\partial z}\left(ss_z\right) - 2c_0\beta s_{zzzz} = 0$$

$$s\left(z, t = 0\right) = s_0\phi\left(z\right). \tag{2.14}$$

Finally, it is convenient to transform to a moving frame coordinate system, defined by the parameters $t' = \varepsilon t$, $y = z - c_0 t$, where the small constant ε is used to express the slowness of the time coordinate. After substitution of these variables and maintaining terms up to order ε, we arrive at terms consisting of only one derivative with respect to the traveling coordinate y. Integrating the resulting expression once, we finally obtain the equation

$$
\begin{aligned}
s_{t'} + \frac{a}{2\rho c_0} s s_y + \beta s_{yyy} &= 0 \\
s(y, t' = 0) &= s_0 \phi(y).
\end{aligned}
\tag{2.15}
$$

This is the well-known Korteweg-de Vries (KdV) equation, describing, for example, the formation of stable wavepackets (solitons) in a narrow water channel.[31] In the experimental configuration in the sapphire crystal, $\alpha < 0$ and $\beta > 0$, resulting in soliton development for $s < 0$, (i.e., for compressive strain pulses).

2.2.3 SOLITON TRAINS

We are interested in the development of an arbitrary initial waveform $s(y, t) = s_0 \phi(y/l_0)$, where s_0 and l_0 are the typical (negative) strain amplitude and the spatial width of the compressive part of the initial strain packet, respectively. After transformation to the coordinates $\eta = s/s_0$, $\xi = y/l_0$, and $\tau = t'\beta/l_0^3$, the initial value problem of Equation 2.15 takes on the form[36]

$$
\begin{aligned}
\eta_\tau + 6\sigma^2 \eta \eta_\xi + \eta_{\xi\xi\xi} &= 0 \\
\eta(\xi, 0) &= \phi(\xi),
\end{aligned}
\tag{2.16}
$$

where σ is a dimensionless parameter defined by

$$
\sigma = l_0 \left(\frac{\alpha s_0}{12 \rho c_0 \beta} \right)^{1/2}.
\tag{2.17}
$$

Equation 2.17 yields identical solutions for initial wavepackets with the same value of σ and $\phi(\xi)$, and therefore σ is called the similarity parameter. The magnitude of σ determines the relation between nonlinearity and dispersion in the wavepacket development. According to Equation 2.17, σ is proportional to the area under the square root of the compressive strain waveform. To compare the similarity parameters for different initial value conditions, it is important to choose l_0 and s_0 so that the residual area of $[\phi(\xi)]^{1/2}$ is normalized to some predefined value, which in Ref. 37 equals π, the area under the square root of the soliton waveform of Equation 2.19.

For the KdV initial value problem Equation 2.16, there exists an associated eigenvalue problem for the Schrödinger equation[38]:

$$\Psi_{\xi\xi} + \left(\lambda + \frac{\sigma^2}{6}\eta(\xi,\tau)\right)\Psi = 0. \tag{2.18}$$

The solutions of this eigenvalue equation can consist of free or bound states, depending on the sign of the initial potential $\phi(\xi)$. The reflection and transmission coefficients can be derived using the time dependence of the eigenfunctions Ψ, as dictated by the KdV equation for the potential $\eta(\xi,\tau)$. A concise overview of the inverse scattering method for the KdV initial value problem can be found in Refs. 36 and 37. We limit our discussion here to the most relevant result for this chapter, namely, the stationary states, when $t \to \infty$, for the potential $\eta(\xi,\tau)$, for a discrete spectrum of eigenmodes $\lambda_n < 0$. It turns out that all bound states of the initial potential $\phi(\xi)$ correspond to soliton pulses determined by the eigenvalue λ_n, according to

$$\eta(\xi,\tau) = \frac{-2\lambda_n}{\sigma^2}\operatorname{sech}^2\left(|\lambda_n|^{1/2}(\xi + 4\lambda_n\tau)\right). \tag{2.19}$$

It can be observed from Equation 2.19 that the amplitude of the nth soliton is $\eta_n = -2\lambda_n/\sigma^2$, or $a_n = -2\lambda_n s_0/\sigma^2$ in normal strain units. The velocity of these solitons is $2\eta_n\sigma^2$ in normalized coordinates, or $c_n = \alpha a_n/6\rho c_0$ in real coordinates. Finally, the width is given by $[2/\eta_n\sigma^2]^{1/2}$, or $l_n = l_0[2s_0/a_n\sigma^2]^{1/2}$ in normal units.

The exact number of solitons developing from an initial perturbation can be found from the depth of the potential well $\phi(\xi)$ and the spacing of the eigenvalues of the energy. An analytical solution of the eigenvalue equation, Equation 2.18 can be obtained in several special cases of the initial waveform $\phi(\xi)$. In particular, the spectrum of eigenvalues for a potential of the form $\phi(\xi) = \operatorname{sech}^2\xi$ can be found in quantum mechanical textbooks[39] as

$$\lambda_n = -\frac{1}{4}\left(1 - 2n + \sqrt{1 + 4\sigma^2}\right)^2. \tag{2.20}$$

However, in our experiments we will be dealing with strain pulses that have not the above form, but a profile given to good approximation by the derivative of a Gaussian. It turns out that the compressive part of this waveform resembles quite well the hyperbolic secant function, so that we can use Equation 2.20 in our estimates for the experimental soliton trains. The corresponding values of σ are to be calculated using Equation 2.17 after the appropriate normalization of the compressive waveform.[2]

Figure 2.3 shows the development of a typical initial waveform in the situations of nonlinear propagation, both in absence (Figure 2.3b) and presence (Figure 2.3c) of the third-order dispersion.

The dispersionless case results in the formation of shock fronts, the compressive phase of the bipolar wavepacket traveling slightly faster and the rarefaction phase slightly slower than the linear velocity of sound. This situation is discussed in Section

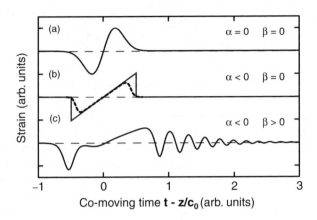

FIGURE 2.3 Development of an initial wavepacket for different combinations of nonlinearity alpha and dispersion beta, showing linear propagation (a), typical shock wave formation without (b, line) and with damping (b, thick dash), and soliton development (c), for a similarity parameter $\sigma \approx 1$.

2.2.4. The trace, including the dispersion, shows the development of a soliton pulse in the leading part and a oscillating tail in the trailing part of the bipolar initial wavepacket. The parameters α, β were chosen so that the similarity parameter for the compressional part equals $\sigma \approx 1$, leading to only one negative eigenvalue λ. The radiative tail is associated with the continuum of positive energy states of the scattering problem, Equation 2.19. Although analytical expressions for this tail exist in the case of linear dispersive waves,[36,37] no simple form is known for the combination of nonlinearity and dispersion in the wavepacket propagation.

Note that, for comparison with experiments, we use a boundary value problem instead of an initial value problem in all numerical simulations that is equivalent to Equation 2.15 after a trivial transformation $t \to z/c_0$, $z \to c_0 t$. The theoretical description was, however, presented as an initial value problem for compatibility with the existing literature.[36–38]

2.2.4 SHOCK WAVE DEVELOPMENT

In practice, except at the lowest temperatures, acoustic waves are attenuated by scattering at thermal phonons during propagation through a crystal. Classically, this scattering can be accounted for by including a viscous term in the wave equation

$$s_t + \frac{\alpha}{2\rho c_0} ss_y - \varepsilon s_{yy} + \beta s_{yyy} = 0$$

$$s(y,t=0) = s_0 \phi(y/l_0). \tag{2.21}$$

Here, ε denotes the viscosity constant of the medium, which can be obtained experimentally from the thermal scattering length of ultrasonic waves.[34] The combination of nonlinearity, dispersion, and viscosity in the wave equation is called the Korteweg-de Vries-Burgers (KdV-Burgers) equation. Unlike the KdV initial value problem, this is not an integrable equation, and a solution can only be obtained numerically. For special situations, however, we can neglect any one of the three constituent terms and recover the KdV-, Burgers-, or a linear wave equation. At this point we consider the problem of a *small* dispersive term relative to the viscous term. Neglecting the dispersion, one recovers the Burgers equation that can be solved in an analytical way. Starting from the Burgers equation in the form $s_t + \frac{1}{2}\left(s^2\right)_y = v s_{yy}$, one can apply the Cole-Hopf transformation $s = -2vu_y/u$ to arrive at

$$u_t = vu_{yy}. \tag{2.22}$$

This is the well-known heat equation, which has solutions that spread out diffusively with time.[36] The development of the initial wavepacket depends on the relative magnitude of the nonlinearity and dispersion, as contained in the parameter $v = c_0\varepsilon/\alpha$. The most relevant case for this work is the condition that nonlinearity dominates wavepacket development. In the extreme case, when viscosity is neglected, the initial shape develops until an infinitely sharp wavefront is formed. For an initially symmetric bipolar waveform, it can be shown that, in the limit $y \to \infty$, it develops into a shape of the form[36]

$$s(t,y) = \begin{cases} y/t & -\sqrt{2A_0 t} < y < \sqrt{2A_0 t} \\ 0 & |y| > \sqrt{2A_0 t} \end{cases}, \tag{2.23}$$

where A_0 denotes the area under one phase of the initial waveform. The solution for a bipolar wavepacket after a long time in the case of nonzero viscosity can be found using the solution of Equation 2.22

$$s(y,t) = \frac{y}{t} \frac{\sqrt{a/t}\exp\left(-y^2/4vt\right)}{1+\sqrt{a/t}\exp\left(-y^2/4vt\right)}. \tag{2.24}$$

The typical form of a so-called N-wave is shown as the dashed line in Figure 2.3b. This expression can be scaled to the initial waveform via the area under one of its phases, $A_0 = 2vR_0$, where we have introduced an initial Reynolds number $R_0 = \log\left(1+\sqrt{a/t_0}\right)$. The characteristic property of Equation 2.24 is the scaling of the main profile (i.e., the N-structure) over $y = \left(2A_0 t\right)^{-1/2}$, combined with a smearing out of the shock front over a distance $\Delta y = R_0^{-1}\log t/t_0$. Further, the integral under each phase of the bipolar wavepacket decays by dissipation through the viscous term as $A(t)/A_0 = \log\left(1+\sqrt{a/t} - \exp R_0\right)$.

2.2.5 Discrete and Multidimensional Models

As an extension of the above derivation, we want to discuss (1) an approach starting from a discrete model and (2) multidimensional corrections to the Korteweg-de Vries equation. The classical theory of finite elasticity resulted in a fully three-dimensional, nonlinear wave equation for arbitrary crystallographic symmetry. After reduction to one dimension, phonon dispersion was added by means of a correction *a posteriori*. For the problem of ultrashort strain packets in a crystalline lattice, it may well be worth the effort to consider a derivation that maintains the discrete character of the atomic lattice. The theory of nonlinear lattices has been developed since the early work of Fermi, Pasta, and Ulam,[40] but has more recently focused on analytical model systems in one dimension (see, e.g., Ref. 41). In these models, dispersion is an inherent feature of the discrete system, and solitary waves have been identified as the normal modes of the nonlinear lattice. It has been demonstrated that the one-dimensional integrable lattice with exponential interactions, known as the Toda chain,[41] reduces to the KdV equation in the long-wavelength limit.

Potapov et al.[42] showed that a two-dimensional microscopic lattice model can be used to derive the nonlinear dispersive wave equation for the atomic displacements. They started from a two-dimensional construction of point masses and considered only nearest-neighbor interactions in the first two configurational shells. Meijer[43] extended this approach to a cubic lattice in three dimensions. In these calculations, the local deformations are again defined by the square of the interatomic distances, similar to Equation 2.1 but now on a discrete lattice. Subsequently, the transition from a discrete to a continuous description is made by maintaining a second-order term in the expansion for the local displacement gradient $\vec{u}(\vec{r}_2,t) - \vec{u}(\vec{r}_1,t)$. This approach results in a wave equation for the displacement that is similar to Equation 2.14 but with microscopic force constants, demonstrating that the dispersive correction indeed has the form of the *a posteriori* term introduced in Equation 2.13. However, as the elastic constants usually are determined experimentally from the strain derivative of the free energy, the resulting expressions must be compared with continuum elasticity if one wants to describe an experimental situation.

We now arrive at the addition of diffraction phenomena in the nonlinear dispersive wave equation. To investigate the effect of weak higher-dimensional propagation effects, one must maintain a higher-order term in the transverse gradients of the displacement vector. This can be done starting from the full three-dimensional equation (Equation 2.9) by the assumption that the transverse derivatives will have an influence of order $\varepsilon^{1/2}$ (the infinitesimal parameter). In this way, the first contribution to the wave equation, which is the Laplacian operator in Equation 2.25, will be of order ε. Assuming, for simplicity, a rotational symmetry around the propagation axis, the resulting correction to the KdV wave equation can be written as[44]

$$\left(s_t + \frac{\alpha}{2\rho c_0} s s_y + \beta s_{yyy} \right)_y = \frac{C_\perp}{2\rho c_0} \Delta_\perp s , \qquad (2.25)$$

where $C_\perp$ is an elastic constant perpendicular to the direction of propagation. For a non-centrosymmetric crystal, the right-hand side consists of two terms in the directions of the symmetry axes with their appropriate elastic constants. In the simulations of Sections 2.3.2 and 2.3.3, along the c axis of sapphire, we will study the effect of the additional transverse gradients on the propagation of narrow strain beams.

2.3 SIMULATIONS

2.3.1 SOLITON TRAINS IN SAPPHIRE

This section presents numerical calculations of the nonlinear dispersive propagation of high-amplitude strain pulses in sapphire. It was discovered in the 1960s by Zabusky and Kruskal[32] that the Korteweg-de Vries model is a nonlinear partial differential equation (PDE) that is integrable, meaning that analytical solutions can be obtained for any initial value or boundary condition imposed on the system. In practice, however, solutions can be written down in concise form only for the initial stage of development of the dispersive shock front and in the limit for the soliton part of the wavepacket. The transition from the self-steepening regime to soliton trains, and the evolution of the rarefaction part of the wavepacket into an oscillating tail, can only be calculated numerically. The wave equation including damping, the KdV-Burgers equation, can also only be evaluated numerically. Therefore, we choose to compute the development of the initial pulse using numerical simulations of the KdV-Burgers equation.

The construction of a numerical solution for a nonlinear, dispersive wave equation with initial conditions is a specialized topic in itself. In particular, the combination of a large nonlinear and small dispersive term poses strong demands on the stability of the numerical algorithm. Of the three approaches[45–47] we have tested in the course of our investigations, the fastest, reasonably stable algorithm appeared to be the fourth-order Runge-Kutta scheme by Driscoll.[46] Its efficiency gain and method of implementation further allowed for a simple extension to the two-dimensional problem of diffraction, which is addressed in Section 2.3.2.

Using the one-dimensional wave equation (Equation 2.21), we have simulated the propagation of a typical experimental wavepacket over several millimeters in a sapphire (Al_2O_3) single crystal. The transition from the shock-wave regime to soliton trains was studied by changing the damping parameter ε in the KdV-Burgers equation, between the value at room temperature and the zero-damping limit. The material constants of the sapphire crystal along the crystallographic c direction [0001] are shown in Table 2.1.

Figure 2.4 shows the wavepacket developed after 1 mm of traveled distance through the sapphire, for an initial strain of 0.13% and for several values of the damping parameter ε. A value of 4.54×10^{-4} Ns/m corresponds to the viscosity of sapphire at room temperature.[34] We clearly observe the transition from an N-shaped shock wave at high viscosity values, via an oscillating shock front at around 2×10^{-5} Ns/m, to soliton train formation at a two times smaller value of ε. At a nonzero viscosity value, dissipation reduces the soliton pulse amplitudes as they propagate

TABLE 2.1
Material Parameters of c-axis Sapphire Relevant
for the Development of Ultrashort Acoustic Soliton Trains

	c_0 km/s	ρ kg/m³	C_{33} GPa	α GPa	β 10^{-17} m³/s	ε (293 K) 10^{-4} Ns/m	$C_\perp$ GPa
Al$_2$O$_3$[0001]	11.23	3.98	502	−1830	3.5	4.54	500

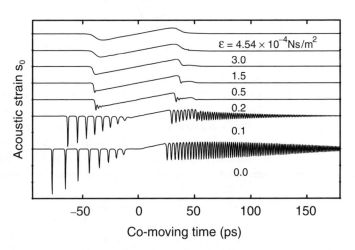

FIGURE 2.4 Strain waveforms after 1 mm propagated distance in sapphire [0001] for several values of viscosity parameter ε, calculated using the KdV-Burgers equation (Equation 2.22) at a strain s_0 of 1.3×10^{-3}. The transition from shock waves to soliton trains takes place for a viscosity parameter between 2 and 5% of the room-temperature value.

and thus leads to a slowing down of the shortest solitons, as can be seen by comparing the two lowermost traces in Figure 2.4. The bottom trace corresponds to the case of zero damping, where the solitons can propagate over long distances without distortion.

The processes of self-steepening and subsequent breakup into soliton trains of the leading part of a typical strain pulse can be distinguished in Figure 2.5a. In the first 40 μm of the propagation distance, the peak of the wavepacket shifts to the left, as its velocity is slightly higher than the linear sound velocity due to the local compression of the lattice. Figure 2.5b shows the frequency spectrum associated with the bipolar waveforms. We observe that the initial frequency spectrum below 100 GHz rapidly redistributes toward the higher harmonics and that the whole spectrum scales toward lower frequencies due to the expansion of the waveform. The strong up-conversion of strain, up to terahertz frequencies, is accompanied by a removal of energy from the lower-frequency part of the spectrum, an effect that we recognize in the experimental traces of Section 2.4. At the point where frequencies

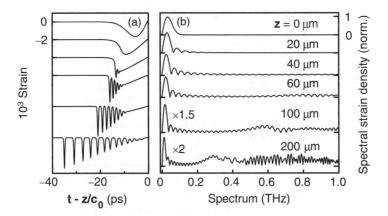

FIGURE 2.5 (a) Calculated development of the compressive part of the wavepacket, using Equation 2.21 with zero damping, for several propagation distances z. (b) Acoustic spectra corresponding to the development stages of (a), but for the complete bipolar wavepacket. Lower two spectra have been multiplied vertically for displaying convenience. (From Muskens, O.L. and Dijkhuis, J.I., *Phys. Rev. B*, 70, 104301, 2004. With permission.)

are high enough for dispersion to play a role (which in Equation 2.21 is proportional to ω^3), the shock front starts to oscillate. Eventually, the whole wave structure breaks up into a number of sharp peaked, half-cycle strain pulses, the acoustic solitons, with increasing amplitudes toward the front of the train. Concomitantly, as the soliton width and velocity scale with its amplitude, the leading soliton in the train is the fastest and the narrowest.

2.3.2 DIFFRACTION AND SOLITON TRAIN FORMATION

This section investigates, by a series of numerical simulations, the influence of diffraction on the formation of acoustic solitons. We consider the typical conditions met in our experiments and take sapphire as the material for our calculations. We assume cylindrical symmetry and a value of $C_{\perp} = 500$ GPa, which is a reasonable approximation around the c axis of sapphire. The ratio between nonlinearity and diffraction is varied by changing the transverse width of the Gaussian strain pencil launched into the material.

Diffraction has been included in the nonlinear wave motion by means of the de Kadomtsev-Petviashvili equation (Equation 2.25).[44] The numerical algorithm used to calculate the propagation of a picosecond strain packet in a cylindrical geometry is an extension of the Runge-Kutta scheme of Section 2.3.1. Computational limitations restrict the calculations to strain amplitudes of 0.05% and to the fist few hundred micrometers of propagation. This is sufficient to follow the initial steps of the nonlinear evolution and the formation of a soliton train and to draw qualitative conclusions about the influence of diffraction.

We calculated the two-dimensional development of a cylindrical beam with a Gaussian radial intensity profile. Figure 2.6a shows radial profiles of the three-dimensional pulses after 0.45 mm of propagation, in the form of two-dimensional

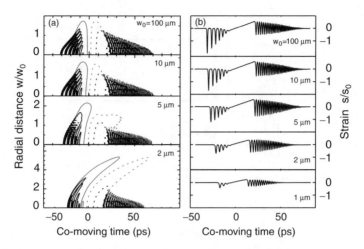

FIGURE 2.6 (a) Contour plots of the developed strain profiles after 0.05 mm in sapphire [0001], calculated for different transverse beam sizes w_0, with a radial distance w normalized to w_0. Solid lines denote negative strain contours, and dashed lines denote positive strains. (b) Strain plots of the on-axis sections of (a), normalized to the initial strain s_0 of 0.05%.

contour plots, for several diameters of the initial acoustic beam. The evolution of the profile with the largest beam waist of 100 μm (upper graph in Figure 2.6a) is completely one-dimensional: diffraction plays no role. As usual, solitons appear at the front and a radiative tail is produced at the rear of the wavepacket, as shown more clearly by the on-axis strain waveforms presented in Figure 2.6b. The appearance of a positive curvature of the solitons is due to the variation in strain amplitude over the profile, which translates directly into a velocity gradient by the properties of the KdV solitons.

For acoustic beam dimensions below 10 μm, an increasing influence of diffraction is observed. For these narrow beams, the transverse gradient of Equation 2.25 results in a significant flow of energy out of the center of the acoustic beam. The nonlinear steepening process and concomitant breakup into solitons is severely frustrated by the reduction of the strain amplitude in the center of the beam. This can be most clearly seen in the on-axis curves presented in Figure 2.6b. Clearly, the number of solitons that develop from the initial packet decreases from seven for the widest acoustic beams, down to three for the narrow, 1-μm wide profile.

2.3.3 STABILITY OF INDIVIDUAL SOLITONS

In the above simulations, the breakup into solitons of a micrometer-wide beam with frequency components around 30 GHz was shown as being strongly influenced by diffraction. This section considers the diffraction of a single terahertz strain soliton. At first glance, the development of a *single*, ultrafast strain soliton is expected to be less disturbed by diffraction. In the linear case, diffraction of a Gaussian beam is governed by the Fraunhofer formula[48]

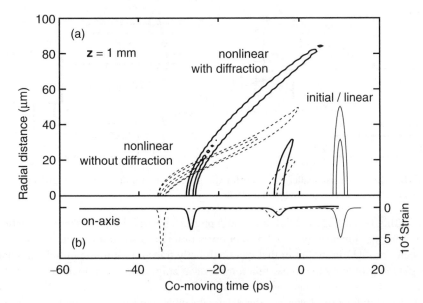

FIGURE 2.7 (a) Radial profiles of an initial picosecond strain pulse (thin line, translated to 10 ps for displaying convenience), in sapphire [0001], in the cases of nonlinear propagation without diffraction (dashed line), and with diffraction (thick solid line). (b) On-axis strain waveforms of the profiles shown in (a).

$$w(z) = w_0 \left(1 + \left(\frac{c_0 z}{\pi v_{ph}(1+2p)w_0^2} \right)^2 \right)^{1/2} \tag{2.26}$$

where c_0 denotes the sound velocity, the acoustic frequency, and p is the phonon focusing parameter. For an initial frequency of 1 THz and an initial acoustic beam width of 10 μm, divergence is of order 10^{-4}, which is negligible over millimeter propagation distances. Hao and Maris[26] used this linear diffraction argument to demonstrate that diffraction can be neglected in their soliton experiments. In the following, we show that this assumption is not valid for nonlinear soliton beams.

The transverse gradient of Equation 2.25, which is responsible for linear diffraction, results also in strong *nonlinear* distortion of the single, ultrashort solitons. This effect is illustrated by the calculations presented in Figure 2.7. Here, we calculate how a narrow cylindrical, high-amplitude wavepacket of 1-ps time duration breaks up into two soliton discs. The initial pulse, with a Gaussian transverse profile of 15 μm width (thin line in Figure 2.7), is followed over 1 mm of propagation in sapphire. We consider three different regimes of propagation: (1) linear propagation with diffraction, (2) nonlinear propagation without diffraction, and (3) nonlinear propagation with diffraction.

A low-amplitude pulse with this shape would not be influenced by diffraction, due to the high-frequency content of the wavepacket (c.f. Equation 2.26). The initial

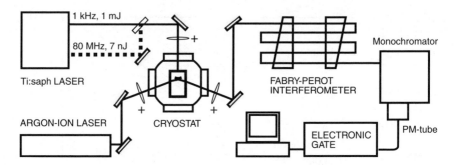

FIGURE 2.8 Brillouin scattering setup for the generation and detection of high-amplitude acoustic wavepackets in a transparent crystal.

pulse (thin line in Figure 2.7a) thus also represents the developed packet for situation (1). The one-dimensional, nonlinear propagation (2) leads to a typical bell-shaped velocity profile (dashed line) that can be directly traced back to the initial Gaussian amplitude profile by the linear relation between amplitude and velocity of the KdV solitons. If, finally, dispersion is added to the nonlinear propagation, the profile changes significantly. Even for the relatively wide acoustic beam under study we observe a significant transverse spreading of the wavepacket (thick solid line), accompanied by a decrease of the on-axis soliton amplitude, as shown in Figure 2.7b.

From the above simulation we conclude that diffraction not only influences the self-steepening part of the wavepacket evolution, but also the propagation of single, three-dimensional soliton objects, which are shown to be unstable in two dimensions. It is therefore impossible in this first-order framework to produce nanometer-sized solitary strain packets that are stable in all spatial dimensions.

2.4 BRILLOUIN-SCATTERING EXPERIMENTS

2.4.1 INTRODUCTION

The previous sections have shown that nonlinear elasticity theory, combined with phonon dispersion, predict the formation of soliton trains from a high-amplitude, picosecond acoustic wavepacket. The experimental verification of this phenomenon is actually not straightforward. One must be able to monitor the strain after large distances of propagation. The spatial dimensions of the solitons are much less than an optical wavelength, making direct observation only possible at sharp interfaces (e.g., at the surface of an opaque film). However, the requirement of a boundary layer prevents a continuous variation of the propagation distance, and interactions at the interface and material roughness may well disturb the soliton pulses before being detected. At much lower strain intensities, and concomitantly wider soliton pulses, it turned out already impossible to extract quantitative information from the reflectivity signal other than the propagation velocity.[49] As we anticipate ultrashort strain solitons of nanometer dimensions, surface detection is expected to become increasingly difficult.

An experimental technique well-suited to the detection of ultrashort strain solitons in *the bulk* of a crystal is Brillouin scattering. This method has a long history in the detection of incoherent and thermal phonons and allows for a sensitive determination of acoustic-phonon sound velocities in crystals. Only very recently, its use has been demonstrated in the field of coherent phonons and picosecond acoustic wavepackets.[48,50–52] As the light scattered from the propagating acoustic waves undergoes a frequency shift corresponding to the absorbed phonon energy, signals can be easily filtered using multipass interferometry. Furthermore, its selectiveness for the momentum ensures that only the directional phonon wavevectors matching with the scattering angle can be selected. In this way, Brillouin spectroscopy yields important information on the development of the gigahertz frequency spectrum of the acoustic wavepackets.

We have chosen to use the spectrometer as a local probe of strain components, by focusing the probe laser tightly through the side windows of an optical cryostat in which the sample is immersed and collecting the directional beam of scattered light. In this way we can obtain information on the development of individual spectral components as the acoustic wavepacket propagates over millimeter distances through the crystal. As we show below, this method yields typical patterns in the distance dependence of the scattered intensity that demonstrate the transition from the shock wave regime to soliton trains.

2.4.2 SETUP

For the propagation experiments, we use a piece of high-quality (<1 ppm impurity ions), single-crystal sapphire of $5 \times 11 \times 10$-mm^3 dimension, with the *c*-axis aligned perpendicular to the 5×11-mm^2 surface, covered with a 100-nm thin chromium transducer. The setup is shown in Figure 2.8. The crystal is mounted into an optical cryostat to perform experiments down to liquid helium temperatures. Strain wavepackets of very short (picosecond) time duration are generated using the conventional method of ultrafast heating of the chromium film that is evaporated onto the sapphire crystal. It is well-known that chromium has a very high electron-phonon coupling constant of 420×10^{15} W/m^{3}K, and thus the electronic excitation is converted to a coherent strain over a distance of the order of the optical absorption length. Acoustic strain pulses in the high-amplitude regime are generated using an amplified Ti:sapphire laser setup, which provides 130-fs laser pulses at 800 nm with an energy of 0.7 mJ/pulse, at a repetition rate of 1.0 kHz. To obtain excitation over a large area, the output of this laser is weakly focused to a spot of several millimeters in diameter onto the sample. The optical pump fluence is varied by controlling the position of this focusing lens, the upper limit being the damage threshold of our transducer at 15 mJ/cm^2. Small-area, low-amplitude acoustic pulses can be generated as well, using the mode-locked output of the Ti-sapphire laser system, carrying 7-nJ pulses at a repetition rate of 80 MHz.

Figure 2.8 shows the experimental configuration for the scattering experiment. The optical probe is a single-mode argon-ion laser operating at 514.5 nm and delivering 60 mW optical output power. The optical beam is tightly focused inside

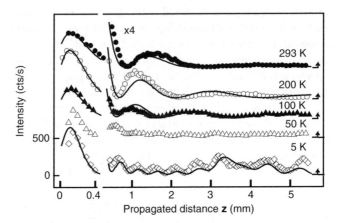

FIGURE 2.9 Brillouin scattering intensity at 22 GHz as a function of propagated distance in sapphire [0001], for several temperatures between 293K and 5K. Symbols indicate experimental data, lines simulation results for the KdV-Burgers model, Equation 2.21. (From Muskens, O.L. and Dijkhuis, J.I, *Phys. Rev. Lett.*, 89, 285504, 2002. With permission.)

the crystal to a waist of several micrometers. The frequency-shifted scattered radiation is analyzed by a quintuple-pass Fabry-Pérot interferometer (Burleigh RC110) and detected using standard photon counting techniques. The same spectrometer setup was used previously to study the propagation of coherent, monochromatic phonon beams and of ultrashort strain pulses in lead molybdate and paratellurite.[48,50–52]

2.4.3 TRANSITION FROM SHOCK WAVES TO SOLITON TRAINS

By measuring the Brillouin scattering intensity in the crystal as a function of distance from the transducer, we monitor one single Fourier component of the strain wavepacket. Figure 2.9 shows this development at an acoustic frequency of 22 GHz for five values of the sample temperatures. Generally, the shape of the traces consists of two distinct parts: (1) within the first hundreds of micrometers, the Brillouin intensity drops dramatically by almost an order of magnitude, followed by (2) oscillations of the intensity as a function of propagated distance. Feature (1) does not appear to depend strongly on the temperature, be it that the decay speeds up slightly towards lower temperatures. However, the oscillations of (2) show a pronounced, qualitative change in behavior below a certain critical point, at a temperature between 50 and 100K. Above this critical temperature, only a few oscillations take place in the propagation length of about 6 mm, with a period that grows longer upon propagation in the crystal.

Below the critical temperature, however, we clearly observe a large number of oscillations with a period that remains constant over the long travel distances of several millimeters. After a distance of several millimeters in the crystal, the patterns become more intricate and can no longer be interpreted as a single oscillation. This behavior of the scattering intensity at the lowest temperature is studied further as a function of the pumping power of the femtosecond laser. These low-temperature

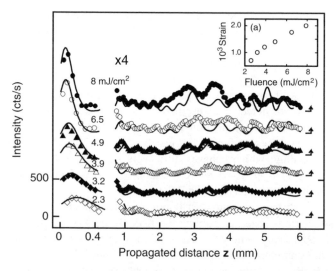

FIGURE 2.10 Brillouin scattering intensity at 22 GHz against propagation distance for several values of pump fluence E, at a temperature of 5K. Symbols indicate experimental data, lines simulations using the KdV model, Equation 2.15. (From Muskens, O.L. and Dijkhuis, J.I, *Phys. Rev. Lett.*, 89, 285504, 2002. With permission.)

data, shown in Figure 2.10, clearly demonstrate that the oscillations in the Brillouin traces become faster with increasing pump fluence.

At this point there are two ways to compare these data with theory: (1) either direct comparison with numerical simulations of the nonlinear wavepacket evolution or (2) analysis of the oscillation frequencies followed by comparison with analytical expressions to recover the essential soliton parameters.

Reproduction of the experimental data by numerical calculations using the KdV-Burgers equations can be performed over all ranges of strain amplitudes and temperatures under study, as was shown by Muskens and Dijkhuis.[1] Together with independent estimates of the initial wavepacket shape from time-resolved experiments[2] and the known material parameters of sapphire (see Table 2.1), we calculated the nonlinear evolution of the pulse over millimeter distances using the numerical schemes of Section 2.3. To arrive at the simulated Brillouin traces, we Fourier-transformed the time-domain simulations of the wavepacket taken at discrete distances z and took the square of the resulting spectral component at the Brillouin frequency. The resulting curves as a function of propagated distance z are plotted as the lines in Figure 2.9 and Figure 2.10. We observe excellent agreement with the experimental data points for realistic values of acoustic strain s_0 (see Figure 2.10a) and the viscosity parameter ε (see Figure 2.4).

The features observed in the Brillouin data can now be traced back via the simulations to the time-domain evolution of the wavepacket. For example, the decay of the Brillouin scattering intensity within the first 100 micrometers is the signature for self-steepening of the acoustic wavepacket. Figure 2.5 showed that the formation of a sharp shock front is accompanied by the formation of many higher harmonics of the fundamental frequency of the bipolar waveform. This redistribution of energy over a spectral window as large as 1 THz inevitably leads to a reduction in amplitude of the lower, GHz strain components that are detected with the Brillouin scattering method.

Subsequently, the spreading of the wavepacket results in a drift of the spectral characteristics toward lower frequencies. As the packet broadens on its journey through the crystal, the harmonic peaks are transported toward the GHz frequency range where they are detected as oscillations in the scattered intensity traces of Figures 2.9 and 2.10. The transition from slow to fast oscillations at a critical temperature in the traces of Figure 2.9 can be explained by the dynamics of the wavepacket spreading, which behaves diffusively (i.e., proportional to $z^{1/2}$) in the case of viscously damped N-waves (c.f. Section 2.2.4) and goes with a constant velocity for soliton trains (c.f. Section 2.2.3).

We conclude that the measured Brillouin intensity can be explained well using numerical simulations based on solving the KdV-Burgers equation. We now arrive at the second approach, namely the explanation of our low-temperature data using exclusively the analytical KdV framework of Section 2.2.3. In this method we interpret the Brillouin scattering intensity in terms of spatial resonance and Bragg reflections of the light scattered from the moving soliton train. This approach was worked out into detail by Muskens and Dijkhuis.[2]

The Brillouin intensity plots of Figure 2.10 consist of an intricate pattern that can be unraveled by performing a spatial Fourier transform of the experimental traces. The highest frequencies in the pattern are due the characteristic oscillations during the first millimeters. By counting the number of these oscillations per millimeter, we can determine the typical frequency contained in the oscillation pattern. This quantity, which we denote as the spatial beating frequency ν_x, can be determined for traces at different pump intensities and for different scattering angles. These two degrees of freedom yield a complete picture of the behavior of ν_x, as can be shown in Figure 2.11. This figure shows the dependence of the typical beating frequency on the selected Brillouin frequency ν_B for some selected pump intensities E. We observe a clear relation between the oscillation frequency in the scattering pattern and the frequency of the selected Brillouin component. The dependence is approximately linear, with a slope increasing with pump fluence. This slope can be used to derive information on the velocities and amplitudes of the strain solitons, as we show below.

The presence of oscillations in the Brillouin scattering traces can be explained quantitatively by a simple model based on optical interference of solitons moving at different velocities.[2] It can be easily deduced that two objects moving with a velocity difference Δc in the moving frame system produce a beating in the scattered intensity against distance with a spatial frequency given by

$$\nu_x = \frac{\Delta c \nu_B}{c_0^2}. \tag{2.27}$$

The dependence of ν_x on the Brillouin frequency in Equation 2.27 agrees well with that observed in Figure 2.11, and we can fit the experimental data to obtain

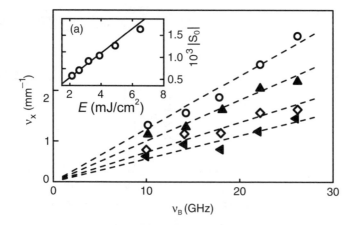

FIGURE 2.11 Maximum spatial frequencies v_x determined from Brillouin traces at frequencies v_B for several values of pump fluence E. Lines (dash) denote fits using Equation 2.27. (a) Initial strain values s_0 obtained from fitting the experimental data. (From Muskens, O.L. and Dijkhuis, J.I., *Phys. Rev. B*, 70, 104301, 2004. With permission.)

values for the velocity difference Δc. Considering that the highest beating frequency is produced by the interference of the fastest solitons and the linear parts of the wavepacket, the expressions relating velocity and amplitude derived below Equation 2.19 can be used to link these velocity differences to the soliton amplitude, which for sapphire gives $\Delta c/c_0 = 0.61|a_1|$. Using further that the amplitude of the first soliton a_1 is two times that of the initial strain pulse s_0 (c.f. Equation 2.20), we can deduce the strain from the slopes of Figure 2.11. The resulting values for s_0 obtained from the fits at different pump fluences are shown in Figure 2.11a and show good agreement with those values obtained from direct numerical reproduction of the Brillouin traces at 22 GHz (c.f. Figure 2.10a).

2.4.4 DIFFRACTION AND THE FORMATION OF SOLITONS

The above discussion considered the nonlinear propagation of millimeter-wide, nanometer-thick acoustic disks. The simulations of Chapter 3 predict negligible diffraction under these conditions. This situation changes as soon as the initial acoustic beam width becomes of the order of several micrometers. We have attempted to investigate experimentally the propagation of an initially narrow strain profile. For this purpose, we focused down the ultrafast pump laser to a measured waist of 12 μm. We have chosen to study a different system, namely lead molybdate [001] covered with a gold transducer, mainly because of the superior acousto-optic properties of the lead molybdate compared to the sapphire used in the one-dimensional experiments.

The sample is a $9 \times 11 \times 10$-mm³ crystal of lead molybdate (PbMoO₄), with the [001]-direction aligned perpendicular to the 9×11-mm² surface. A 500-nm gold film is deposited onto this surface. Small-area, low-amplitude acoustic pulses can be generated using the mode-locked output of the system ("oscillator" in Figure 2.12

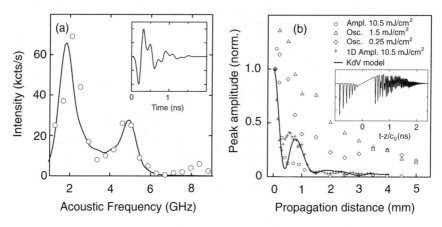

FIGURE 2.12 (a) Acoustic power spectrum of a strain packet generated in a gold film with a 70-nm chromium overlayer. (o) denotes experimental points; line is fit using the wavepacket shown in (inset). (b) Peak amplitude of the small-area (12-μm wide) strain pencils against propagated distance in the lead molybdate crystal, both for high-power ("amplifier") and low-power ("oscillator") excitation, together with an evolution trace of a large-area acoustic beam at high power. Inset shows developed wavepacket corresponding to the one-dimensional fit (line) in (b).

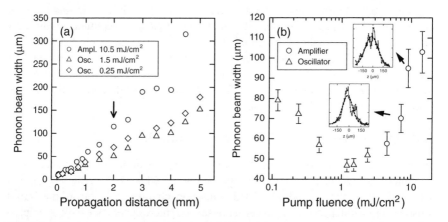

FIGURE 2.13 (a) Acoustic beam 1/e halfwidth against propagated distance for the three pump fluences 0.25, 1.5, and 10.5 mJ/cm². Arrow denotes position where the dependence of (b) is determined. (b) Phonon beam width against pump fluence at $z = 2$ mm. (Insets) typical transverse scans of the profile at high pump fluences.

and Figure 2.13), carrying 7 nJ/pulse at a repetition rate of 80 MHz. The amplified laser pulses can also be focused to the same 12-μm waist after severe attenuation to study the regime of high-intensity, small-area excitation.

Figure 2.12a shows a typical power spectrum of the initial acoustic wavepacket in the crystal, after a 100-μm traveling distance. The spectrum can be fitted with a bipolar waveform consisting of multiple reflections of 45% from the gold-lead molybdate interface. The large temporal width of the pulses implies that soliton formation will only start up after a significant amount of self-steepening of the packet has taken place. In the following we concentrate on a single frequency component of the wavepacket, which we have chosen to be the sub-maximum at 5 GHz. The propagation and diffraction of this component for a small-area initial packet was followed through the crystal at three settings of the pump-laser fluence, covering the entire dynamic range of our experiment. The peak amplitude in the center of the acoustic beam and the width of the strain profiles are shown in Figure 2.12b and Figure 2.13a, respectively. At the highest pump fluence under study, the evolution of a large-area, nondiffracting packet was also measured for comparison. From the figures we can observe a significant difference in evolution at the three pump fluences under study. We know from simulations based on the KdV equation that, in absence of diffraction, the initial decay of the power at low frequencies can be attributed to self-steepening of the wave packet.[3] Comparison of the two traces at 10 mJ/cm² shows, for both, a fast decrease of peak intensity, but for the "small-area" trace followed by an oscillation of lower amplitude than in the "large-area" curve. The decrease of the amplitude of oscillation makes us believe that part of the initial spectral power is diffracting away from the central part of the beam. Evidence of this can also be found in Figure 2.13a. We observe an increase in beam-width of almost an order of magnitude within the first millimeters of propagation. Further, the diffraction is observed to depend on pump fluence. This point is investigated further in Figure 2.13b, which shows the dependence of the beam width after 2 mm propagated distance against pump fluence. Clearly the behavior can be separated into two regimes of decreasing and increasing beam divergence, respectively, with increasing pump fluence. The two regimes are distinguished by a minimum in beam divergence found at a fluence of 1.5 mJ/cm². At the highest intensities, we have further observed structures of smaller dimensions inside the broad "envelope" [insets of Figure 2.13 (right)], which indicate that the profiles consist of different contributions in the radial distribution of the wave packet.

The behavior observed in Figure 2.13b can be explained qualitatively by the influence of nonlinear propagation of the total packet on the diffraction of individual frequency components. As we show below, two regimes can be identified: namely, (1) suppression of diffraction due to self-steepening and frequency up-conversion, and (2) defocusing of nonlinear shock fronts and the solitons. This is made clear by a series of simulations presented in Figure 2.14, showing the calculated propagation of a typical initial wave packet (upper graph, Figure 2.14a) that is consistent with the observed frequency spectrum in lead molybdate. We used the estimated values of the acoustic nonlinearity and phonon dispersion for lead molybdate estimated by Muskens and Dijkhuis.[51] Simulations were limited to a maximum traveling distance of 1.5 mm and to an initial beam profile of 15-μm halfwidth. Qualitatively, however, the results can be compared to the above diffraction experiment.

Figure 2.14a shows the developed state of the two-dimensional wavepacket for three values of the initial acoustic strain s_0. Transverse profiles of the spectral strain component at 5 GHz were obtained from the time-domain data by performing Fourier transforms at each radial distance. The resulting profiles, shown in Figure 2.14b, demonstrate that the radial distribution of individual strain components depends

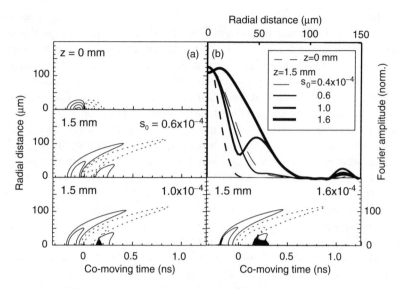

FIGURE 2.14 Calculated diffraction of an acoustic wavepacket in lead molybdate [001] after 1.5 mm of propagation. (a) Radial profiles of the strain waveforms: solid lines denote negative strain contours and dashed lines denote positive strains. (b) Absolute Fourier component at 5 GHz of the wavepackets of (a) against radial distance.

strongly on the nonlinear propagation of the total wavepacket. The profile at the lowest strain amplitude corresponds to the form obtained by linear Fraunhofer diffraction of the 5-GHz component (c.f. Equation 2.26) and shows a factor of two increase in beam width relative to the initial profile. With increasing strain amplitude we find initially a *decrease* of the beam width. However, at slightly higher strain values, a strong *increase* of the profile width appears. This increase of the width is accompanied by a typical oscillating structure in the transverse profile that resembles the observed behavior at the highest pump fluences in Figure 2.13b.

Comparison with the corresponding time-domain traces of Figure 2.14a relates the behavior to the different stages in nonlinear and dispersive development of the wavepacket. The two lowest strain values, which showed the reduction of divergence with strain amplitude, differ only in the amount of self-steepening. At these intensities, no dispersive features such as solitons and an oscillating tail have yet developed. We therefore conclude that self-steepening leads to a reduction in beam divergence.

In the simulations at higher strain amplitudes, we do observe signatures of dispersive development. An oscillating tail emerges and we can even distinguish one or two solitons separating at the front. The Fourier-transform profiles taken from these simulations show a broadening of the beam width and the appearance of some structure in the transverse profile. We thus conclude that the combined nonlinear dispersive behavior is accompanied by an enhanced divergence of the acoustic beam.

These simulation results can be understood more intuitively from the behavior of Equation 2.25. At moderate strain amplitudes, the initial effect of nonlinearity is to weakly couple the spectral modes of the strain wavepacket. A mutual up- and down-conversion of frequency components is responsible for self-steepening and the generation of higher harmonics, as was observed in the one-dimensional simulations of Figure 2.5. As the higher frequencies diffract less, the divergence of lower-frequency components will also be reduced due to the nonlinear coupling of modes. At higher strain amplitudes, the appearance of appreciable nonlinear velocity differences and breakup into dispersive structures will result in the conversion of amplitude variations into a velocity profile, resulting in enhanced (positive) curvature of the wavefront. The concomitant appearance of strong transverse gradients then results in a flow of energy from the center of the acoustic beam toward the edges of the profile, as was observed in the simulations of Section 2.3.2.

The presented work gives a qualitative idea of the influence of nonlinearity and dispersion on the diffraction of a narrow strain beam. On the other hand, the simulations presented in Section 2.3.2 have explored the influence of diffraction of the soliton development. The two together provide a first impression on the stability of nonlinear waves against transverse disturbances. This could be of key importance for future applications of nano-ultrasonic waves in imaging and manipulation of condensed matter. More detailed experimental verification of these complex interactions — achievable, for example, using time-resolved, two-dimensional imaging — will have to be performed to obtain deeper insight into the development of acoustic solitons in more dimensions.

2.5 CONCLUSIONS AND PROSPECTS

We have investigated, both theoretically and experimentally, the nonlinear development of high-intensity acoustic pulses through a single crystal of macroscopic dimensions. Brillouin spectroscopy turned out to be a useful technique for studying shock wave and soliton development in the bulk of a transparent crystal. We used the decay of the scattered intensity as a signature of self-steepening and analyzed the subsequent oscillation pattern in terms of spatial resonances of a moving soliton train. The spatial beating frequencies found in the oscillation pattern can be used as a measure for the different velocities in the train, which translate directly into estimates for the number and width of the solitons using either numerical simulations or the analytical framework of the Korteweg-de Vries model.

Further, we have explored the influence of diffraction on the nonlinear development of a narrow strain pencil. Comparison with simulations shows that self-steepening tends to decrease divergence due to the formation of nondiffracting, higher-frequency components. However, as the wavefronts curve due to the nonlinear velocity difference between the center and exterior regions of the beam, the influence of diffraction on the nonlinear solitons is enhanced. This effect results in a transverse instability and eventual the decay of narrow soliton discs upon propagation in the crystal.

One might argue for another application of Brillouin scattering in the analysis of soliton trains. In principle, it would be possible to determine the different velocities in the wavepacket directly from the frequency shifts of the light scattered from the individual solitons. For the backscattering condition, this shift amounts to several gigahertz for the typical solitons found in this chapter. The condition for spectral

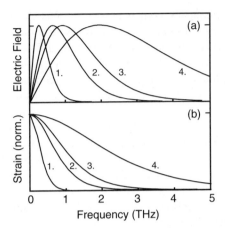

FIGURE 2.15 (a) Predicted terahertz electromagnetic spectrum from the reflection of a single-strain soliton at a piezoelectric interface, calculated for the four substrate materials of (b). (b) Calculated single-soliton spectra for 0.4% strain solitons generated in (1) silicon [100], (2) quartz [0001], (3) sapphire [0001], and (4) MgO [100].

separation of the backscattered radiation for each *individual* soliton can be easily fulfilled by the spectral resolution of our interferometer. This allows, in principle, the detection of the Brillouin peaks for the individual solitons in the train. The detection of these individual soliton contributions will be a topic for future experiments.

At this point we want to speculate about some possible new avenues and perhaps applications for ultrafast high-amplitude strain pulses in solids.

Of course, one can ask the question whether it if feasible to generate transverse soliton trains. One possibility, perhaps, is to inject the strain pulses at an angle to the symmetry axis of the crystal. This will lead to both longitudinal and transverse components of the injected strain packets that will quickly separate during propagation and may develop into trains of longitudinal and transverse solitons. Alternatively, one can think of mode-converting a developed longitudinal soliton train by reflection at a surface and choosing a suitable angle between the soliton beam and the normal to the surface.

Another promising direction might be to generate terahertz solitons in piezoelectric crystals and convert them, by reflection at a crystal surface, into intense electromagnetic terahertz pulses. Figure 2.15 shows the typical terahertz electromagnetic spectrum that could be obtained using the strain solitons formed in some typical materials. These terahertz electric fields should be detectable by electro-optic sampling. Conversion of ultrasound into microwaves is well known in the field of ultrasonics but is never demonstrated in the terahertz regime simply because coherent terahertz sound waves were not available until now. On the one hand, such schemes might allow for a direct analysis of the time-dependent strain in the soliton train and on the other hand permit the generation of strong terahertz fields at surfaces that can be instrumental in studying electrons in nanostructures on surfaces, or perhaps to address elements in micro-photonic devices.

A quite fundamental question is to what extent a strain soliton in a resonant medium can transform to a resonant soliton. From soliton experiments in optically excited ruby, it appeared that the coupling is strong enough that the typical Bloch angle for the $2\bar{A}(^2E) \rightarrow \bar{E}(^2E)$ phonon transition at 870 GHz reaches values of a few degrees.[3] In case the two-level systems are prepared inverted, coherent amplification will take place, as in a laser amplifier, and the Bloch angle might exponentially grow until angles exceeding π are reached and stable resonant 2π-solitons can be formed. Here one could create a *soliton phonon laser* at terahertz frequencies. Such pulses can be used to study *self-induced transparency* of phonons or to produce *phonon echoes* in resonant media. Again, such phenomena are already demonstrated in ultrasonic experiments at microwave frequencies but have not been shown in the terahertz domain.

A fundamental limit of soliton propagation can be reached in materials with a low sound velocity, like lead molybdate; here it is possible to create solitons with a width approaching the lattice constant.[52] As a result, dispersion becomes particularly large and the description in terms of discrete lattice model becomes a prerequisite.

Finally, it is of interest to devise schemes in which the soliton reaches pressures close to the internal pressure of the crystal and study properties of matter under extreme conditions and create *nano-explosions*. Thanks to the high-frequency content of the soliton pulses, focusing strain pulses to extremely small spots might be possible. To examine these effects, it is relevant to study the nonlinear propagation of spherical wave fronts employing the numerical techniques described in this chapter to see to what extent geometrical focusing can enhance the strains in the pulses to the desired levels. Experimentally, spherical wave fronts can be generated by evaporating the transducer on a surface of the crystal that is prepared with a spherical shape.

We expect high-amplitude nano-ultrasonics to develop further the direction of technological applications. Possibly, the concept of ultrashort strain solitons will play an important role in this. The ability to convert energy from an initially broad packet into sharp, localized pulses is a unique feature that may find its way into manipulation and characterization of condensed matter. The feasibility of generating soliton pulses up to more elevated temperatures will certainly increase the range of potential applications in science and technology.

Acknowledgments

The authors wish to thank P. Jurrius and C.R. de Kok for their technical assistance. This work was supported by the Netherlands Foundation Fundamenteel Onderzoek der Materie (FOM) and the Nederlandse Organisatie voor Wetenschappelijk Onderzoek (NWO).

References

1. Muskens, O.L. and Dijkhuis, J.I., High amplitude, ultrashort, longitudinal strain solitons in sapphire, *Phys. Rev. Lett.*, 89, 285504, 2002.
2. Muskens, O.L. and Dijkhuis, J.I., Inelastic light scattering by trains of ultrashort acoustic solitons in sapphire, *Phys. Rev. B*, 70, 104301, 2004.

3. Muskens, O.L., Akimov, A.V., and Dijkhuis, J.I., Coherent interactions of terahertz strain solitons and electronic two-level systems in photoexcited ruby, *Phys. Rev. Lett.*, 92, 335503, 2004.

4. Tucker, J.W. and Rampton, V.W., *Microwave Ultrasonics in Solid State Physics*, 1st ed., Elsevier, New York, 1972.

5. Grill, W. and Weis, O., Excitation of coherent and incoherent terahertz phonon pulse in quartz using infrared laser radiation, *Phys. Rev. Lett.*, 35, 588, 1975.

6. Shiren, N..S., Measurement of signal velocity in a region of resonant absorption by ultrasonic paramagnetic resonance, *Phys. Rev.*, 128, 2103, 1962.

7. Jacobsen, E.H. and Stevens, K.W.H., Interaction of ultrasonic waves with electron spins, *Phys. Rev.*, 129, 2036, 1963.

8. Tucker, E.B., Amplification of 9.3-kMc/sec ultrasonic pulses by maser action in ruby, *Phys. Rev. Lett.*, 6, 547, 1961.

9. Bron, W.E. and Grill, W., Stimulated phonon emission, *Phys. Rev. Lett.*, 40, 1459, 1978.

10. Hu, P., Stimulated emission of 29 cm^{-1} phonons in ruby, *Phys. Rev. Lett.*, 44, 417, 1980.

11. Sox, D..J., Rives, J.E., and Meltzer, R.S., Stimulated emission of 0.2-THz phonons in $LaF_3:Er^{3+}$, *Phys Rev. B*, 25, 5064, 1982.

12. Van Miltenburg, J.G.M. et al., Stimulated emission and decay of phonons resonant between the Zeeman states of $E(^2E)$ in ruby, in *Phonon Scattering in Condensed Matter*, Eisenmenger, W., Labman, K., and Döttinger, S., Eds., Springer, New York, 1984, p. 130.

13. Overwijk, M.H.F., Dijkhuis, J.I., and de Wijn, H.W., Superfluorescence and amplified spontaneous emission of 29-cm^{-1} phonons in ruby, *Phys. Rev. Lett.*, 65, 2015, 1990.

14. Thomsen, C. et al., Coherent phonon generation and detection by picosecond light pulses, *Phys. Rev. Lett.*, 53, 989, 1984; Thomsen, C. et al., Surface generation and detection of phonons by picosecond light pulses, *Phys. Rev. B*, 34, 4129, 1986.

15. Wright, O.B. and Kawashima, K., Coherent phonon detection from ultrafast surface vibration, *Phys. Rev. Lett.*, 69, 1668, 1992.

16. Bosco, C.A.C., Azevedo, A., and Acioli, L.H., Laser-wavelength dependence of the picosecond ultrasonic response of a NiFe/NiO/Si structure, *Phys. Rev. B*, 66, 125456, 2002.

17. Baumberg, J.J., Williams, D.A., and Köhler, K., Ultrafast acoustic phonon ballistics in semiconductor heterostructures, *Phys. Rev. Lett.*, 78, 3908, 1997.

18. Wright, O.B. et al., Ultrafast carrier diffusion in gallium arsenide probed with picosecond acoustic pulses, *Phys. Rev. B*, 64, 081202, 2001.

19. Yahng, J.S. et al., Probing strained InGaN/GaN nanostructures with ultrashort acoustic phonon wave packets generated by femtosecond lasers, *Appl. Phys. Lett.*, 80, 5223, 2002.

20. Sugawara, Y., et al., Watching ripples on crystals *Phys. Rev. Lett.*, 88, 185504, 2002.

21. Vertikov, A. et al., Time-resolved pump-probe experiment with subwavelength lateral resolution, *Appl. Phys. Lett.,* 69, 2465, 1996.

22. Duquesne, J.-Y. and Perrin, B., Ultrasonic attenuation in a quasicrystal studied by picosecond acoustics as a function of temperature and frequency, *Phys. Rev. B*, 68, 134205, 2003.

23. Daly, B.C. et al., Imaging nanostructures with coherent phonon pulses, *Appl. Phys. Lett.*, 84, 5180, 2004.

24. Hao, H.-Y. and Maris, H.J., Study of phonon dispersion in silicon and germanium at long wavelengths using picosecond ultrasonics, *Phys. Rev. Lett.*, 84, 5556, 2000.

25. Hao, H.-Y. and Maris, H.J., Dispersion of the long-wavelength phonons in Ge, Si, GaAs, quartz and sapphire, *Phys. Rev. B*, 63, 224301, 2001.
26. Hao, H.-Y. and Maris, H.J., Experiments with acoustic solitons in crystalline solids, *Phys. Rev. B*, 64, 064302, 2001.
27. Bradley, C.C., *High pressure Methods in Solid State Research*, 1st ed., Plenum Press, New York, 1969.
28. Trunin, R.F., *Shock Compression of Condensed Materials*, 1st ed., Cambridge University Press, Cambridge, 1998.
29. Love, A.E.H., *Treatise on the Mathematical Theory of Elasticity.*, 1st ed., Cambridge University Press, Cambridge, 1937.
30. Murnaghan, F.D., *Finite Deformations of an Elastic Solid*, 1st ed., John Wiley & Sons, New York, 1951.
31. Korteweg, D.J. and de Vries, G., On the change of form of long waves advancing in a rectangular canal, and on a New Type of Long Stationary Waves, *Philos. Mag.*, 39, 422, 1895.
32. Zabusky, N.J. and Kruskal, M.D., Interaction of "solitons" in a collisionless plasma and the recurrence of initial states, *Phys. Rev. Lett.*, 15, 240, 1963.
33. Wallace, D.C., in *Solid State Physics*, Ehrenreich, H., Seitz, F., and Turnbull, D., Eds., Academic, New York, Vol. 25, p. 301, 1970.
34. Auld, B.A., *Acoustic Fields and Waves in Solids*, Vol. 1., 2nd ed., Robert E. Krieger Publishing Company, Malabar, FL, 1990.
35. Kittel, C., *Introduction to Solid State Physics*, 6th ed., John Wiley & Sons, New York, 1986.
36. Whitham, G.B., *Linear and Nonlinear Waves*, 1st ed., Wiley, New York, 1974.
37. Karpman, V.I., *Non-linear Waves in Dispersive Media*, 1st ed., Pergamon Press, New York, 1975.
38. Gardner, C.S. et al., Method for solving the Korteweg-deVries equation, *Phys. Rev. Lett.*, 19, 1095, 1967.
39. Landau, L.D. and Lifschitz, E.M., *Quantum Mechanics, Non-relativistic Theory*, 2nd ed., Pergamon Press, New York, 1965.
40. Fermi, E., Pasta, J., and Ulam, S., Technical Report LA-1940, Los Alamos, NM, 1955.
41. Toda, M., *Theory of Nonlinear Lattices*, 1st ed., Springer-Verlag, Berlin, 1981.
42. Potapov, A.I. et al., Nonlinear interactions of solitary waves in a 2D lattice, *Wave Motion*, 34, 83, 2001.
43. Meijer, H.G.E., Master's thesis, University of Utrecht, 2003.
44. Kadomtsev, B.B. and Petviashvili, V.I., On the stability of solitary waves in weakly dispersive media, *Soviet Phys. Dokl.*, 15, 539, 1970.
45. Infeld, E., Senatorski, A., and Skorupski, A.A., Numerical simulations of KP soliton interactions, *Phys. Rev. E*, 51(4), 3183, 1995.
46. Driscoll, T.A., A composite Runge-Kutta method for the spectral solution of semilinear PDE, *J. Comp. Phys.*, 182, 357, 2002.
47. Balogh, A. and Krstic, M., Boundary control of the KdV-Burgers equation, *IEEE Trans. Autom. Contr.*, 45(9), 1739, 2000.
48. Damen, E.P.N. et al., Generation and propagation of coherent phonon beams, *Phys. Rev. B*, 64, 174303, 2001.
49. Singhsomroje, W. and Maris H.J., Generating and detecting phonon solitons in MgO using picosecond ultrasonics, *Phys. Rev. B*, 69, 174303, 2004.
50. Damen, E.P.N, Arts, A.F.M., and De Wijn, H.W., Experimental verification of Herring's theory of anharmonic phonon relaxation: TeO_2, *Phys. Rev. B*, 59(1), 349, 1999.

51. Muskens, O.L. and Dijkhuis, J.I., Propagation of ultrashort acoustic wave packets in PbMoO$_4$ studied by Brillouin spectroscopy, *Physica B*, 316-317, 373, 2002.
52. Muskens, O.L. and Dijkhuis, J.I., Trains of ultrashort acoustic solitons, *Phys. Stat. Sol.*, *Phys. Stat. Sol. (b)*, 241(15), 3469, 2004.

3 Nonlinear Optical Properties of Artificial Dielectrics in the Nano-Scale

H. Grebel

CONTENTS

3.1 Preface 49
3.2 Introduction 50
3.3 Theoretical Considerations 53
3.4 Experiments and Results..61
3.5 Discussion 66
3.6 Conclusions 67
References..67

3.1 PREFACE

Artificial dielectrics constitute a class of man-made materials; the effective permittivity and permeability of a given dielectric material are altered by embedding small clusters of another material, typically nano-size metal clusters. Additional dielectric properties are achieved as a result of local interactions between the material components. Such composition was used over many centuries as a method to stain glass: in the case of gold-embedded glass, the result is a red, winelike color. This concept was extended to include nano-size semiconductor embedded dielectrics, conditional artificial dielectrics,[1] to emphasize the dependence on control parameters such as light. It is tempting to take this concept one step further and attempt to alter the capacitance and inductance of dielectric materials by geometrical means. Such features should be of sizes much smaller than the propagating electromagnetic wavelength to give the material new effective properties.[2] For example, ordered arrays of nano-size spheres offer a practical route to develop periodic nanostructures on a sub-wavelength scale; by using these structures as templates for other, nonlinear materials, one opts to achieve novel optical characteristics.

We follow the advent of artificial dielectrics through the nonlinear optical properties of periodic nanostructures and, in particular, those embedded with nano-size semiconductor nano-materials ("nano within nano"). Such a concept can also be employed indirectly to include vibration states; for example, when embedding Raman-active molecules on metal-coated periodic nanostructures, invoking surface-enhanced Raman spectroscopy (SERS).

3.2 INTRODUCTION

The optical properties of a coherent array of spheres have gained increased interest in recent years (Figure 3.1a and Figure 3.1b for millimeter-size and nano-size arrangements, respectively). The combined scattering effect from individual spheres and the overall coherent scattering effect of the periodic three-dimensional structure strongly contribute to its nonlinear optical transmission properties.[3] These spheres are typically made of glass or polymers, which by themselves do not exhibit strong nonlinear characteristics. The nonlinear optical properties of the structure are therefore attained by incorporating into it other, relatively strong nonlinear optical materials. Of particular interest are low-dimensional nano-size materials such as quantum dots or quantum wires, where the optical nonlinearity is enhanced by electronic wave confinement. We call this concept "nano within nano." One can show that the periodic structure substantially contributes to the enhancement of nonlinear optical properties of the embedded material. For example, one could compare the nonlinear coefficients of quantum silicon dots embedded in either a flat silica film or in a film made of a coherent array of silica spheres; we demonstrated an enhancement by more than four orders of magnitude when these quantum dots were incorporated into the structured film.[4,5]

In view of the above, photonic crystals[6] comprise a class of three-dimensional periodic structures that frustrate wave propagation in either all or some directions. In essence, the composite structures portray the combined effects of Mie and Bragg scattering. Such strong scattering leads to highly dispersive characteristics, namely, the relationship between the frequency and the wavevector is nonlinear and enables us to design and realize linear imaging elements thinner than the propagating wavelength.[7] As previously suggested, the optically confining environment of these structures is particularly attractive when it is implanted with nonlinear materials such as semiconductors,[8] erbium ions,[9] or embedded with single-wall carbon nanotubes (SWCNTs).[10] The quantum dot-based structure is achieved by ion implantation of semiconductor materials into the self-assembled structure of a silica-based array of spheres. Upon annealing, the ion-implanted material precipitates into small (on the order of a few nanometers) nano-crystallites, hence the name "nano within nano." The combination of nonlinear properties of the quantum dots with the strong dispersion of the matrix leads to controllable short pulse broadening or even pulse compression.[11]

Tapping into a particular nonlinear effect depends on the nonlinear nano-material of choice. One can form quantum dots of semiconductor by ion implantation followed by annealing. The size of the dots is on the order of a few nanometers, with a reasonable 20% distribution about the mean. An important size measure is the Rydberg length, below which a nano-crystallite can be considered amorphous. This

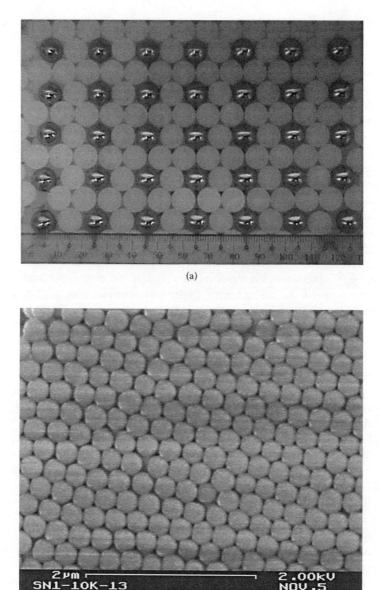

(a)

(b)

FIGURE 3.1 Coherent array of spheres: (a) macroscopic version of a composite made of millimeter-size polymer and metal spheres; (b) nano-size version of silica-based spheres.

size is 3 nm for silicon and 8 nm for germanium. Erbium, on the other hand, remains in ion form surrounded by silica molecules.

Nano-size silicon clusters (Si-nc) exhibit strong nonlinearity around 2.4 eV, which can be attributed to indirect transitions to surface states formed as a result of the interface between the silicon nano-crystal and the surrounding oxide. In contrast, Ge nanoclusters (Ge-nc) feature several direct transitions in the wavelength region between 0.5 and 1.8 μm. Of particular interest is the transition around 2.2 eV, which is size independent owing to mid-Brillouin transitions.[12] In the case of erbium, the ion surrounded by oxide has a transition in the green and in the near-infrared (near-IR) (most notable are the transitions at 0.98 and 1.54 microns). When pumped in the green, the overall transmission of artificial dielectric composite is changed via the Kerr effect; there exist nonlinear loss and nonlinear refraction components. The loss is due to absorption of the nonlinear material within the composite. Yet, in esu units, the overall nonlinear refraction of the composite is about 10 times larger than the nonlinear loss owing to the light-confining effect of the periodic matrix. Photon localization in the periodic structure in addition to near resonance transition result in an index change of $\Delta n = 0.25$ at relatively modest intensities of a few tens of MW/cm^2.

Single-wall carbon nanotubes (SWCNTs) are one-dimensional crystals formed by a rollover of single grapheme sheets. The result is a tubular structure whose electronic wavefunction is, on one hand, confined by the monolayer thickness and on the other hand by the circumference of the tube. The electrons are free to move along the axis of the tube. The way the grapheme sheet is rolled about its axis (chirality) determines the conductive or semiconductive nature of the tube. Chiral tubes, which roll at an angle with respect to the tube axis, are generally semiconductors; namely, they portray an energy gap for all possible wavevectors. In addition, the one-dimensional nature of the tube results in singularities in the density of states. Transitions are made between these singularities at discrete wavelengths. Because the widths of the singular point of density of states are quite narrow, one can expect band-filling effects as the intensity of the pump light increases. The perfect nature of the one-dimensional crystal and the absence of large electron-electron or electron-phonon interaction results in an almost ballistic conductance. In contrast to the case of semiconductor quantum dots-embedded composites, the nonlinear properties of SWCNT-embedded structures are mostly determined by nonlinear absorption rather than by nonlinear refraction.[13] As a function of intensity, the nonlinear curve portrays a small nonlinear absorption followed by a large bleaching effect. The bleaching occurs at a particular intensity value and the entire transmission curve has a step-like characteristic. Such behavior may well be employed in optical switching or optical limiting devices.

When the periodic assembly of dielectric spheres is coated with metal, couplings can be made to charge propagating waves. In general, periodic structures are useful when mode-coupling between two wavevectors is required. Examples are coupling between an optical mode in free space and a guided mode, or the coupling of a free space mode to a surface charge mode. The latter process is used in surface-enhanced Raman spectroscopy (SERS). SERS is a modified version of Raman spectroscopy; the usually weak signal of a nonresonant, spontaneous Raman line is amplified via coupling of the Raman-active, optical phonons to localized electric fields. The probed molecule is adsorbed onto a rough metallic silver, gold, or copper surface to enable such amplification.[14] One can argue that in the case of strong coupling between the nonlinear properties of the embedded material and the scattering properties of the

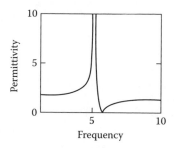

FIGURE 3.2 Permittivity vs. normalized frequency for 5% clusters volume in glass media. The resonance frequency was set to 5. Note the asymmetrical curve.

periodic matrix may define the composite as an artificial dielectric of sorts. Such an artificial dielectric is interrogated by light and exhibits gain provided by the feedback mechanism of the structure. The Raman spectra of SWCNT are now well understood to be resonance-enhanced, near density-of-state singularities. Conventional SERS result for SWCNTs adsorbed on aggregates of silver particles.[15–18] Yet, and as we will see below, when placed on periodic metallic surface, a *selective* amplification of vibration state is achieved.[19]

3.3 THEORETICAL CONSIDERATIONS

Artificial dielectrics have been associated with two basic effective models: (1) the Garnett theory[20,21] and (2) the Lorentz approach.[22,23] The Garnett theory is an effective medium theory that deals with low dispersion of clusters, mostly metal clusters, in a dielectric host material. The linear polarizability of the cluster is written as $\alpha = R^3(\varepsilon_m - \varepsilon_d)/(\varepsilon_m + 2\varepsilon_d)$, with R being the cluster radius. The subscripts m and d are used for metal and dielectric host material, respectively. Using the Clausius-Mossotti relation, $(\varepsilon_{eff} - \varepsilon_d)/(\varepsilon_{eff} + 2\varepsilon_d) = 4N\alpha/3$, with N being the number density of clusters, we write the effective permittivity, ε_{eff}, as

$$\varepsilon_{eff} = \varepsilon_d + 3f\varepsilon_d(\varepsilon_m - \varepsilon_d)/(\varepsilon_m + 2\varepsilon_d). \tag{3.1}$$

Here, f is the fill factor, $f = 4NR^3/3 \ll R^3/\alpha$. For metals, ε_m is a large negative number and a resonance occurs whenever $Re\{\varepsilon_m + 2\varepsilon_d\} = 0$. The latter condition heavily depends on the optical frequency used, as seen from Figure 3.2, in which $\varepsilon_m = \varepsilon_0[1 - (\omega_p/\omega)^2]$ with ω_p being the plasma frequency. The nonlinear permittivity change is related to the local field E_i by, $\delta\varepsilon_m = 12\chi^{(3)}|E_i|^2$. The local field is related to the external field E_0 via $E_i = 3\varepsilon_d/(\varepsilon_m + 2\varepsilon_d) \equiv gE_0$, where g is the "gain" coefficient. The nonlinear polarization is written as

$$P_{NL}^{(3)}(\omega) = 3fg^2|g|^2\chi^{(3)}|E_0|^2E_0. \tag{3.2}$$

Any intensity dependence is entering this equation through the factor g. Frequency dependence is made through g (because $\varepsilon_m = \varepsilon_m(\omega)$) and the nonlinear

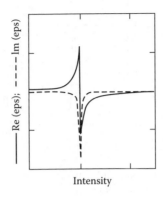

FIGURE 3.3 Real and imaginary values of the nonlinear permittivity as a function of intensity.

susceptibility $\chi^{(3)}(\omega)$. The nonlinear model is applicable to SWCNTs due to its high conductance and may be applicable to semiconductor clusters at high levels of induced carriers.

The Lorentz model assumes that the polarization at any given point is arising from couplings between neighboring dipoles — those of the matrix and those of other clusters. With the total electric field being the sum of the external and internal electric fields, respectively, $E_T = E_0 + E_i$; the internal field E_i is related to the polarization of the cluster p as $E_i = Cp/\varepsilon$; the polarization p is related to the electric polarizability, α_e, as, $p = \alpha_e \varepsilon E_T$. The total polarization, $P = Np$, is written in a self-consistent manner as $P = N\alpha_e \varepsilon (E_0 + Cp/\varepsilon) = N\alpha_e \varepsilon E_0/(1 - N\alpha_e C)$. Because the displacement field is related to the total field as $D = \varepsilon_{eff} E_T$, we get,

$$\varepsilon_{eff} = \varepsilon(1 + N\alpha_e E_0/(1 - N\alpha_e C). \tag{3.3}$$

The nonlinear polarization can be developed if we let the coupling constant and the polarization be nonlinear; thus, $C(E) = C_0 + C_1|E_0|^2$ and $p = \alpha_e^{(1)} \varepsilon E_T + \alpha_e^{(3)} \varepsilon|E_T|^2 E_T$. In a self-consistent manner, $P = N_e^{(1)} \varepsilon E_0/[1 - N\alpha_e C(E)] + 3\alpha_e^{(1)} N\alpha_e^{(3)} \varepsilon|E_0|^2 E_0/[1 - N\alpha_e^{(1)} C(E)]^4$. To a first order of approximation,

$$P_{NL}(\omega) \sim N\alpha_e \varepsilon E_0/(1 - N\alpha_e(C_0 + C_1|E_0|^2). \tag{3.4}$$

A plot of normalized P_{NL} is shown in Figure 3.3. The experimental nonlinear index of refraction change is shown in Figure 3.4. The nonlinear absorption of Figure 3.3 is very narrow and could be easily missed (see, for example, Ref. 5).

The transmission through an ordered array of silica spheres is a combined effect of resonance scatterings from individual and from strongly coupled spheres (that is, Mie and Bragg scatterings combined). To gain intuitive insight into this complicated three-dimensional problem, it might be useful to consider the propagation of a wave through a stack of two-dimensional screens (Figure 3.5a). Each screen possesses a (transverse) resonance[24,25] that is related to the resonance of the individual feature (pixel, Figure 3.5b) and the periodic structure as a whole, and can be also viewed as a generalized case of transverse Bragg waveguides.[26] The coupling between

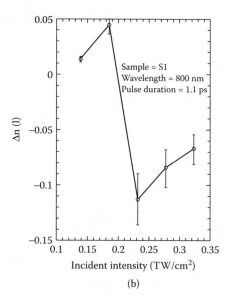

(b)

FIGURE 3.4 Nonlinear index of refraction change when nano-Si is embedded in a flat silica matrix. We use a Ti:sapphire laser at $\lambda = 800$ nm with pulse duration of 1.1 ps.[3] The nonlinear absorption was flat and may have missed the narrow characteristics of Figure 3.3.

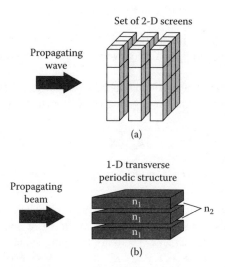

FIGURE 3.5 Representation of photonic crystal as a set of two-dimensional screens. Each box represents a unit cell (pixel) of the structure.

screens results in resonances along the direction of propagation, otherwise known as Fabry-Perot modes. Such a simplified approach is valid as long as the propagation of the wave does not deviate much from normal incidence.[27] In short, the screen is represented by a matrix, the screen's matrix; the propagation between screens is represented by another matrix; and the stack is represented by the block of these two matrices raised to the power of the number of screens. We note that there are two phase shifts associated with this setting: one is the phase delay as the beam transverses each screen[27] and the other is the obvious delay associated with the propagation between screens.

This concept is applied to our opalline structure as follows: First we assume that the three-dimensional structure is made of stacked, two-dimensional monolayers of spheres. The spheres in each "screen" form two-dimensional gratings (Figure 3.1). Such a picture holds accurate if the Fabry-Perot resonance modes are far from the transverse (lateral) ones. There exist "magic angles" where lateral propagation is frustrated, as shown in Figure 3.6, resulting in maximized transmission values. Because the lateral component in the structure is a standing wave, the corresponding longitudinal component *within the composite*, β_z, is $\beta_z = k_0 n_{eff} cos(\theta')$. On the other hand, momentum conservation dictates that $(\beta_z)^2 = (k_0 n_{eff})^2 - (\beta_x)^2$, with $2\beta_x = K = 2\pi/\Lambda$ and Λ is the pitch of the transverse grating. Empirically, one may write (note that θ' is the refracted angle inside the composite and θ is the angle of incidence):

$$k_0 n_{eff} sin(\theta) \cong K. \qquad (3.5)$$

Here, k_0 is the propagation constant in vacuum, n_{eff} is the effective refractive index of the media, and K is the crystallographic wavenumber. The effective refractive index is typically estimated using the relationship

$$n^2_{eff} = (1 - F)n^2_{cl} + Fn^2_d, \qquad (3.6)$$

with n^2_{cl} and n^2_d being the permittivity of the cladding material (often air) and the dielectric material (often glass), respectively. Here, F is the fill factor; $F = 0.74$ for an fcc structure.

Another look at the same problem is as follows. The gratings, with a wavenumber K, affect the dispersion of the propagating optical beam with a wavenumber k_n. At resonance,

$$|k_n - K| \sim k_n. \qquad (3.7a)$$

or

$$(k_{nx} \pm K_x)^2 + (K_{n2} \pm K_2)^2 = k_n^2 \qquad (3.7b)$$

Thus, a resonance occurs at specific incident angles. Again, and to simplify matters, we treat the cross-section of the ordered array of silica spheres as composed of stacks of two-dimensional screens. There are two perpendicular gratings: the longitudinal periodicity is $m_z 0.816D = 2\pi/K_z$, due to close packing along the $\langle 111 \rangle$ direction (the z-direction), with D being the diameter of each sphere. The transverse periodicity, the t-direction, has two gratings in directions $\langle 100 \rangle$ and $\langle 110 \rangle$. The transverse periodicity is either $m_t D/2 = 2\pi/K_t$ where $m_t = 1, 2, 3, ...$, or $m_t 0.816D = 2\pi/K_t$. One combination contributes to a transmission-only mode; for example, with $D = 300$ nm, $n_{eff} = 1.22$, $m_z = 1$ (for a monolayer of Er-implanted silica spheres) and $m_t = 2$ in the $\langle 100 \rangle$ direction. Two "magic" incident angles exist: $\theta = 7.7°$ and $\theta = 48.8°$. The latter is in excellent agreement with experiments.[9]

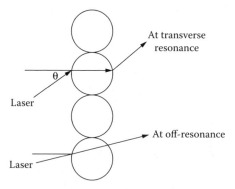

FIGURE 3.6 Schematics of propagation at on and off transverse resonance.

Such approaches can be easily extended to the nonlinear regime by letting the refractive index be intensity dependent. Specifically, $n(I) = n_0 + n_2 I$, where $n(I)$ is the intensity-dependent refractive index and n_0 and n_2 are constants, respectively. The structure is treated in a thick hologram limit, namely, $2\pi\lambda L/n_{eff}\Lambda^2 \gg 1$. Here, L is the screen thickness and Λ is the pitch. The pitch value Λ depends on the crystallographic orientation (either [100] or [110] planes). In the experiments, we launched light into the resonance mode bounded by the (100) planes. The resonance occurs when the pitch equals an integral number of the opal sphere's radius, D, or, $\Lambda = sD$, with $s = 1, 2, 3, ...$ Each local mode was launched with a relative phase shift determined by the incidence angle θ. Thus, the output radiation is a result of a coherent sum of all local modes. The output radiation will propagate in the same direction as the input radiation if the relative phase shift between the local modes is preserved. The far-field transmission of this phase screen is proportional to the square of the Bessel function of the first kind, $J_0^2(M(I))$, with an intensity-dependent modulation function $M(I)$. At transverse resonance conditions, the transverse propagation is frustrated, as shown in Figure 3.6. The modulation function is the difference between phases of the local mode propagating in the medium and the local mode propagating in its surrounding air: $M(I)^{resonance} = (\beta_z(n,I) - \beta_z(1,0))L$, which manifests the localization characteristics of photons in this structure. The nonlinear index is approximated by $n(I)^{resonance} = n_{sphere} + A \cdot tanh(B \cdot I)$, due to saturation in the nonlinear absorption of the nano-clusters.

The modulation function at off-resonance conditions is written again as the difference between the phases of waves propagating in the media and in air. Thus, $M(I)^{off-resonance} = k(n(I)^{off-resonance} - 1)L/cos(\theta)$. Here, θ is the incident angle. The corresponding nonlinear refractive indices, $n(I)^{resonance}$ and $n(I)^{off-resonance}$ differ only by their gain coefficient B, which is larger at resonance conditions and thus lowering the corresponding intensity level. Their saturation level A is related to the average properties of each individual cluster and is therefore independent of the incident angle.

The theoretical results are presented in Figure 3.7. In this particular example, we examined an ion-implanted Ge opalline structure. Typically, the implanted region is thin and occupies a thickness on the order of a sphere. Here, the screen thickness

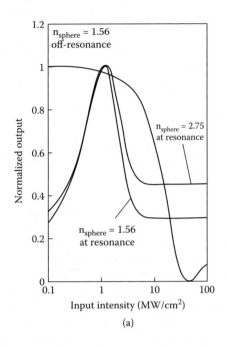

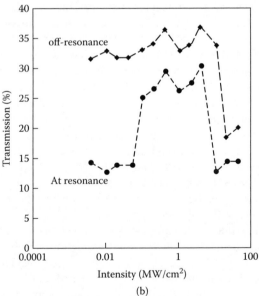

FIGURE 3.7 (a) Theoretical transmission curves as a function of peak laser intensity. (b) Experimental data at on and off-resonance conditions.

was taken as $L = 2D$ because some implantation of the next row of spheres took place through the voids of the opalline structure. The grating pitch is taken as $\Lambda = D = 0.3\ \mu\mathrm{m}$. The saturation level of the nonlinear refractive index, $A = 0.25$, is taken as a free parameter whose value can be justified by nonlinear measurement at other wavelengths. The remainder of the opalline structure contributes little to the nonlinear process and will result in an overall multiplicative factor to the transmission value. Because this constant is unknown (it depends on the overall sample thickness and on scattering from structural defects), the graphs are normalized to unity.

Such a modal approach comes in handy as we consider the effect of structural dispersion. The propagation of an electromagnetic wave in an opalline structure is affected by three main dispersion factors: (1) intramodal dispersion, whereby the various frequency components of the pulse propagate at different phase velocities. This dispersion is considered negative; namely, the high-frequency components of the pulse precede the lower ones. (2) Intermodal dispersion, whereby the same frequency component of different modes propagate at different phase velocities. Linear intermodal dispersion leads to self-imaging effects.[7] This dispersion type is also considered negative. (3) Nonlinear material and structural dispersion, whereby the nonlinearity adds frequency chirp to the propagating pulse.

The nonlinear effect in a grating system affects the temporal as well as the spatial dispersion. Owing to the large linear index of refraction difference, Δn, between the silica spheres in the opalline structure and its surrounding air, $\Delta n/n = 1/3$, the transverse Bragg condition is maintained for a large spectral bandwidth, $\Delta \omega = 2\pi \cdot 8 \cdot 10^{13}$ rad·sec^{-1}, larger than the frequency width of the pulse itself, $\Delta \omega_p = 2\pi \cdot 0.53 \cdot 10^{13}$ rad·sec^{-1}. Thus, the transverse propagation constant of the entire spectra of the qth mode can be made to be at resonance with the periodic structure, namely, $\beta_x^{(q)} \sim k_0 n_0 \sin(\theta) = q(\pi/\Lambda)$, where $\Lambda = s\Lambda_0$, $s = 1, 2, 3, ...,$ is the corresponding transverse pitch and θ is the incident angle. The close-packed fcc structure contains another wavevector along the $\langle 011 \rangle$ direction and for which the transverse pitch is $\sqrt{2}\Lambda$.

The transverse coupling strength for a nonimplanted sample is $|\kappa_t| = \pi \Delta n/\lambda_0 \cos\theta$. With $\Delta n = 0.5$, we calculate, $\kappa_t \sim 2.7$ at $\lambda = 0.8$ μm. This means that the linear transverse coupling process is very efficient; a typical coupling process, which obeys the condition $\kappa L \sim \pi$ occurs within less than a transverse distance of 1.2 μm or four silica spheres. This value also implies that only the first few transverse planes (namely, $s = 1$ and $s = 2$) contribute to the propagation.

The propagation constant in the direction of propagation is $\beta_z^{(q)} = \{k_0^2 n_0^2 - [\beta_x^{(q)}]^2\}^{1/2}$. The time of arrival per unit length of a spectral line, ω, of mode number q is

$$\tau(\omega,q) = \partial\beta_z^{(q)}/\partial\omega = (n_0/c)/[1 - (q\lambda_0/2n_0\Lambda)^2]^{1/2}. \tag{3.8}$$

Here, we used n_0, the effective linear refractive index, and $k_n = \omega n_0/c = 2\pi/\lambda_n$. In Figure 3.8 we show the intermodal time delay per unit length of a spectral line λ of mode $q = 1$ for $n_{eff} = 1.35$ and 1.5, respectively. The maximum delay between all spatial modes is $\delta\tau_{Q0} = (n_0/c)([1 - (q_{max}\lambda_0/2n_0 s\Lambda_0)^2]^{-1/2} - 1)$, where q_{max} is the maximum mode number.

The intermodal time delay between the zeroeth and first spatial modes is shown in Figure 3.8. These time delays are, in general, sources of pulse broadening. However, one may note the larger effect of the intramodal dispersion for the higher-order spatial mode.

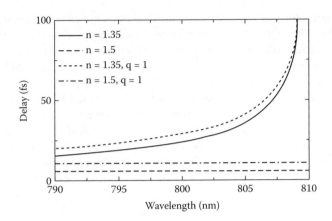

FIGURE 3.8 Intermodal delay time as a function of propagating wavelength for effective refractive indices $n = 1.35$ and $n = 1.5$.

Assume now that the pulse is Gaussian; that is, it can be written as $I(t) = I_0 exp[-2(t/\tau)^2]$. The instantaneous phase of the wave along a distance L is $\phi = \omega t - \beta_z^{(q)}L$. Assume also that the nonlinear refractive index can be written as $n(t) = n_0 + \gamma I(t) \equiv n_0 + \Delta n(t)$. The instantaneous frequency change as a function of time (frequency chirp) in a nonlinear homogeneous medium, or self-phase modulation, is ($\beta_z^{(q=0)} = k_0 n(t)$),

$$\partial\phi/\partial t = \Omega_0 = 2k_0\Delta n(t)Lt/\tau^2. \tag{3.9}$$

The frequency chirp in the presence of transverse gratings where $\beta_z^{(q)} = \{[k_0 n(t)]^2 - (q\pi/\Lambda)]^2\}^{1/2)}$, is

$$\partial\phi/\partial t = \Omega_q = [2k_0\Delta n(t)Lt/\tau^2]/[1 - (q\pi/\Lambda k_0 n(t))^2]^{1/2}. \tag{3.10}$$

Frequency chirp, normalized to the spectral pulse width, as a function of time for the first two spatial modes is plotted in Figure 3.9 for positive and negative values of $\Delta n(t)$, respectively. As seen from the figures, the higher-order mode has a larger impact on the frequency chirp. In addition, a negative nonlinear material has a larger impact than the positive one. The increase in the spectral width results in a larger delay times for the lower frequencies of the pulse. In Figure 3.14 we show the broadening of the pulse as it propagates through Si-implanted samples. The implanted region of the 3- to 4-nm silicon clusters is narrow, occupying a width of approximately 0.3 µm with $n_{eff} = 1.5$, compared with the total sample thickness of 10 µm with $n_{eff} = 1.35$. Thus, the temporal broadening of the pulse for the implanted samples can be attributed to the nonlinear frequency chirp in combination with the dispersive medium of the opalline structure.

Surface Charge Waves: Surface charge waves are excited according to the momentum conservation equation: $k_t + K + k_p = 0$, where k_t is the transverse wavevector of the excitation light, K is the wavevector of the crystallographic structure, and k_p is the surface plasmon wavevector. The excitation incident angle can be calculated as

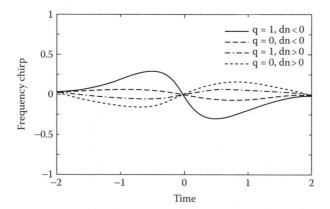

FIGURE 3.9 Frequency chirp as a function of normalized time for the first two modes of a grating medium. Parameters are: $k_0 L/\tau^2 = 1$; $|\Delta n(0)| = 0.075$.

$$\sin(\theta) = \pm\{\lambda/\Lambda - [\varepsilon_1\varepsilon_2/(\varepsilon_1 + \varepsilon_2)]^{1/2}\}, \tag{3.11}$$

where θ is the incident angle of incidence, λ/Λ is the ratio between the optical and structural wavelengths, ε_1 is the dielectric constant of air, and ε_2 is the dielectric constant of the metal. We used $\varepsilon_2 = -16.4 - j0.54$ and $\varepsilon_2 = -10.28 - j1.03$ for silver and gold, respectively. There are two major transverse crystallographic wavevectors for the face-centered-cubic array of the silica spheres along: (a) the $\langle 100 \rangle$ and (b) $\langle 110 \rangle$ directions. Based on Equation 3.11, we predict, $\theta = 25°$ for the (110) planes of silver- or gold-coated array made of 250-nm spheres. We also predict $\theta = 0°$ for (1/2,0,0) planes. The excited radiation wavelength also follows Equation 3.11 by replacing the excitation wavelength λ with the excited wavelength λ_e. Note that the direction of propagation being reversed to indicate back-reflection mode.

The perpendicular component of reflected wavenumber at surface plasmons resonance is

$$k_n = [k^2 - (k_t - sK)^2]^{1/2} = [k^2 - (k_p)^2]^{1/2} \tag{3.12}$$

with $s = \dots 3, 2, 1, 1/2, 1/4, \dots$ The resonance condition has been extended to superstructure, including every other or, more planes of reflection. The coefficient k_n is usually imaginary; the propagation is exponentially decaying away from the metallic recess region and most of the excitation energy is specularly reflected. However, when the metal is a good conductor, the surface charge wavenumber $k_p = k_0[\varepsilon_1\varepsilon_2/(\varepsilon_1 + \varepsilon_2)]^{1/2} \sim k_0$ and $k_n \ll 1$ — almost zero. This enables excitation of a standing wave within the structure voids.

3.4 EXPERIMENTS AND RESULTS

The substrate, composed of a three-dimensionally ordered array of silica spheres (commonly referred to as opals), was prepared according to a well-known procedure.[5] It

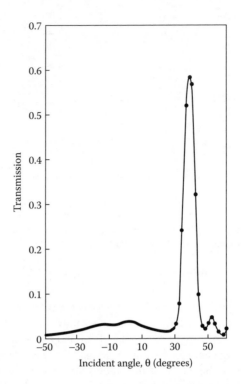

FIGURE 3.10 Enhanced linear transmission for SWCNT embedded in an ordered array of 275-nm size silica spheres. Data taken with a Nd:YAG laser at 532 nm.

was annealed at 600°C prior to embedding the nonlinear materials or, deposition of silver or gold layers on top of the silica spheres. Turbo pumps were used to evacuate these chambers to avoid formation of a carbon layer on top of the metallic surface.

Opals of various sphere sizes, ranging from 100 to 1000 nm, were prepared for the nonlinear experiments. Ion implantation of these structures was made at various energies and fluences. For example, ion implantation of the Ge ions was made at 150 keV and fluence levels ranged between 2×10^{14} and 2×10^{15} cm^{-2}. The corresponding Ge peak concentrations were estimated as 3×10^{20} cm^{-3} and 3×10^{21} cm^{-3}, respectively. At this energy, the range of implantation (depth below surface) in flat samples is 925 ± 266 Å; it may be a bit larger in opalline systems due to the local curvature of the silica spheres.

As suggested by the previous discussion, one expects a peak in the transmission curve as a function of incident angle whenever the propagating beam is transversely confined by the periodic structure of the opalline matrix. An example is shown in Figure 3.10 for SWCNT embedded in the voids of a 275-nm size sphere opalline structure. Clearly seen is the peak transmission around incident angle of $\theta = 40°$ in addition to a much smaller peak at $\theta = 7°$. The curve is asymmetric with respect to normal incidence, $\theta = 0°$, due to the relatively small coherent grain of the opalline structure (on the order of 20 to 30 μm) comparable to the spot size.

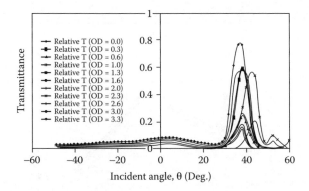

FIGURE 3.11 Nonlinear transmission of SWCNT-embedded, ordered array of 275-nm size silica spheres. Data was taken by varying the intensity of a Nd:YAG laser at 532 nm using neutral density filters.

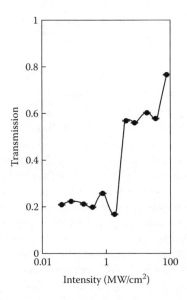

FIGURE 3.12 Intensity-dependent nonlinear transmission of SWCNT embedded in opalline structure. The plot is made of intensity values at the magic angle, $\theta \sim 40°$.

The nonlinear characteristics are measured simply by attenuating the incident beam with neutral density (ND) filters. Changing the laser intensity through the provided current control might result in beam profile change and should be avoided. The results are presented in Figure 3.11. The corresponding intensity-dependent transmission is plotted in Figure 3.12 with its steplike characteristics. A comparison with Ge implanted quantum dots is given in Figure 3.13.

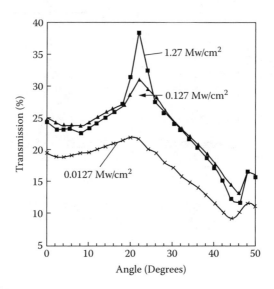

FIGURE 3.13 Nonlinear transmission for a sample of Ge-implanted, ordered array of 300-nm size silica spheres, near the "magic angle" of $\theta = 22°$ at various intensities. The average size of the Ge-nc was approximately 2 nm.

As suggested, a large broadening of ultrashort pulses is expected at resonance conditions by transverse confining planes. The pulse is broadened due to intermodal dispersion. Thus, one should also expect that the broadening of the pulse will nonlinearly depend on the pulse intensity as index changes affect the distribution of propagating modes. Indeed, this is the case for 3- to 4-nm size crystallites of Si embedded in an opalline matrix.[11] The change in the pulse broadening as a function of pulse peak intensity is shown in Figure 3.14. The initial broadening of approximately 60 fs is almost eliminated when a large, intensity-dependent index change of $\Delta n = -0.1$ removes the transverse confinement condition.

As argued, the nonlinear interaction between a periodic metallic surface and adsorbed molecules on it may constitute an artificial dielectric material and, specifically, if there is a selective amplification process as a result of the modified matrix structure. Such is the case of Raman spectroscopy of SWCNT on nano-patterned metallic surfaces. The Raman spectroscopy data depend on the crystallographic orientation of the ordered array of the "crystal" substrate with respect to the polarized laser beam. Shown in Figure 3.15 is the Raman signal at various incident angles. The typical signature of SWCNTs appears only at resonance angles for which the surface waves are excited. In Figure 3.16 we show a comparison between the data at resonance with the (110) planes of opal ($\theta = 25°$) and (1/2,0,0) planes ($\theta = 0°$). This behavior is also true for the scattered (frequency-shifted) Raman signal: signal amplification for the scattered light is detected at the corresponding angles (see Equation 3.11) when λ is replaced by the scattered (excited) wavelength$_e$.

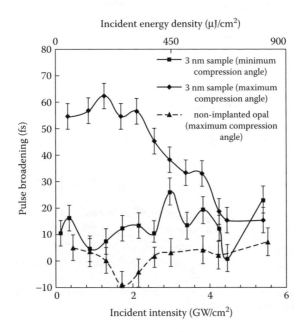

FIGURE 3.14 Nonlinear pulse broadening in 3- to 4-nm size Si crystallites embedded in an opalline matrix. As intensity increases, the confinement condition of the matrix is removed.

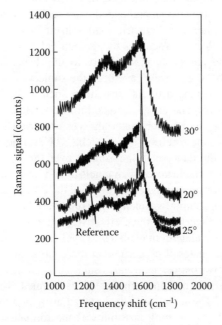

FIGURE 3.15 Raman signal from a single-wall nanotube excited at on- and off-resonance conditions. The resonance condition occurs at $\theta = 25°$.

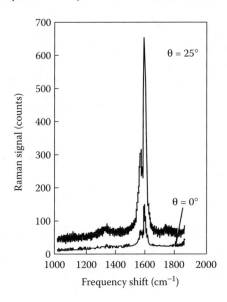

FIGURE 3.16 Comparison between signals obtained at resonances with the (110) planes (θ = 25°) and (1/2,0,0) planes (θ = 0°).

3.5 DISCUSSION

Artificial dielectrics exhibit novel optical characteristics when we invoke interactions between a dielectric matrix and an embedded nano-scale nonlinear material. Further development occurs when the dielectric matrix itself is shaped in the nano-scale level to change the dielectric properties by geometrical means. As such, the dispersion properties of the photon bear similarity to the dispersion properties of the electron. In fact, artificial materials have been the subject of investigation for many years.[28] To this class belong artificial structures such as multiple quantum wells (MQWs), quantum wires, and arrays of quantum dots. Most of these elements are based on inorganic semiconductor materials, yet improved synthesis has put the organic counterparts in the forefront of research.[29,30] Nevertheless, the key issues in device fabrication at the nano-scale are control over the position of the elements and the ability to form reliable array of channels to and from the elements. The issue is somewhat relaxed for optical materials because the critical length, determined by the corresponding propagating wavelength, is relatively large — on the order of a few hundreds of nanometers and, thus, effective properties on a sub-micron scale, yet, smaller than the wavelength may also find application (e.g., in optical limiters). Thus, control over the material properties on the nano-scale extends our ability to fabricate new and novel optical materials. Such an approach is enabled by nano-scale templates: a periodic opalline template is first formed. The template voids are filled with material A. The template is then etched away and the resultant voids are filled with material B. An exciting possibility is the formation of a distributed p-n junction, with material A being p-doped and material B being n-doped SWCNT.[31] Such an arrangement extends the two-dimensional concept of MQW to three dimensions.

Overall, the effective dielectric properties of materials can be altered significantly by geometrically controlling the dielectric matrix in which a nonlinear material is embedded. This has proven useful for quantum Si and Ge dots, and one-dimensional materials such as SWCNT. There exist angles for which the nonlinear transmission peaks. This occurs when the transverse propagation constant of the propagating beam couples to the opalline periodic structures at resonance conditions. Two such conditions exist in photonic crystals: (1) the wave becomes *transversely* confined by the periodic structure; and (2) the wave is locally confined to a resonant local mode of individual sphere (Mie scattering). The second condition requires a large index of refraction ratio between each sphere and the surrounding cladding (air in our case). Using these considerations, one might achieve a sizeable nonlinear effect: index changes as large as 0.25 have been demonstrated at modest peak power intensities.

To this class of artificial materials we now add a metallic periodic structure on the nano-scale. Coherent effects, such as surface charge waves, extend local characteristics to relatively large distances. Using scatterings from sub-wavelength structures, an almost noise-free signal can be obtained. We have demonstrated these effects in back-reflected SERS experiments on SWCNT bundles deposited on metal-coated opal substrates. Such experiments might help realize a new family of nano-patterned substrates for SERS characterization and detection of low concentration levels of nano-particles.

3.6 CONCLUSIONS

In summary, we may realize useful three-dimensional nanostructures by a combination of chemical and physical synthesis. Two applications for these structures have been described: (1) novel substrates for surface-enhanced Raman spectroscopy (SERS) to detect a small number of molecules, and (2) a new class of nonlinear optical structures that can be used for all-optical switches. The latter can also be fabricated at the tip of an optical fiber or on the top of an optical waveguide.

References

1. H. Grebel and P. Chen, *Opt. Lett.*, 15, 667, 1990.
2. C. Wu and H. Grebel, *J. Physics D: Appl. Phys.*, 28, 437, 1995.
3. M. Ajgaonkar, Y. Zhang, H. Grebel, and C.W. White, *Appl. Phys. Lett.*, 75, 1532, 1999.
4. S. Vijayalakshmi, H. Grebel, G. Yaglioglu, R. Pino, R. Dorsinville, and C.W. White, *J. Appl. Phys.*, 88, 6415, 2000.
5. Y. Zhang, S. Vijayalakshmi, M. Ajgaonkar, H. Grebel, and C.W. White, *J. Opt. Soc. Am. (JOSA) B*, 17, 1967, 2000.
6. E. Yablanovich, *Phys. Rev. Lett.*, 58, 2059, 1987.
7. J.M. Tobias and H. Grebel, *Opt. Lett.*, 24, 1660, 1999.
8. M. Ajgaonkar, Y. Zhang, H. Grebel, and R.A. Brown, *JOSA B,* 19, 1391–1395, 2002.
9. M. Ajgaonkar, Y. Zhang, H. Grebel, M. Sosnowski, and D. Jacobson, *Appl. Phys. Lett.*, 76, 3876, 2000.

10. AiDong Lan, Zafar Iqbal, Abdelaziz Aitouchen, Mattew Libera, and Haim Grebel, *Appl. Phys. Lett.*, 81, 433–435, 2002.
11. S. Vijayalakshmi, H. Grebel, G. Yaglioglu, R. Dorsinville, and C.W. White, *Appl. Phys. Lett.*, 78, 1754, 2001.
12. J. R. Heath, J.J. Shiang, and A.P. Alivisatos, *J. Chem. Phys.*, 101, 1607, 1994.
13. H. Han, S, Vijayalakshmi, A. Lan, Z. Iqbak, H. Grebel, E. Lalanne, and A.M. Johnson, *Appl. Phys. Lett.*, 82, 1458–1460, 2003.
14. M. Fleischmann, P.J. Hendra, and A.J. McQuillan, *Chem. Phys. Lett.*, 26, 163, 1974.
15. A.M. Rao, E. Richter, S. Bandow, B. Chase, P.C. Eklund, K.A. Williams, S. Fang, K.R. Subbaswamy, M. Menon, A. Thess, R.E. Smalley, G. Dresselhaus, and M. Dresselhaus, *Science,* 275, 187, 1997.
16. J. Azoulay, A. Debarre, A. Richard, P. Techenio, S. Bandow, and S. Iijima, *Chem. Phys. Lett.,* 331, 347, 2000.
17. K. Kneipp, H. Kneip, P. Corio, S.D.M. Brown, K. Shafer, J. Motz, L.T. Perelman, E.B. Hanlon, A. Marucci, G. Dresselhause, and M.S. Dresselhaus, *Phys. Rev. Lett.*, 84, 3470, 2000.
18. G.S. Duesberg, W.J. Blau, H.J. Byrne, J. Muster, M. Burghard, and S. Roth, *Chem. Phys. Lett.,* 310, 8, 1999.
19. H. Grebel, Z. Iqbal, and A. Lan, *Chem. Phys. Lett.*, 348, 203–208, 2001.
20. J.C.M. Garnett, *Philos. Trans. R. Soc. London, Ser. B,* 203, 385, 1904.
21. R. Landauer, in *Electrical Transport and Optical Properties of Inhomogeneous Media*, edited by J.C. Garland and D.B. Tanner, AIP Conf. Proc. No. 40 ~AIP, New York, 1978.
22. H.A. Lorentz, *Theory of Electrons*, Note 55, B.G. Teubner, Leipzig, 2nd edition, 1916.
23. R.E. Collin, *Field Theory of Guided Waves*, Wiley-IEEE Press, 2nd edition, 1990.
24. R. Ulrich, *Infrared Phys.*, 7, 37–55, 1967.
25. J.S. McCalmont, M.M. Sigalas, G. Tuttle, K.-M. Ho, and C.M. Soukolis, *Appl. Phys. Lett.*, 68, 2759–2761, 1996.
26. J.M. Tobias, M. Ajgaonkar, and H. Grebel, *J. Opt. Soc. Am. (JOSA) B*, 19, 285–291, 2002.
27. J. Shah, D. Moeller, H. Grebel, O. Sternberg, and J.M. Tobias, Three-dimensional metallo-dielectric photonic crystals with cubic symmetry as stacks of two-dimensional screens, accepted by *J. Opt. Soc. Am. (JOSA)* A, 2004.
28. P.N. Butcher and D. Cotter, *The Elements of Nonlinear Optics*, Cambridge University Press, 1990.
29. D.S. Chemla and J. Zyss, *Nonlinear Optical Properties of Organic Molecules and Crystals*, Academic Press, 1987.
30. A.J. Heeger, S. Kivelson, J.R. Schrieffer, and W.P. Su, *Rev. Mod. Phys.,* 60, 781, 1988.
31. H. Han, J. Chen, Y. Diamant, H. Grebel, M. Etienne, and R. Dorsinville, Novel nonlinear three-dimensional distributed nano-structures with single-wall carbon nanotubes, paper, *W Non Linear Optics Topical Meeting,* Hawaii, 2004.

4 Optical Properties of Hexagonal and Cubic GaN Self-Assembled Quantum Dots

Yong-Hoon Cho and Le Si Dang

CONTENTS

4.1 Introduction 70
4.2 Physical Properties of GaN Self-Assembled Quantum Dots......................71
4.3 Growth of Hexagonal and Cubic GaN Self-Assembled Quantum Dots 73
    4.3.1 MBE Growth of Hexagonal Phase GaN Self-Assembled
        Quantum Dots 73
    4.3.2 MOCVD Growth of Hexagonal Phase (In)GaN
        Self-Assembled Quantum Dots 74
    4.3.3 Growth of Cubic Phase GaN Self-Assembled Quantum Dots 74
 4.4 Optical Properties of GaN Self-Assembled Quantum Dots..........................75
    4.4.1 Optical Properties of Hexagonal and Cubic GaN
        Self-Assembled Quantum Dots 75
    4.4.2 Optical Properties of GaN Quantum Dots Doped with Rare
        Earth Materials..79
4.5 Time- and Space-Resolved Optical Studies on GaN Self-Assembled
    Quantum Dots 80
    4.5.1 Time-Resolved Optical Properties of GaN Self-Assembled
        Quantum Dots 80
    4.5.2 Space-Resolved Optical Properties of GaN Self-Assembled
        Quantum Dots 82
    4.5.3 Single Quantum Dot Spectroscopy Using Micro-PL and
        Micro-CL on GaN-Based Quantum Dots...84
4.6 Prospects of GaN Self-Assembled Quantum Dots 87
    4.6.1 Growth Challenges: Control of Quantum Dot Nucleation
        Sites and Sizes 87
    4.6.2 Control of Electronic Properties...89
    4.6.3 UV Bipolar Opto-Electronic Devices: Interband Transitions
        in GaN Quantum Dots 89

4.6.4   IR Unipolar Opto-Electronic Devices: Intersubband Transitions
in GaN Quantum Dots 90
4.6.5   White LEDs: Self-Assembled GaN Quantum Dots Doped
with Rare Earth Materials 91
4.7   Conclusion 91
4.8   Acknowledgments 92
References 92

4.1   INTRODUCTION

Quantum dots (QDs) are the ultimate three-dimensional confinement structures of carriers in semiconductor heterostructures, which can be regarded as artificial solid-state atoms, with discrete energy levels and reduced density of states (DOS). Their unusual electronic properties, mostly due to the nanometer scale of the confinement potential, have attracted increasing interest; a recent review of their physical studies and device applications can be found in the literature.[1-4] For example, it has been discussed that low-dimensional QDs play an important role in achieving lower threshold current, enhanced differential gain, reduced temperature dependence of threshold current, and no chirping and/or high spectral purity in semiconductor laser diodes (LDs).[5-9]

GaN-based QDs are attractive for both basic physics and novel device applications because they combine many unique features of zero-dimensional QDs and wide bandgap materials. With respect to other QD systems, nitride-based QDs exhibit special interests because of the following features: Nitrides are wide bandgap compound semiconductors that have specific electronic properties, such as stronger coupling to LO phonons, heavier effective masses, stronger excitonic (Coulomb) effects, etc. In addition, nitride materials are good hosts for carrier-mediated energy transfer to rare earth ions, particularly at elevated temperatures.

Due to a large bandgap difference in InN/GaN and GaN/AlN (>2.5 eV) heterostructures, III-nitride systems can be applicable for not only visible-UV photonic device applications based on interband transitions, but also near-IR applications for optical telecommunications using intersubband transitions. Because of large amounts of spontaneous and piezoelectric polarizations of nitride-based heterostructures in the wurtzite crystalline phase, the quantum confined Stark effect (QCSE) plays a dominant role instead of confinement effects. Because QDs with both wurtzite (hexagonal) and zinc-blende (cubic) phases can be grown, one can have a direct comparison of QCSE and confinement effects. Most of all, due to the lack of adapted substrate for epitaxial growth of nitride nanostructures, GaN-based QDs are an "ideal" system to develop emitters with high radiative quantum efficiency and better performance.

This chapter reviews the physics and optical properties of GaN-based self-assembled QDs, which are much less well-known than their II-VI (CdTe/ZnTe, CdSe/ZnSe) or III-V (InAs/GaAs) semiconductor counterparts. The chapter is organized in the following way. The unique physical properties and applications of GaN-based QDs are described in Section 4.2. Then, recent progresses in the growth of hexagonal and cubic phase GaN self-assembled QDs by molecular beam epitaxy (MBE) and metal-organic chemical vapor deposition (MOCVD) are described in Section 4.3. Sections 4.4 and 4.5 discuss the optical properties and carrier dynamics of hexagonal and cubic GaN self-assembled QDs. Section 4.6 discusses some growth and application prospects of GaN-based QDs. Finally, Section 4.7 summarizes the main points.

4.2 PHYSICAL PROPERTIES OF GAN SELF-ASSEMBLED QUANTUM DOTS

In a QD, the electronic states are quantized and the energy levels become discrete. The spatial confinement of electrons and holes along all three dimensions leads to a discrete energy-level structure with sharp optical absorption lines. In this sense, a semiconductor QD can be regarded as an artificial solid-state atom. On the other hand, this naive picture is oversimplified because, by contrast to atoms, electrons and holes in QDs are coupled to their environment through interaction with lattice phonons. As a result, QD optical transitions should become broader at high temperature and quenched eventually by thermal activation of carriers.

Synthesis of group-III nitride nanostructures has attracted considerable attention because of their potential applications in high-performance opto-electronics and information processing.[9–11] With respect to other QD systems such as III–V arsenides or II–VI tellurides and selenides, the GaN/AlN QDs present some unique features, which derive from the wide bandgap of this material system and from the fact that their thermodynamically stable crystalline phase is of wurzite symmetry. First, these wide bandgap QDs are particularly well adapted for any carrier-mediated energy transfer process, such as the one involved in the excitation of rare earth ions in semiconductors. Next, the bandgap difference between GaN and AlN exceeds 2.5 eV. This ensures not only exceptional stability for carrier confinement at elevated temperatures, but also intersubband electronic transitions at near-IR wavelengths of interest for optical fiber telecommunications.

It has been reported that the built-in internal macroscopic polarizations, which consist of the spontaneous polarization caused by charge accumulations between two constituent material interfaces and the piezoelectric polarization due to lattice-mismatch-induced strain, play an important role in carrier recombination in the hexagonal-phase (wurzite) InGaN/GaN and AlGaN/GaN QW structures.[12–16] For hexagonal GaN/AlN QDs grown on (0001) substrates, this polarization induces a large built-in internal electric field of the order of 7 MV/cm along the height direction of QDs ([0001] or c-axis).[17] Because of the quantum confined Stark effect (QCSE), the interband transition energy can be varied from the visible to UV spectral range, depending on the QD size. Moreover, the electron and hole wavefunctions are well separated along the direction of the c-axis, resulting in strong modifications of electronic and optical properties.

Figure 4.1 shows the low-temperature photoluminescence (PL) spectra of two samples of hexagonal GaN/AlN QDs of large size ("with ripening," QD height of 4.1 nm) and small size ("without ripening," QD height of 2.3 nm). The emissions appear below and above the GaN bandgap energy, respectively, and the energy

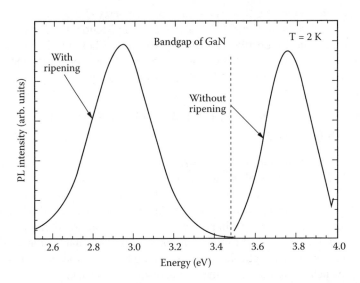

FIGURE 4.1 Low-temperature PL spectra of two samples of hexagonal GaN/AlN QDs of large (with ripening) and small sizes (without ripening). The PL peaks appear below and above GaN bandgap energy, respectively.

difference is as much as 0.8 eV. By contrast, the PL spectra of cubic QDs are always observed at higher energy than the GaN bandgap energy. The origin of this remarkable spectral variation in hexagonal QD interband transitions is the QCSE, due to the presence of a giant internal electric field (F) of ~7 MV/cm induced by the spontaneous and piezoelectric polarizations along the QD height (c-axis). Neglecting electron–hole Coulomb interaction, the interband transition energy is well-described by $E(d) = E_g + E_e + E_h - Fd$; where E_g is the bandgap energy of GaN, E_e (E_h) is the electron (hole) confinement energy, and d is the QD height. In practice, E_e and E_h do not vary much with the QD size for $d \geq 2$ nm because electrons and holes are repelled by the strong electric field at opposite interfaces and confined in a (similar) triangular profile potential. Therefore, the interband transition energy $E(d)$ would linearly decrease with increasing d. A theoretical study of electron-hole electronic states and recombination in hexagonal truncated pyramidal-shaped GaN/AlN QDs has also been reported.[18] It is based on a multiband $k \cdot p$ model that takes into account the three-dimensional strain and the built-in internal field distributions. It shows that *both* the PL emission wavelength and the radiative lifetime (τ_{rad}) for the ground-state transition strongly increase with QD size due to the influence of the strong built-in internal field. The calculated value of τ_{rad} changes from the order of a few nanoseconds to microseconds by a small variation of QD height from about 2 to 4 nm. The total exciton lifetime (τ_{tot}) of about 1 to 3 ns is estimated by $\tau_{tot} = \tau_{rad} \cdot \tau_{non-rad}/(\tau_{rad}+\tau_{non-rad})$ with a nonradiative lifetime ($\tau_{non-rad}$) of 3.6 ns extracted from the experimental data of Ref. 19.

The spatial separation of electrons and holes by the internal electric field in hexagonal QDs should strongly modify their other electronic properties as well. For example, Coulomb repulsion effects should be important because carriers of the same sign will be localized at the same interface. In fact, negative binding energy of biexciton has been reported recently in single QD spectroscopy of GaN/AlN QDs.[20] The dipolelike distribution of excitons in QDs should enhance their Fröhlich coupling interaction with LO phonons. One would also expect an amplified spectral diffusion induced by spatial fluctuations of nearby charges,[21] which is a drawback for single QD spectroscopy. Finally, of great interest in nitride QDs is the possibility of also growing QDs in the cubic phase. The latter do not exhibit any spontaneous polarization, which allows one to compare their electronic properties to those of hexagonal QDs and single out confinement effects from QCSE.

4.3 GROWTH OF HEXAGONAL AND CUBIC GAN SELF-ASSEMBLED QUANTUM DOTS

The lattice mismatch between GaN and AlN is about 2.4%. The strain-driven Stranski-Krastanow (SK) growth mode, in which elastic energy stored in the two-dimensional GaN layer is released by the spontaneous formation of three-dimensional islands, and other growth modes, such as those making use of surface energy, have been demonstrated for GaN-based QDs on Al(Ga)N layers. Several epitaxial growth methods have been extensively used to fabricate GaN-based QDs, including MBE,[13,22–39] MOCVD,[40–51] and other techniques.[52–54]

4.3.1 MBE GROWTH OF HEXAGONAL PHASE GaN SELF-ASSEMBLED QDS

Self-assembled GaN/AlN QDs by the SK growth mode have been grown for the first time by Daudin et al.[22] using MBE and a radio-frequency plasma cell for active N. Under N-rich conditions and a growth temperature around 730°C, the SK growth mode transition is induced after the growth of about 2.5 monolayers of GaN on AlN. QDs are truncated pyramids exhibiting (10–13) facets, with a base/height aspect ratio of about 5. Depending on the growth conditions, the QD heights are typically 2 to 5 nm, and their densities range from 5×10^{10} to 5×10^{11}/cm². Figure 4.2 shows a TEM image of a stack of hexagonal GaN QD layers.[55] The existence of a wetting layer, specific to the SK growth mode, is clearly visible. A marked "vertical" correlation is observed between the nucleation sites of QDs located on different layers. This self spatial organization is due to the propagation of the strain field generated by QDs in the surrounding barrier layers.[56]

Recently, self-assembled GaN QDs were also grown on AlN under Ga-rich conditions.[57,58] It is shown that for a certain Ga flux excess, the GaN layer is covered by a self-regulated two-monolayer-thick Ga layer.[59] This Ga bilayer completely modifies the surface energy of the GaN epilayer and inhibits the SK transition toward the spontaneous formation of QDs. Instead, a layer-by-layer growth is maintained for the GaN epilayer, well beyond two or three monolayers, with a progressive strain relaxation by generation of misfit dislocations. Then, growth interruption under vacuum will result in the evaporation of the surface Ga layer and the subsequent SK transition to form self-assembled QDs. It is interesting to note that growth under very high Ga excess, in fact under the Ga droplet regime, allows for a drastic reduction in the QD density, down to 3×10^8/cm².[57]

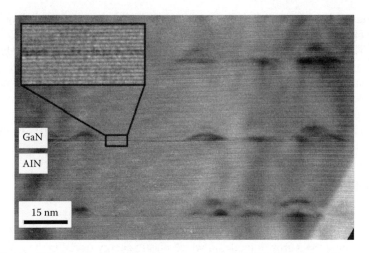

FIGURE 4.2 TEM image of a stack of hexagonal phase GaN QD layers. QDs are formed on top of a wetting layer and exhibit a strong vertical correlation effect.

The size control of QDs has been demonstrated with MBE growth of GaN on AlN using NH_3 as the nitrogen active source.[25] It was shown that, by playing with the GaN coverage and growth interruption, it is possible to increase the QD size to obtain light emission from blue to green and orange through the QCSE.

4.3.2 MOCVD GROWTH OF HEXAGONAL PHASE (IN)GAN SELF-ASSEMBLED QDs

The SK growth mode with and without "anti-surfactant" has been achieved by MOCVD. The first MOCVD-grown nitride QDs were obtained using Si atoms as anti-surfactants.[11,40,41,60] To grow QDs, Si is first supplied onto the AlGaN surface, before a short supply of Ga and N. Si acts as a random nano-mask so that QDs will be grown on the other regions without the nano-mask. GaN or InGaN QDs were successfully obtained with the SK growth mode without using the anti-surfactants. Structural and optical properties of the GaN QDs and InGaN QDs by the SK growth mode were characterized using transmission electron microscopy (TEM), Raman scattering, PL, and micro-PL techniques.[21,43,44,48,61-64] Stimulated emission and lasing action were demonstrated by an optical pumping method.[41,65] Low-pressure MOCVD has also been used to grow self-assembled GaN QDs on an AlN layer.[50,66] It requires a very low V/III flux ratio, much smaller than for conventional two-dimensional GaN growth. The QD base/height aspect ratio is about 10, with a typical height of 2 nm, and densities could be as low as $5 \times 10^9/cm^2$. An interesting result is the strong vertical correlation effect observed for stacked QD layers, which tends to increase the QD size and reduce its distribution.[66]

4.3.3 GROWTH OF CUBIC PHASE GAN SELF-ASSEMBLED QDs

Concerning the GaN QDs in zinc blende (cubic) phase, there are much fewer growth studies. The most significant results have been obtained by MBE using radio plasma

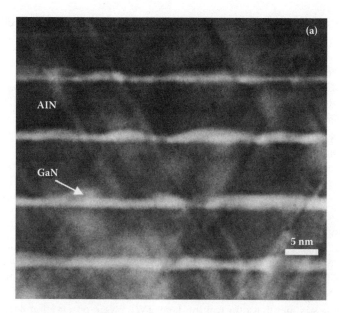

FIGURE 4.3 TEM image of a stack of cubic phase GaN QD layers. QDs are formed on top of a wetting layer and surrounded by stacking faults.

source for active nitrogen and cubic 3C-SiC pseudo-substrates grown by CVD on (001) Si substrates.[29,36] Growth was carried out at 720°C under stoichiometric conditions for III and V fluxes. After depositing about three monolayers of GaN on AlN, a clear SK transition could be observed, corresponding to the formation of self-assembled QDs. TEM images reveal that QDs are formed on top of a wetting layer of about two monolayers thick and exhibit a base/height aspect ratio of about 8 and rather small heights of 1.6 nm. By contrast to the wurtzite (hexagonal) case, TEM images show that these QDs are surrounded by {111} stacking faults (see Figure 4.3).[29,36] The high stacking fault densities could induce a broad distribution of QD size because they can lead to local relaxation and, thus, to an inhomogeneous strain distribution on the epilayer surface. Nevertheless, the strong PL observed at room temperature suggests that the structural quality should be high inside cubic QDs.

4.4 OPTICAL PROPERTIES OF GAN SELF-ASSEMBLED QUANTUM DOTS

4.4.1 OPTICAL PROPERTIES OF HEXAGONAL AND CUBIC GaN SELF-ASSEMBLED QDs

Understanding the optical properties and carrier recombination mechanisms in the GaN self-assembled QDs is important not only for the physical interest in atomic-like confined systems, but also in designing practical visible and UV light-emitting applications with better performance and quantum efficiency. Comparisons between the optical properties of hexagonal-phase, wurtzite GaN (*h*-GaN) and cubic-phase, zinc-blende GaN (*c*-GaN) QDs allows us to discriminate giant built-in polarization field effect in *h*-GaN QDs from the pure three-dimensional confinement effect in both *h*-GaN and *c*-GaN QDs.

The *h*-GaN and *c*-GaN QD samples presented here were grown in SK mode on a 1.0-μm-thick AlN buffer layer by plasma-assisted MBE.[35,37] The 6H-SiC and chemical vapor-deposited 3C-SiC on Si were used as the substrates for the growth of *h*-GaN and *c*-GaN QDs, respectively. For some *h*-GaN QD samples, separate confinement heterostructures were fabricated, consisting of five stacks of GaN QDs with 15-nm-thick $Al_{0.25}Ga_{0.75}N$ spacers, sandwiched by 45-nm-thick $Al_{0.25}Ga_{0.75}N$ optical waveguide layers, and capped by a 0.12-μm-thick AlN layer. The QDs studied here have a typical height of 3 nm, with a diameter-to-height aspect ratio of 5 and 10 for the *h*-GaN and *c*-GaN QDs, respectively. The typical QD density was in the $10^{11}/cm^2$ range.

Figure 4.4 shows 10K PL spectra of the *h*-GaN QD separate confinement heterostructure. In this study, the PL spectra were measured at excitation wavelengths of 266 and 325 nm, using a xenon lamp to excite above and below the bandgap of $Al_{0.25}Ga_{0.75}N$ layers, respectively. The near-band-edge emission of the $Al_{0.25}Ga_{0.75}N$ waveguide layers and that of a wetting layer (WL) were observed at 4.07 and 3.87 eV, respectively, for the 266-nm excitation. The peak energies of GaN QD emission were 3.57 and 3.55 eV at the excitation wavelength of 266 and 325 nm, respectively. The relative red shift for 325-nm excitation is primarily due to the weaker absorption of smaller QDs for this excitation wavelength.

Figure 4.5 shows the temperature-dependent PL spectra of the same *h*-GaN QD sample excited by the 266-nm line of a fourth harmonic generated Nd:YAG laser over the temperature range from 10 to 300K. The total PL intensity from the GaN QDs is reduced by only one order of magnitude from 10 to 300K, while that from $Al_{0.25}Ga_{0.75}N$ layers almost disappears at 300K, so the PL efficiency of GaN QDs is much better than that of 2D $Al_{0.25}Ga_{0.75}N$ layers due to strong carrier localization in GaN QDs. Keeping in mind that, in this experiment, carriers are mostly created in $Al_{0.25}Ga_{0.75}N$ layers by the 266-nm excitation, part of the PL quenching of QDs at room temperature could be due to a reduced efficiency of carrier capture in QDs with increasing temperature. If this effect is neglected, then the temperature dependence of the QD PL intensity can be attributed to the thermal activation of carriers trapped in QDs. A curve fitting for the data in Figure 4.5 gives an activation energy of about 126 meV at T > 175K for GaN QDs. By considering the energy difference between GaN QD emissions and the WL (more than 300 meV), the thermal quenching of the GaN QD-related emission may be due to the thermal activation of electrons and/or holes from the QD ground states into the excited states (or band tail states of the WL) followed by nonradiative recombination processes. For comparison, we also measured temperature-dependent PL spectra for the *h*-GaN QW sample that was carefully grown under the same growth condition as that of the *h*-GaN QDs except for the growth time of the GaN active region. It was found that the PL intensity of the *h*-GaN QW emission is reduced by two orders of magnitude from 10 to 300K, thus indicating stronger carrier localization in *h*-GaN QDs compared to *h*-GaN QWs.[33,35] This behavior can be explained by (1) the QD density (in the $10^{11}/cm^2$

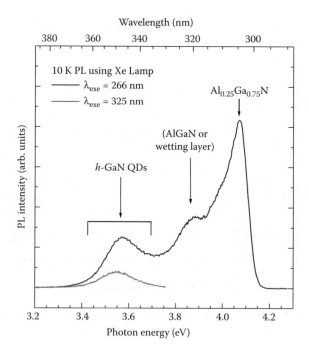

FIGURE 4.4 PL spectra at 10K of a separate confinement heterostructure consisting of 5 stacks of *h*-GaN QDs sandwiched by $Al_{0.25}Ga_{0.75}N$ optical waveguide, excited by a Xe lamp at 325 nm and 266 nm.

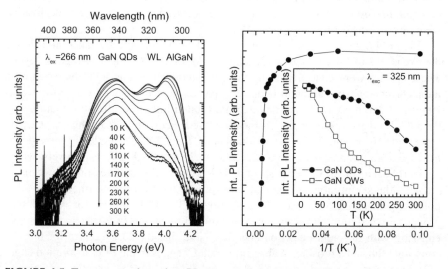

FIGURE 4.5 Temperature-dependent PL spectra of the same separate confinement heterostructure consisting of 5 stacks of *h*-GaN QDs sandwiched by $Al_{0.25}Ga_{0.75}N$ optical waveguide, excited by the 266-nm line of a fourth harmonic generated Nd:YAG laser over the temperature range from 10 to 300K.

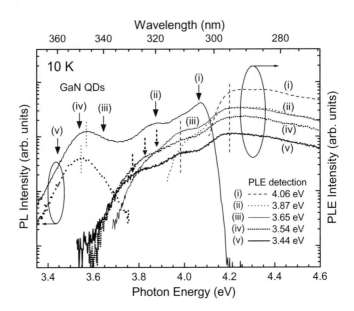

FIGURE 4.6 1K low-excitation-power PL and PLE spectra of the same separate confinement heterostructure consisting of 5 stacks of *h*-GaN QDs sandwiched by $Al_{0.25}Ga_{0.75}N$ optical waveguide.

range) being much higher than the defect density, and (2) carriers in GaN QDs being efficiently confined and thermally stable due to the record difference in bandgap energies between GaN (3.6 eV) and AlN (6.2 eV). On the other hand, carriers in QWs are more subject to nonradiative defects due to their two-dimensional degree of freedom. Similar experimental results demonstrating strong carrier localization were also obtained for the *c*-GaN QDs.[35,38]

Figure 4.6 shows low-excitation-power PL and PLE spectra taken at 10K for the five stacks of *h*-GaN QDs sandwiched by $Al_{0.25}Ga_{0.75}N$ optical waveguide layers. These spectra were measured using the quasi-monochromatic light dispersed by a monochromator from a xenon lamp. PL spectra were taken with excitation wavelengths of 266 nm (solid line) and 325 nm (dotted line) to excite above and below the AlGaN layers, respectively. The near-band-edge emissions from the $Al_{0.25}Ga_{0.75}N$ waveguide layers and the WL were found at 4.07 and 3.87 eV, while the corresponding PLE absorption edges were observed at ~4.20 and ~3.97 eV, respectively, which is clearly seen for all the PLE spectra. This kind of Stokes-like shift between the AlGaN-related emission and the PLE absorption edge has been ascribed to the alloy potential fluctuation in ternary AlGaN alloys.[67] The PLE spectra for several subgroups with different QD sizes can be monitored by taking PLE spectra for the QD emission with different detection energies. Note that the PLE spectra for the QD emissions showed a broadened absorption tail. Changes in the PLE absorption edgelike slope were observed near 3.87, 3.82, and 3.77 eV for the detection energies of 3.65, 3.54, and 3.44 eV, so the energy differences correspond to ~220, 280, and 330 meV, respectively. Therefore, the apparent Stokes-like shift in the GaN QD emissions increases with decreasing emission energy (i.e., increasing QD size), which can be

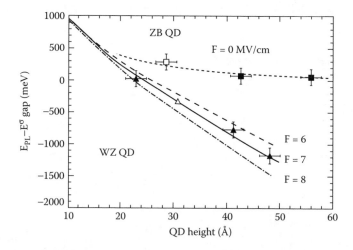

FIGURE 4.7 PL peak shift as a function of QD height, measured for hexagonal (WZ) and cubic (ZB) GaN/AlN QDs. Dotted lines are calculations taking into account QCSE with the electric field F aligned along the dot height.

explained by the separation of wavefunction overlap due to the built-in internal field present in the QDs.

The influence of the built-in internal field on optical emission energy was also studied from the PL peak shift as a function of QD height, as shown in Figure 4.7.[38] The linear decrease in the PL peak with increasing QD height is a clear indication of the dominant role of the QCSE with respect to confinement effects, which is not so surprising considering the magnitude of the internal electric field (about 7 MV/cm). Spectacular screening of this field has been reported recently in both cw and pulsed excitation experiments by Kalliakos et al.[68] These authors showed that optical excitation, even at moderate power densities such as 10 W/cm² in cw experiments, can induce sizeable spectral shifts of the PL as a result of internal field screening.

4.4.2 Optical Properties of GaN QDs Doped with Rare Earth Materials

Rare earth doped semiconductors have attracted much attention due to the prospect of opto-electronic applications based on their intra-f shell transitions, such as the atomic-like transition at 1540 nm of trivalent erbium ions for silica-based optical fibers. The energy transfer mechanism from the host to the rare earth ions most probably involves the trapping of carriers by a rare earth-related level and subsequent rare earth excitation via an Auger-like process induced by the recombination of carriers.[69] This complex carrier-mediated energy transfer strongly depends on the bandgap of the host materials. Quenching of rare earth PL at room temperature was found to be more important in small bandgap semiconductors, which drives current research efforts on rare earth doping in wider bandgap hosts such as nitrides (see, for example, Ref. 70).

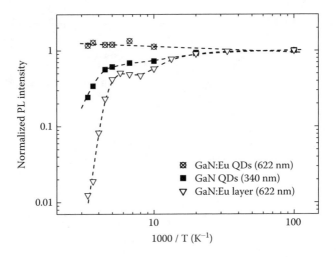

FIGURE 4.8 Temperature dependence of the red PL at 622 nm of Eu^{3+} ions in h-GaN/AlN QDs and in a thick GaN layer, and of exciton recombination in undoped h-GaN/AlN QDs. The superior radiative quantum efficiency of Eu-doped QDs is clearly seen with respect to Eu-doped bulk materials.

Recently, it was pointed out that carrier-mediated energy transfer between host materials and rare earth ions should be greatly enhanced with rare earth doped QDs because the latter are known to be particularly efficient carrier traps.[71] Figure 4.8 shows the temperature dependence of the red PL at 622 nm of Eu^{3+} ions in QDs and in a thick GaN layer. The superior radiative quantum efficiency of Eu-doped QDs with respect to Eu-doped bulk materials is clearly demonstrated. It is even better than for exciton recombination in undoped QDs, which is partially quenched at room temperature. From this remarkable result one can deduce that the energy transfer to rare earth ions in QDs should be much faster than the exciton radiative recombination time, which is on the nanosecond scale.[17,35,38]

4.5 TIME- AND SPACE-RESOLVED OPTICAL STUDIES ON GAN SELF-ASSEMBLED QUANTUM DOTS

4.5.1 Time-Resolved Optical Properties of GaN Self-Assembled QDs

A better understanding of electronic properties can be gained by investigating the radiative recombination dynamics of carriers. In the case of QDs, PL spectra are usually broadened by size distribution. Thus, time-resolved PL spectra allow one to probe the dynamics of several QD subgroups with different QD sizes. Figure 4.9 shows the temporal evolution of PL intensity at 10K for various emission wavelengths (thus, different QD sizes), for the (a) h-GaN and (b) c-GaN QD samples.[35,37] In this study, PL measurements were carried out using a picosecond pulsed Ti:sapphire laser system for sample excitation and a streak camera system for detection. The output laser pulses from the pulsed laser were frequency-tripled into the UV spectral region by a nonlinear crystal and the overall time resolution of the system

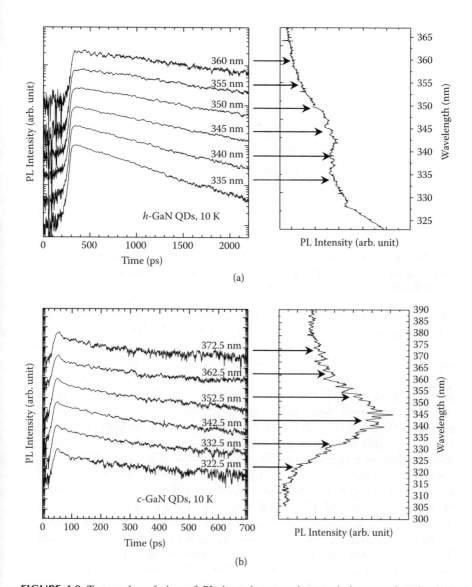

FIGURE 4.9 Temporal evolution of PL intensity at various emission wavelengths (thus different QD sizes) for the (a) *h*-GaN and (b) *c*-GaN QD samples.

is about 5 ps. The *h*-GaN QD emission persists well within a few nanosecond time range and the decay time of the emission becomes longer with increasing emission wavelength, and hence with QD size. This behavior is attributed to the reduction of the electron and hole wavefunction overlap with increasing QD height. This is a direct consequence of the QCSE due to the built-in polarization internal fields in *h*-GaN quantum structures. On the other hand, the measured lifetimes for the *c*-GaN QDs are on the order of 100 ps and almost constant with varying QD emission wavelength. This is attributed to the absence of a built-in internal field in the *c*-GaN

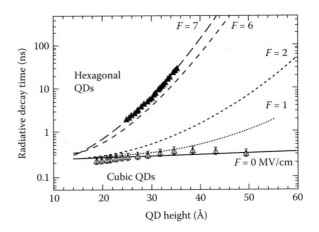

FIGURE 4.10 Variation of h-GaN and c-GaN QD lifetimes as a function of GaN QD height. Dotted lines are fits to data taking into account QCSE with the electric field F aligned along the dot height.

QDs, in contrast to h-GaN QDs. These results demonstrate that the built-in internal field strongly affects the carrier recombination in h-GaN QDs.

Figure 4.10 shows a comparative study of the low-temperature radiative decay time of h-GaN and c-GaN QDs.[38] The variation in h-GaN and c-GaN QD lifetimes with QD height was fitted by a model calculation assuming the existence of a built-in internal field of ~7 and 0 MV/cm, respectively. For cubic QDs, the decay time is shorter by one order of magnitude on average because there is no QCSE as for hexagonal QDs. Moreover, it is only weakly dependent on the QD size in the range 2 to 5 nm, which indicates that size effects on carrier wavefunctions are much less important than on carrier confinement energies. For hexagonal QDs, the fit is in good agreement with the results from the QD emission energy shift as a function of QD height shown in Figure 4.7.[17] From the time-resolved PL (Figure 4.9 and Figure 4.10), PL (Figure 4.7), and PLE (Figure 4.6) results, we conclude that the influence of the built-in electric field plays an important role in the carrier recombination in h-GaN QDs.

In fact, this built-in field can be screened, at least in part, by optical pumping. This is most clearly seen in time-resolved PL experiments. Figure 4.11 shows PL spectra at room temperature of a stack of GaN/AlN QD layers measured for (a) cw excitation and (b) pulsed excitation.[68] A strong spectral red shift of the PL band is observed with increasing time following the pulsed excitation. This can be related to the screening of the internal field by photo-injected carriers. As time goes by, the carrier density becomes smaller, and so does the screening effect.

4.5.2 SPACE-RESOLVED OPTICAL PROPERTIES OF GaN SELF-ASSEMBLED QDs

Another convincing illustration of the superiority of GaN QDs as efficient radiative recombination centers is investigated by a wavelength-resolved cathodoluminescence (CL) mapping technique. The sample used was a separate confinement heterostructure consisting of five-stacks GaN QDs sandwiched by $Al_{0.25}Ga_{0.75}N$ optical

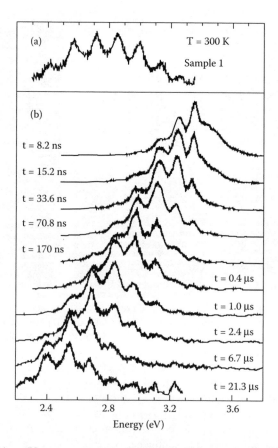

FIGURE 4.11 (a) cw PL spectrum taken at 300K. The excitation power density is 1 W/cm². (b) TRPL spectra at different time delays at 300K, illustrating temporal evolution of internal electric field screening. The energy density per pulse is 2×10^{-4} J/cm². Even for the lowest excitation power density in a cw PL experiment, the spectrum is clearly blue-shifted compared with the TRPL spectra at large time delays.

waveguide layers. The CL experiments were performed at 80K using a MonoCL™ system installed on a scanning electron microscope (SEM) with a low-temperature cryostat sample holder. Similar CL results taken at 5K appear elsewhere.[35] Figure 4.12 shows (a) the SEM image of the investigated sample area, (b) 80K CL spectrum for a given spot of this area, and 80K CL images of the same area, taken at wavelengths selected for emissions of (c) $Al_{0.25}Ga_{0.75}N$ waveguide, (d) wetting layer, (e) smaller QDs, and (f) larger QDs. The CL image taken at the $Al_{0.25}Ga_{0.75}N$ emission clearly shows the dark line defects acting as nonradiative recombination centers in the $Al_{0.25}Ga_{0.75}N$ layer (Figure 4.12c). Such line defects with boundaries at 60° or 120° are currently observed and result from the tendency for columnar growth in these materials. On the other hand, the CL images taken at the smaller and larger QD emissions do not show the luminescence degradation in these line defects, indicating good carrier localization in the QDs, even near the line defects (Figure

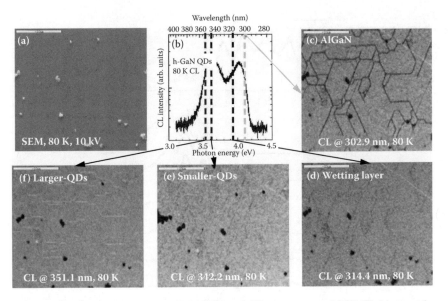

FIGURE 4.12 (a) SEM image, (b) 80K broad area CL spectrum, and 80K CL images taken at the (c) $Al_{0.25}Ga_{0.75}N$, (d) wetting layer, (e) smaller QDs, and (f) larger QDs emissions of a separate confinement heterostructure consisting of 5 stacks of GaN QDs sandwiched by $Al_{0.25}Ga_{0.75}N$ optical waveguide.

4.12e and Figure 4.12f). This is in good agreement with the transmission electron microscopy results that the presence of the threading edge dislocations next to GaN QDs does not degrade the optical emission of the GaN QDs, although GaN QDs preferentially nucleate very close to the dislocations propagating through the AlN matrix layer.[26] In addition, we observed less influence of the dark line defects for the CL image taken at the wetting layer emission (320 nm) than that obtained at the $Al_{0.25}Ga_{0.75}N$ emission (Figure 4.12d). Therefore, benefits of strong carrier localizations within GaN QDs for radiative recombination are demonstrated from the results of the temperature-dependent PL (Figure 4.5) and the wavelength-resolved CL (Figure 4.12).

An interesting reverse contrast in the CL image of the larger QD emission (Figure 4.12f) was found along some line defects compared to that obtained for the $Al_{0.25}Ga_{0.75}N$ emission (Figure 4.12c). This fact is explained by assuming that QD formation preferentially occurs near the grain boundaries due to partial elastic strain relaxation of AlN on free edges. As a result, grain boundaries can be viewed as local minima of the elastic potential for diffusing Ga and N adatoms. This leads to the observation of larger QDs formed near the grain boundaries, which is consistent with the results from atomic force microscopy shown in Figure 4.13.[72]

4.5.3 SINGLE QD SPECTROSCOPY USING MICRO-PL AND MICRO-CL ON GAN-BASED QDs

Single QD spectroscopy is a powerful tool for studying the characteristics of a single QD from the ensemble of QDs. The micro-PL technique is the most popular experimental method to investigate single QD spectroscopy. The QD density is usually

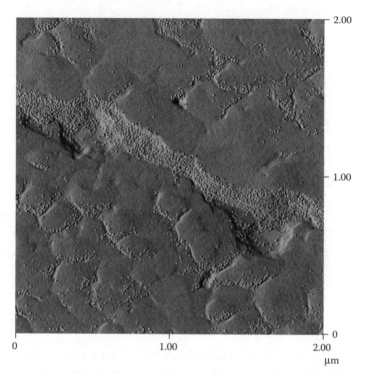

2.00

1.00

0

0   1.00   2.00
                                                              μm

FIGURE 4.13 AFM image of the early stage of the MBE growth of *h*-GaN QDs. QDs tend to form preferentially near the line defect boundaries.

of the order of $10^{10}/cm^2$. Because the focused excitation beam size through the micro-objective lens is typically micrometers (μm), postgrowth processing is needed to reduce the number of excited QDs. It consists of using electron-beam lithography and (dry or chemical) etching either to deposit a thin metallic mask with submicron (a few hundred nanometers) size apertures on the sample surface or to fabricate mesas of submicron sizes. The latter option allows a better collection of emitted photons, but could be more sensitive to fluctuating charge effects due to defects on mesa sidewalls. Recently, many single QD spectroscopic studies on single QDs have been reported using the micro-PL technique combined with metal-coated, nanoscale-apertured sample preparation on the QD samples.[1-4]

Up to now, however, only a few studies on single QD spectroscopy and imaging for GaN-based QDs have been performed, as compared to other III–V and II–VI QDs (e.g., InAs QDs, InP QDs, CdSe QDs, etc.). Single-dot PL spectroscopic studies using a micro-PL setup were performed on metal-masked QD samples with small submicron apertures for single InGaN QDs[21,62–64] and single GaN/AlN QDs.[20] By time-resolved micro-PL experiments on MOCVD-grown InGaN QDs, a single expo-nential decay profile was observed from single InGaN QDs, in contrast to the nonexponential decay behavior observed from the two-dimensional wetting layer.[64] The decay time of single InGaN QD was found to be in the nanosecond range at 4K, similar to radiative decays from an ensemble of GaN QDs (see Figure 4.9 and

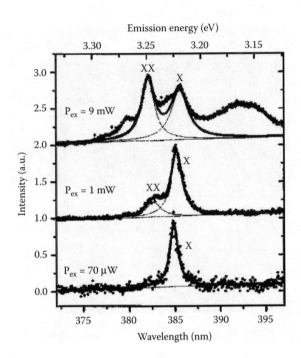

FIGURE 4.14 Micro-PL spectra of single h-GaN/AlN QD at excitation powers of 70 μW, 1 mW, and 9 mW. The circles show experimental data, and the solid lines show fitting results using a Lorentzian lineshape function. X and XX are assigned to exciton and biexciton recombination, respectively.

Figure 4.10). However, it becomes shorter at 10 to 20K as a result of the thermal activation of carriers, which indicates a rather weak localization of carriers in these QDs.

It has been observed that very sharp lines [with a linewidth of approximately microelectron volts (μeV)] typically seen from single QDs can suffer temporal fluctuations in their peak energy, intensity, and linewidth on the time scale of seconds, because of interactions with the local environment.[73,74] For single InGaN QDs, temporal variations in peak position and linewidth were also observed in optical transitions by monitoring PL spectra with time, which result in a rather broad linewidth of time-integrated PL spectra. This was explained by the Stark effect in emission peaks of InGaN QDs caused by a randomly generated local electric field.[21]

For self-assembled h-GaN/AlN QDs, several isolated emission peaks from individual GaN QDs were observed through submicron mesa structures.[20] An additional peak showing a quadratic power dependence appeared on the higher energy side of the ground state (Figure 4.14). This peak was attributed to the emission from a biexciton state with a negative binding energy of 30 meV.[20] This negative binding energy, which is rather unusual, is a direct consequence of the QCSE in nitride QDs. It results from the repulsive Coulomb interaction between carriers of the same sign that are repelled to the same interface of the QD by the internal electric field. Recently, a single-dot spectroscopic study of isolated c-GaN/AlN QDs was also carried out using low-temperature CL technique on submicron mesa structures.[75]

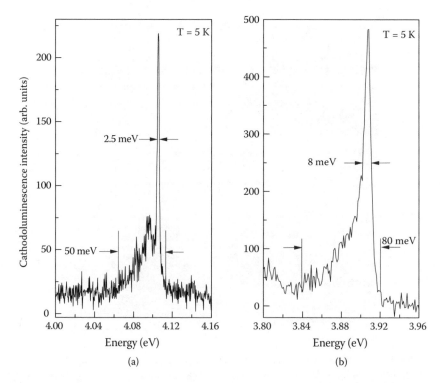

FIGURE 4.15 Micro-PL spectra of single c-GaN/AlN QDs. Sideband at low-energy side of sharp line is due to interaction with acoustic phonons.

Several CL peaks with a typical linewidth of 2 to 8 meV were observed for individual QDs (see Figure 4.15). This linewidth, much larger than what could be expected from the radiative decay time of cubic QDs, was explained by charge fluctuations around individual QDs. Another feature is the sideband extending up to 50 to 80 meV on the low-energy side of the sharp line. It was attributed to interaction with acoustic phonons, which is much stronger than in other QD systems such as CdTe/ZnTe[76] or InAs/GaAs.[77]

4.6 PROSPECTS OF GAN SELF-ASSEMBLED QUANTUM DOTS

4.6.1 GROWTH CHALLENGES: CONTROL OF QD NUCLEATION SITES AND SIZES

The unique properties of QDs as "artificial atoms" have aroused novel opto-electronic applications (e.g., low-threshold laser or single photon source). The development of practical devices requires, however, a high degree of control over QD nucleation sites, size uniformity, as well as densities. It is clear that these critical issues cannot be fully addressed with the SK growth mode. For nitrides, attempts to control QD sites have been carried out using selective growth on patterned substrates,[47,78] by an electron-beam-induced CVD technique,[79] or by using vicinal substrates.[80] In the latter case, it was shown that AlN layers with stepped surfaces can be grown on vicinal

FIGURE 4.16 AFM image of *h*-GaN QDs grown on a vicinal AlN (0001) surface. The upper inset gives the details of the two-dimensional -FFT spectrum and the lower inset gives details of the center of the autocorrelation picture.

4H-SiC (0001) substrates to promote the nucleation of GaN QDs along the step edges (Figure 4.16).

Very recently, remarkable developments in the QD technology, including selective carrier injection, have been achieved by Kapon et al.[81–83] for arsenide QDs. Using MOCVD growth on prepatterned substrates, they demonstrated full control over QD nucleation sites and size uniformity. In contrast to the strain-driven SK mode, dot formation is based on growth rate anisotropy, which allows for much better control of the QD size. For example, the inhomogeneous broadening of the InGaAs/AlGaAs QD ground state was found to be less than 8 meV for a set of 120 QDs emitting at 1.55 eV. It will be a challenging task to transfer this technology to nitrides.

Another promising alternative to the SK growth mode is to grow quantum disks embedded in nanorods (i.e., QW structures along the growth of nanorods). Because the growth axis of GaN nanorods has been succeeded by several growth methods,[84,85] the formation of heterostructures during nanorod growth has attracted a lot of attention for the various applications (such as high-brightness LEDs and one-dimensional resonant tunneling diodes). Recently, current-injected p-n junction LEDs using InGaN quantum disks were successfully demonstrated by forming InGaN/GaN multiple QWs during the growth of GaN nanorod arrays by metal organic-hydride vapor phase epitaxy. Due to the lack of dislocations and the large surface areas provided by the sidewalls of nanorods, both internal and extraction efficiencies are significantly enhanced.[84] GaN quantum disks embedded in AlGaN nanorods were grown on Si (111) substrates by plasma-assisted MBE under highly N-rich conditions.[85]

4.6.2 Control of Electronic Properties

We discussed in Sections 4.4 and 4.5 that the optical properties of (0001)-oriented GaN/AlN QDs are dominated by the presence of an internal electric field up to ~7 MV/cm oriented along the QD height (that is, along the c-axis). The radiative recombination rate is strongly reduced and is accompanied by a dramatic increase in the inhomogeneous spectral broadening induced by size fluctuations. These effects are detrimental for opto-electronic applications, which explains important efforts to develop the growth of cubic GaN QDs. Another alternative is to use nonpolar substrate surfaces to force the electric field to lie in the plane of the epitaxial layer. Indeed, recent reports on the nonpolar GaN QWs grown on r-plane (10–12) sapphire and a-plane (11–20) SiC substrates tend to support the absence of any electric field aligned along the well width.[86–89] For (11–20)-oriented QDs, the internal electric field will be in the plane of the QD base, but with a much reduced value. This is because, by contrast to (0001)-oriented QDs, the spontaneous and piezoelectric fields are expected to be in opposite directions. Preliminary reports show that (11–20) QDs emit at shorter wavelengths than the bulk GaN bandgap. Moreover, their radiative recombination time is found to be as short as for cubic QDs,[90] indicating that the quantum confinement effects should be dominant over the QCSE. These promising studies open the way for more effective applications of nitride QDs.

4.6.3 UV Bipolar Opto-Electronic Devices: Interband Transitions in GaN QDs

Nitride semiconductors are well-suited for UV bipolar opto-electronic devices such as UV LDs and LEDs based on interband transitions. For LDs, the threshold current depends on various material parameters, and it increases with increasing effective mass of electrons or the ratio of the effective masses of holes and electrons. The effective masses in GaN-based semiconductors are, however, much heavier than those of smaller bandgap semiconductors such as GaAs and InP. Therefore, conventional nitride QW LDs would require much higher thresholds because of their higher two-dimensional DOS: the minimal threshold current density of GaN-based lasers is ~1 kA/cm^2, while that of GaAs-based lasers is ~100 A/cm^2. However, if GaN QDs with small QD size are used in the active medium for LDs, GaN QDs are good candidates for low-threshold UV LDs due to discrete energy levels and the reduced zero-dimensional DOS of QDs.[9]

Another major advantage of using QDs instead of QWs is the benefit of carrier localization in zero-dimensional QDs. Because high-quality substrates are still being developed for nitride epitaxial growth, the use of QDs enhances optical properties and device performances dramatically compared to those of QWs grown on typical substrates, as shown in Section 4.4. This is because the carriers localized in QDs can be restrained from being trapped at nonradiative recombination centers caused by dislocations and other defects. For InGaN-based QW structures, it also has been reported that the carrier localization caused by indium composition fluctuations formed in the plane of the layers enhances the quantum efficiency by suppressing lateral carrier diffusion, thereby reducing the probability that carriers will be captured by nonradiative recombination centers.[91–98] Again, any practical application of nitride QDs will need improved control of QD size uniformity and density.

4.6.4 IR Unipolar Opto-Electronic Devices: Intersubband Transitions in GaN QDs

Transitions of electrons (or holes) in quantum structures between two confined quantum states in the same conduction (or valence) band are called intersubband transitions. An intersubband transition is unipolar because it involves only one type of carrier (electron or hole) making transition between two of its levels within the same (conduction or valence) band, which is useful for other kinds of QD-based device applications such as IR photodetectors (QDIPs) and quantum cascade lasers. These unipolar photonic devices with the use of intersubband transitions have been fabricated from InGaAs/AlInAs on InP or GaAs/AlGaAs on GaAs in the wavelength range of 3 to 20 μm.

For optical communication applications operating at wavelengths of 1.3 to 1.55 μm, other material systems that provide a sufficiently large band offset have been searched, because the shortest possible wavelength achievable for a certain material system is primarily determined by its band offset. Among them, GaN-based hetero-structures are receiving new attention due to their large conduction-band offsets (e.g., ~2 eV for GaN/AlN and InN/GaN), which allow record tuning of electron intersubband transitions to cover the near-IR region of interest for telecommunication applications based on optical fibers.[99–102] Despite the poor crystalline quality of GaN-based materials, caused mainly by the lack of suitable substrates, visible LEDs and blue-violet LDs based on interband transitions have been already commercialized with excellent device performance and reliability. This fact makes us expect that the GaN-based QD system can also be a good choice for realizing such unipolar devices using intersubband transitions as photodetectors, modulators, and quantum cascade lasers for high-bit-rate fiber-optic telecommunication applications.

One of the important factors in the use of the intersubband transition is its fast relaxation process. For GaAs and InGaAs MQWs, intersubband transition relaxation times of 0.65 to 2.7 ps have been reported.[103–105] GaN-based heterostructures are of particular interest due to their large effective electron mass and large LO phonon energy, which are key parameters in achieving ultrafast electron relaxation at large transition energies. Therefore, the absorption recovery time is expected to be fast in these GaN-based materials, and in fact, the electron relaxation time has been observed as 150 and 370 fs for the intersubband transition at 4.5 and 1.7 μm, respectively.[106,107]

Room-temperature intersubband absorptions in the whole optical communica-tion wavelength range of 1.3 to 1.55 μm have recently been observed in GaN/AlGaN QW structures.[108,109] Intersubband transition wavelengths as short as 1.08 μm have also been demonstrated using $(GaN)_m/(AlN)_n$ superlattice structures.[110] Intersubband transitions have also been observed in GaN/AlN QW structures grown by MOCVD[111] and MBE.[112]

Recent progress in epitaxial growth technology, such as MBE and MOCVD, allows us to fabricate high-quality nitride QDs using the SK growth mode. The dot density as high as the order of $10^{11}/cm^2$, and size fluctuations comparable to that of the well-known InAs/GaAs QD system are now achievable, which are preferable for developing efficient QD-based optical devices. Furthermore, the three-dimensional confinement of electrons provides the intersubband transitions with TE (in the plane of growth) and TM (along the growth axis) polarizations, in contrast to the case of QWs. Recently, intersubband transitions from MBE-grown GaN/AlN QDs have been demonstrated in the wavelength range of 1.27 to 2.4 μm.[113] From the aspect of practical device applications, the following two factors might favor QDs: (1) normal incidence is forbidden for electron intersubband transitions in QWs (dipole aligned along the confinement axis) but not in QDs; and (2) the electron scattering time between subbands in nitride QWs is found to be in the sub-picosecond range,[106,107] which is attractive for high-speed photodetector and modulator applications but could be a limiting factor for laser devices. To our knowledge, the electron scattering time has not been measured in nitride QDs yet, but could be significantly longer due to the discrete energy level structure.

4.6.5 WHITE LEDs: SELF-ASSEMBLED GaN QDs DOPED WITH RARE EARTH MATERIALS

Hexagonal GaN/AlN QDs can emit light in the visible spectral region due to the strong QCSE. Thus, by growing a QD sample of three different sizes to emit in the blue, green, and orange, Damilano et al.[25] have succeeded in generating white light emission without any postgrowth process. Recently, it was suggested that white light emission can also be obtained by combining QDs doped with rare earth ions emitting in the blue (Tm), green (Er, Tb, or Ho), and red (Eu, Sm, or Pr).[71] The main advantage of this option is the extremely high temperature stability of rare earth ion intra-f shell transitions. High radiative quantum efficiency could be expected also, provided that the rare earth related trap level, required for the carrier-mediated energy transfer to rare earth ions, can be accommodated in QDs. This seems to be the case for Eu (red), Er (green), and Tm (blue) doping in GaN/AlN QDs.[114] More insight into the energy transfer mechanism can be gained by PLE spectroscopy as well as dynamical studies of the PL. Another challenging task for practical applications of GaN/AlN QDs is the issue of carrier injection, which has yet to be realized.

4.7 CONCLUSION

We have presented a review of recent research progresses on the growth and optical properties of GaN-based, self-assembled QDs. First, we briefly introduced the unique physical properties and importance of these wide bandgap semiconductor QDs. Second, we reviewed various advanced growth techniques, such as MBE and MOCVD, for the growth of GaN-based QDs using the Stranski-Krastanow growth mode.

We discussed the optical properties and carrier dynamics of hexagonal and cubic phase GaN-based QDs, which were investigated by various optical spectroscopic and imaging techniques, including PL, PLE, time-resolved PL, micro-PL, CL, etc. The decay time of GaN h-QD emission was found to become longer for smaller emission energy (i.e., larger QD size), indicating a strong influence of the built-in internal field on the optical properties of h-GaN QDs. Other evidence of this internal field was observed with the large Stokes-like shift between GaN QD emissions and PLE absorption edges induced by the QCSE, or with the time-dependent red shift of QD emission induced by screening effects in time-resolved PL experiments.

Strong carrier localization in QDs was confirmed by temperature-dependent PL and wavelength-resolved CL imaging experiments. It was observed that GaN QDs are remarkable hosts for carrier-mediated energy transfer to rare earth ions. We have presented a few studies on single QD spectroscopy of both h-GaN and c-GaN QDs.

Finally, we reviewed some alternatives to the Stranski-Krastanow growth mode to control the QD nucleation site, size uniformity, and density, which are crucial for any practical applications. Prospects for UV bipolar and IR unipolar applications or white lighting were discussed also.

The potentialities of GaN-based QDs in terms of opto-electronic applications extending from the IR to the UV are quite fascinating. From this brief overview it is clear that this wide bandgap QD system is still in its early developmental stage, and more work is needed to reach the level of profound understanding of other QD systems, such as InAs/GaAs, CdSe/ZnSe, or CdTe/ZnTe.

4.8 ACKNOWLEDGMENTS

We are particularly grateful to all collaborators at Chungbuk National University in Korea and at CEA-CNRS-UJF Nanophysics and Semiconductors group in France, including B. J. Kwon, H. S. Kwack, J. S. Hwang, B. Daudin, H. Mariette, J. Barjon, J. Brault, N. Gogneau, F. Enjalbert. Y. Hori, and E. Monroy. Special thanks to A. Gokarna for her patient editorial assistance. We acknowledge support by the Korea Science and Engineering Foundation (KOSEF) through the National Research Laboratory program and by the KOSEF–CNRS exchange program.

References

1. Y. Masumoto and T. Takagahara (Eds.). *Semiconductor Quantum Dots: Physics, Spectroscopy and Applications.* New York: Springer-Verlag, 2002.
2. M. Grundmann (Ed.). *Nano-Optoelectronics: Concepts, Physics and Devices.* New York: Springer-Verlag, 2002.
3. Y. Arakawa and S. Tarucha (Eds.). *Proceedings of the 2nd International Conference on Semiconductor Quantum Dots (QD2002), Nara, Japan, 2002. Phys. Stat. Sol. (c)* 0(4) (2003, Wiley-VCH).
4. *Proceedings of the 3rd International Conference on Semiconductor Quantum Dots (QD 2004), Banff, Canada, 2004. Physica E,* 26, 1–499, 2005.
5. Y. Arakawa and H. Sakaki. Multidimensional quantum well laser and temperature dependence of its threshold current, *Appl. Phys. Lett.,* 40, 939–941, 1982.

6. Y. Arakawa, K. Vahala, and A. Yariv. Quantum noise and dynamics in quantum well and quantum wire lasers, *Appl. Phys. Lett.,* 45, 950–952, 1984.

7. M. Asada, Y. Miyamoto, and Y. Suematsu, Gain and the threshold of three-dimensional quantum-box lasers, *IEEE J. Quantum Electron.,* 22, 1915–1921, 1986.

8. A Yariv. Scaling laws and minimum threshold currents for quantum-confined semiconductor lasers, *Appl. Phys. Lett.,* 53, 1033–1035, 1988.

9. Y. Arakawa, T. Someya, and T. Tachibana. Progress in growth and physics of nitride-based quantum dots. *Phys. Stat. Sol. (b),* 224, 1–11, 2001.

10. I. D'Amico, E. Biolatti, F. Rossi, S. Derinaldis, R. Rinaldis, and R. Cingolani. GaN quantum dot based quantum information/computation processing. *Superlattices Microstruct.,* 31, 117–125, 2002.

11. K. Kawasaki, D. Yamazaki, A. Kinoshita, H. Hirayama, K. Tsutsui, and Y. Aoyagi. GaN quantum-dot formation by self-assembling droplet epitaxy and application to single-electron transistors. *Appl. Phys. Lett.,* 79, 2243–2245, 2001.

12. D.L. Smith and C. Mailhiot. Optical properties of strained-layer superlattices with growth axis along [111]. *Phys. Rev. Lett.,* 58:1264-1267, 1987.

13. F. Bernardini and V. Fiorentini. Spontaneous polarization and piezoelectric constants of III-V nitrides. *Phys. Rev. B,* 56, R10024–R10027, 1997.

14. T. Takeuchi, S. Sota, M. Katsuragawa, M. Komori, H. Takeuchi, H. Amano, and I. Akasaki. Quantum-confined Stark effect due to piezoelectric field in GaInN strained QWs. *Jpn. J. Appl. Phys.,* 36, L382–L385, 1997.

15. J.S. Im, H. Lollmer, J. Off, A. Sohmer, F. Scholz, and A. Hangleiter. Reduction of oscillator strength due to piezoelectric fields in $GaN/Al_xGa_{1-x}N$ QWs. *Phys. Rev. B,* 57, R9435–R9438, 1998.

16. R. Langer, J. Simon, O. Konovalov, N. Pelekanos, A. Barski, and M. Leszczynski. X-ray reciprocal lattice mapping and photoluminescence of GaN/GaAlN multiple QWs; strain induced phenomena. *MRS Internet J. Nitride Semicond. Res.,* 3, 46, 1998.

17. F. Widmann, J. Simon, B. Daudin, G. Feuillet, J.L. Rouvière, N.T. Pelekanos, and G. Fishman. Blue-light emission from GaN self-assembled quantum dots due to giant piezoelectric effect. *Phys. Rev. B,* 58, R15989–R15992, 1998.

18. A.D. Andreev and E.P. O'Reilly. Optical transitions and radiative lifetime in GaN/AlN self-organized quantum dots, *Appl. Phys. Lett.,* 79, 521–523, 2001.

19. B. Daudin, F. Widmann, J. Simon, G. Feuillet, J.L. Rouvière, N.T. Pelekanos, and G. Fishman. Piezoelectric properties of GaN self-organized quantum dots. *MRS Internet J. Nitride Semicond. Res.,* 4S1, G9.2, 1999.

20. S. Kako, K. Hoshino, S. Iwamoto, S. Ishida, and Y. Arakawa. Exciton and biexciton luminescence from single hexagonal GaN/AlN self-assembled quantum dots. *Appl. Phys. Lett.,* 85, 64–66, 2004.

21. J.H. Rice, J.W. Robinson, A. Jarjour, R.A. Taylor, R.A. Oliver, G.A.D. Briggs, M.J. Kappers, and C.J. Humphreys. Temporal variation in photoluminescence from single InGaN quantum dots. *Appl. Phys. Lett.,* 84, 4110–4112, 2004.

22. B. Daudin, F. Widmann, G. Feuillet, Y. Samson, M. Arlery, and J.L. Rouvière. Stranski-Krastanov growth mode during the molecular beam epitaxy of highly strained GaN. *Phys. Rev. B,* 56, R7069–R7072, 1997.

23. X.Q. Shen, S. Tanaka, S. Iwai, and Y. Aoyai. The formation of GaN dots on $Al_xGa_{1-x}N$ surfaces using Si in gas-source molecular beam epitaxy. *Appl. Phys. Lett.,* 72, 344–346, 1998.

24. F. Widmann, B. Daudin, G. Feuillet, Y. Samson, J.L. Rouvière, and N. Pelekanos. Growth kinetics and optical properties of self-organized GaN quantum dots. *J. Appl. Phys.,* 83, 7618–7624, 1998.

25. B. Damilano, N. Grandjean, F. Semond, J. Massies, and M. Leroux. From visible to white light emission by GaN quantum dots on Si(111) substrate. *Appl. Phys. Lett.*, 75, 962–964, 1999.

26. J.L. Rouvière, J. Simon, N. Pelekanos, B. Daudin, and G. Feuillet. Preferential nucleation of GaN quantum dots at the edge of AlN threading dislocations. *Appl. Phys. Lett.*, 75, 2632–2634, 1999.

27. B. Damilano, N. Grandjean, S. Dalmasso, and J. Massies. Room-temperature blue-green emission from InGaN/GaN quantum dots made by strain-induced islanding growth. *Appl. Phys. Lett.*, 75, 3751–3753, 1999.

28. C. Adelmann, J. Simon, G. Feuillet, N.T. Pelekanos, and B. Daudin, Self-assembled InGaN quantum dots grown by molecular-beam epitaxy. *Appl. Phys. Lett.*, 76, 1570–1572, 2000.

29. E. Martinez-Guerrero, C. Adelmann, F. Chabuel, J. Simon, N.T. Pelekanos, G. Mula, B. Daudin, G. Feuillet, and H. Mariette. Self-assembled zinc blende GaN quantum dots grown by molecular-beam epitaxy. *Appl. Phys. Lett.*, 77, 809–811, 2000.

30. J. Gleize, J. Frandon, F. Demangeot, M.A. Renucci, C. Adelmann, B. Daudin, G. Feuillet, B. Damilano, N. Grandjean, and J. Massies. Signature of GaN-AlN quantum dots by nonresonant Raman scattering. *Appl. Phys. Lett.*, 77, 2174–2176, 2000.

31. J. Gleize, F. Demangeot, J. Frandon, M.A. Renucci, M. Kuball, B. Damilano, N. Grandjean, and J. Massies. Direct signature of strained GaN quantum dots by Raman scattering. *Appl. Phys. Lett.*, 79, 686–688, 2001.

32. B. Daudin, G. Feuillet, H. Mariette, G. Mula, N. Pelekanos, E. Molva, J.L. Rouvière, C. Adelmann, E. Martinez-Guerrero, J. Barjon, F. Chabuel, B. Bataillou, and J. Simon. Self-assembled gan quantum dots grown by plasma-assisted molecular beam epitaxy. *Jpn. J. Appl. Phys.*, 40, 1892–1895, 2001.

33. J. Simon, E. Martinez-Guerrero, C. Adelmann, G. Mula, B. Daudin, G. Feuillet, H. Mariette, and N.T. Pelekanos. Time-resolved photoluminescence studies of cubic and hexagonal GaN quantum dots. *Phys. Stat. Sol. (b)*, 224, 13–16, 2001.

34. J. Brown, C. Elsass, C. Poblenz, P.M. Petroff, and J.S. Speck. Temperature dependent photoluminescence of MBE grown gallium nitride quantum dots. *Phys. Stat. Sol. (b)*, 228, 199–202, 2001.

35. Y.H. Cho, B.J. Kwon, J. Barjon, J. Brault, B. Daudin, H. Mariette, and L.S. Dang. Optical characteristics of hexagonal GaN self-assembled quantum dots: strong influence of built-in electric field and carrier localization. *Appl. Phys. Lett.*, 81, 4934–4936, 2002.

36. E. Martinez-Guerrero, F. Chabuel, B. Daudin, J.L. Rouvière, and H. Mariette. Control of the morphology transition for the growth of cubic GaN/AlN nanostructures. *Appl. Phys. Lett.*, 81, 5117–5119, 2002.

37. Y.H. Cho, B.J. Kwon, J. Barjon, J. Brault, B. Daudin, H. Mariette, and L.S. Dang. Optical properties of hexagonal and cubic GaN self-assembled quantum dots. *Phys. Stat. Sol. (c)*, 0, 1173–1176, 2003.

38. J. Simon, N.T. Pelekanos, C. Adelmann, E. Martinez-Guerrero, R. Andrè, B. Daudin, L.S. Dang, and H. Mariette. Direct comparison of recombination dynamics in cubic and hexagonal GaN/AlN quantum dots. *Phys. Rev. B*, 68, 035312, 2003.

39. N. Gogneau, D. Jalabert, E. Monroy, T. Shibata, M. Tanaka, and B. Daudin. Structure of GaN quantum dots grown under "modified Stranski–Krastanow" conditions on AlN. *J. Appl. Phys.*, 94, 2254–2261, 2003.

40. S. Tanaka, S. Iwai, and Y. Aoyagi. Self-assembling GaN quantum dots on $Al_xGa_{1-x}N$ surfaces using a surfactant. *Appl. Phys. Lett.*, 69, 4096–4098, 1996.

41. S. Tanaka, H. Hirayama, Y. Aoyagi, Y. Narukawa, Y. Kawakami, and S. Fujita. Stimulated emission from optically pumped GaN quantum dots. *Appl. Phys. Lett.,* 71, 1299–1301, 1997.

42. P. Ramvall, S. Tanaka, S. Nomura, P. Riblet, and Y. Aoyagi. Observation of confinement-dependent exciton binding energy of GaN quantum dots. *Appl. Phys. Lett.,* 73, 1104–1106, 1998.

43. K. Tachibana, T. Someya, and Y. Arakawa. Nanometer-scale InGaN self-assembled quantum dots grown by metalorganic chemical vapor deposition. *Appl. Phys. Lett.,* 74, 383–385, 1999.

44. A. Petersson, A. Gustafsson, L. Samuelson, S. Tanaka, and Y. Aoyagi. Cathodoluminescence spectroscopy and imaging of individual GaN dots. *Appl. Phys. Lett.,* 74, 3513–3515, 1999.

45. J. Wang, M. Nozaki, M. Lachab, Y. Ishikawa, R.S. Qhalid Fareed, T. Wang, M. Hao, and S. Sakai. Metalorganic chemical vapor deposition selective growth and characterization of InGaN quantum dots. *Appl. Phys. Lett.,* 75, 950–952, 1999.

46. P. Ramvall, S. Tanaka, S. Nomura, P. Riblet, and Y. Aoyagi. Confinement induced decrease of the exciton-longitudinal optical phonon coupling in GaN quantum dots. *Appl. Phys. Lett.,* 75, 1935–1937, 1999.

47. K. Tachibana, T. Someya, S. Ishida, and Y. Arakawa. Selective growth of InGaN quantum dot structures and their microphotoluminescence at room temperature. *Appl. Phys. Lett.,* 76, 3212–3214, 2000.

48. M. Kuball, J. Gleize, S. Tanaka, and Y. Aoyagi. Resonant Raman scattering on self-assembled GaN quantum dots. *Appl. Phys. Lett.,* 78, 987–989, 2001.

49. J. Zhang, M. Hao, P. Li, and S.J. Chua. InGaN self-assembled quantum dots grown by metalorganic chemical-vapor deposition with indium as the antisurfactant. *Appl. Phys. Lett.,* 80, 485–487, 2002.

50. M. Miyamura, K. Tachibana, and Y. Arakawa. High-density and size-controlled GaN self-assembled quantum dots grown by metalorganic chemical vapor deposition. *Appl. Phys. Lett.,* 80, 3937–3939, 2002.

51. M. Miyamura, K. Tachibana, T. Someya, and Y. Arakawa. Stranski-Krastanow growth of GaN quantum dots by metalorganic chemical vapor deposition. *J. Cryst. Growth,* 237–239, 1316–1319, 2002.

52. C.W. Hu, A. Bell, F.A. Ponce, D.J. Smith, and I.S.T. Tsong. Growth of self-assembled GaN quantum dots via the vapor–liquid–solid mechanism. *Appl. Phys. Lett.,* 81, 3236–3238, 2002.

53. T.J. Goodwin, V.J. Leppert, S.H. Risbud, I.M. Kennedy, and H.W.H. Lee. Synthesis of gallium nitride quantum dots through reactive laser ablation. *Appl. Phys. Lett.,* 70, 3122–3124, 1997.

54. O.I. Mićić, S.P. Ahrenkiel, D. Bertram, and A.J. Nozik. Synthesis, structure, and optical properties of colloidal GaN quantum dots. *Appl. Phys. Lett.,* 75, 478–480, 1999.

55. C. Adelmann, N. Gogneau, E. Sarigiannidou, J.L. Rouvière, and B. Daudin. GaN islanding by spontaneous rearrangement of a strained two-dimensional layer on (0001) AlN. *Appl. Phys. Lett.,* 81, 3064–3066, 2002.

56. J. Tersoff, C. Teichert, and M.G. Lagally. Self-organization in growth of quantum dot superlattices. *Phys. Rev. Lett.,* 76, 1675–1678, 1996.

57. J. Brown, F. Wu, P.M. Petroff, and J.S. Speck. GaN quantum dot density control by rf-plasma molecular beam epitaxy. *Appl. Phys. Lett.,* 84, 690–692, 2004.

58. N. Gogneau, F. Fossard, E. Monroy, S. Monnoye, H. Mank, and B. Daudin. Effects of stacking on the structural and optical properties of self-organized GaN/AlN quantum dots. *Appl. Phys. Lett.,* 84, 4224–4226, 2004.

59. G. Mula, C. Adelmann, S. Moehl, J. Oullier, and B. Daudin. Surfactant effect of gallium during molecular-beam epitaxy of GaN on AlN (0001). *Phys. Rev. B,* 64, 195406–195408, 2001.

60. H. Hirayama, S. Tanaka, P. Ramvall, and Y. Aoyagi. Intense photoluminescence from self-assembling InGaN quantum dots artificially fabricated on AlGaN surfaces. *Appl. Phys. Lett.,* 72, 1736–1738, 1998.

61. P. Ramvall, P. Riblet, S. Nomura, Y. Aoyagi, and S. Tanaka. Optical properties of GaN quantum dots. *J. Appl. Phys.,* 87, 3883–3890, 2000.

62. O. Moriwaki, T. Someya, K. Tachibana, S. Ishida, and Y. Arakawa. Narrow photoluminescence peaks from localized states in InGaN quantum dot structures. *Appl. Phys. Lett.,* 76, 2361–2363, 2000.

63. R.A. Oliver, G.A.D. Briggs, M.J. Kappers, C.J. Humphreys, S. Yasin, J.H. Rice, J.D. Smith, and R.A. Taylor. InGaN quantum dots grown by metalorganic vapor phase epitaxy employing a post-growth nitrogen anneal, *Appl. Phys. Lett.,* 83, 755–757, 2003.

64. J.W. Robinson, J.H. Rice, A. Jarjour, J.D. Smith, R.A. Taylor, R.A. Oliver G.A.D. Briggs, M.J. Kappers, C.J. Humphreys, and Y. Arakawa. Time-resolved dynamics in single InGaN quantum dots. *Appl. Phys. Lett.,* 83, 2674–2676, 2003.

65. K. Tachibana, T. Someya, Y. Arakawa, R. Werner, and A. Forchel. Room-temperature lasing oscillation in an InGaN self-assembled quantum dot laser. *Appl. Phys. Lett.,* 75, 2605–2607, 1999.

66. K. Hoshino, S. Kako, and Y. Arakawa. Formation and optical properties of stacked GaN self-assembled quantum dots grown by metalorganic chemical vapor deposition. *Appl. Phys. Lett.,* 85, 1262–1264, 2004.

67. Y.H. Cho, G.H. Gainer, J.B. Lam, J.J. Song, W. Yang, and W. Jhe. Dynamics of anomalous optical transitions in $Al_xGa_{1-x}N$ alloys. *Phys. Rev. B,* 61, 7203–7206, 2000.

68. S. Kalliakos, T. Bretagnon, P. Lefebvre, T. Taliercio, B. Gil, N. Grandjean, B. Damilano, A. Dussaigne, and J. Massies. Photoluminescence energy and linewidth in GaN/AlN stackings of quantum dot planes. *J. Appl. Phys.,* 96, 180–185, 2004.

69. J.H. Shin, G.N. van den Hoven, and A. Polmn. Direct experimental evidence for trap-state mediated excitation of Er^{3+} in silicon. *Appl. Phys. Lett.,* 67, 377–379, 1995.

70. J. Zavada and D. Zhang. Luminescence properties of erbium in III–V compound semiconductors. *Solid State Electron.,* 38, 1285–1293, 1995.

71. Y. Hori, X. Biquard, E. Monroy, D. Jalabert, F. Enjalbert, Le Si Dang, M. Tanaka, O. Oda, and B. Daudin. GaN quantum dots doped with Eu. *Appl Phys Lett.,* 84, 206–208, 2004.

72. F. Widmann, Ph.D. thesis, University Joseph Fourier, Grenoble-I, France, 1998.

73. J. Seufert, R. Weigand, G. Bacher, T. Kümmell, A. Forchel, K. Leonardi, and D. Hommel. Spectral diffusion of the exciton transition in a single self-organized quantum dot. *Appl. Phys. Lett.,* 76, 1872–1874, 2000.

74. N. Panev, M.E. Pistol, S. Jeppesen, V.P. Evtikhiev, A.A. Katznelson, and E.Y. Kotelnikov. Spectroscopic studies of random telegraph noise in InAs quantum dots in GaAs. *J. Appl. Phys.,* 92, 7086–7089, 2002.

75. J.P. Garayt, J.M. Gérard, F. Enjalbert, L. Ferlazzo, S. Founta, E. Martinez-Guerrero, F. Rol, D. Araujo, R. Cox, B. Daudin, B. Gayral, L.S. Dang, and H. Mariette. Study of isolated cubic GaN quantum dots by low temperature cathodoluminescence. *Proceedings of the 3rd International Conference on Semiconductor Quantum Dots (QD 2004), Physica E,* 26, 203–206, 2005.

76. L. Besombes, K. Kheng, L. Marsal, and H. Mariette. Acoustic phonon broadening mechanism in single quantum dot emission, *Phys. Rev. B,* 63, 155307, 2001.

77. P. Borri, W. Langbein, S. Schneider, U. Woggon, R.L. Sellin, D. Ouyang, and D. Bimberg. Ultralong dephasing time in InGaAs quantum dots. *Phys. Rev. Lett.,* 87, 157401, 2001.

78. K. Tachibana, T. Someya, S. Ishida, and Y. Arakawa. Uniform array of GaN quantum dots in AlGaN matrix by selective MOCVD growth. *Phys. Stat. Sol. (b),* 228, 187–190, 2001.

79. P.A. Crozier, J. Tolle, J. Kouvetakis, and C. Ritter, Synthesis of uniform GaN quantum dot arrays via electron nanolithography of D_2GaN_3. *Appl. Phys. Lett.,* 84, 3441–3443, 2004.

80. J. Brault, S. Tanaka, E. Sarigiannidou, J.L. Rouvière, B. Daudin, G. Feuillet, and H. Nakagawa. Linear alignment of GaN quantum dots on AlN grown on vicinal SiC substrates. *J. Appl. Phys.,* 93, 3108–3110, 2003.

81. M.H. Baier, S. Watanabe, E. Pelucchi, and E. Kapon. High uniformity of site-controlled pyramidal quantum dots grown on prepatterned substrates. *Appl. Phys. Lett.,* 84, 1943–1945, 2004.

82. M.H. Baier, C. Constantin, E. Pelucchi, and E. Kapon. Electroluminescence from a single pyramidal quantum dot in a light-emitting diode. *Appl. Phys. Lett.,* 84, 1967–1969, 2004.

83. S. Watanabe, E. Pelucchi, B. Dwir, M.H. Baier, K. Leifer, and E. Kapon. Dense uniform arrays of site-controlled quantum dots grown in inverted pyramids. *Appl. Phys. Lett.,* 84, 2907–2909, 2004.

84. H.M. Kim, Y.H. Cho, H. Lee, S.I. Kim, S.R. Ryu, D.Y. Kim, T.W. Kang, and K.S. Chung, High-brightness light emitting diodes using dislocation-free indium gallium nitride/gallium nitride multiquantum-well nanorod arrays. *Nano. Lett.,* 4, 1059–1062, 2004.

85. J. Ristic, E. Calleja, M.A. Sanchez-Garcia, J.M. Ulloa, J. Sanchez-Paramo, J.M. Calleja, U. Jaha, A. Trampert, and K.H. Ploog. Characterization of GaN quantum discs embedded in $Al_xGa_{1-x}N$ nanocolumns grown by molecular beam epitaxy. *Phys. Rev. B,* 68, 125305, 2003.

86. H.M. Ng. Molecular-beam epitaxy of $GaN/Al_xGa_{1-x}N$ multiple quantum wells on R-plane (10-12) sapphire substrates. *Appl. Phys. Lett.,* 80, 4369–4371, 2002.

87. M.D. Craven, P. Waltereit, F. Wu, J.S. Speck, and S.P. DenBaars. Characterization of a-plane GaN/(Al,Ga)N multiple quantum wells grown via metalorganic chemical vapor deposition, *Jpn. J. Appl. Phys.,* 42, L235–L238, 2003.

88. M.D. Craven, A. Chakraborty, B. Imer, F. Wu, S. Keller, U.K. Mishra, J.S. Speck, and S.P. DenBaars. Structural and electrical characterization of a-plane GaN grown on a-plane SiC. *Phys. Stat. Sol. (c),* 0, 2132–2135, 2003.

89. A. Chakraborty, H. Xing, M.D. Craven, S. Keller, T. Mates, J.S. Speck, S.P. DenBaars, and U.K. Mishra, *J. Appl. Phys.,* 96, 4494–4499, 2004.

90. S. Founta, F. Rol, E. Bellet-Amalric, J. Bleuse, B. Daudin, B. Gayral, and H. Mariette. Optical properties of GaN quantum dots grown on non-polar (11-20) SiC by molecular beam epitaxy. *Appl. Phys. Lett.* 86, 171901, 2005.

91. S. Chichibu, T. Azuhata, T. Sota, and S. Nakamura. Spontaneous emission of localized excitons in InGaN single and multiQW structures. *Appl. Phys. Lett.,* 69, 4188–4190, 1996.

92. E.S. Jeon, V. Kozlov, Y.K. Song, A. Vertikov, M. Kuball, A.V. Nurmikko, H. Liu, C. Chen, R.S. Kern, C.P. Kuo, and M.G. Craford. Recombination dynamics in InGaN QWs. *Appl. Phys. Lett.,* 69, 4194–4196, 1996.

93. P. Perlin, V. Iota, B.A. Weinstein, P. Wiśniewski, T. Suski, P.G. Eliseev, and M. Osiński. Influence of pressure on photoluminescence and electroluminescence in GaN/InGaN/AlGaN QWs. *Appl. Phys. Lett.,* 70, 2993–2995, 1997.

94. Y. Narukawa, Y. Kawakami, M. Funato, S. Fujita, S. Fujita, and S. Nakamura. Role of self–formed InGaN quantum dots for exciton localization in the purple laser diode emitting at 420 nm. *Appl. Phys. Lett.*, 70, 981–983, 1997.

95. Y. Narukawa, Y. Kawakami, S. Fujita, S. Fujita, and S. Nakamura. Recombination dynamics of localized excitons in $In_{0.20}Ga_{0.80}N$-$In_{0.05}Ga_{0.95}N$ multiple QWs. *Phys. Rev. B*, 55, R1938–R1941, 1997.

96. Y.H. Cho, G.H. Gainer, A.J. Fischer, J.J. Song, S. Keller, U.K. Mishra, and S.P. Denbaars. "S-shaped" temperature-dependent emission shift and carrier dynamics in InGaN/GaN multiple QWs. *Appl. Phys. Lett.*, 73, 1370–1372, 1998.

97. P. Lefebvre, J. Allègre, B. Gil, A. Kavokine, H. Mathieu, W. Kim, A. Salvador, A. Botchkarev, and H. Morkoç. Recombination dynamics of free and localized excitons in $GaN/Ga_{0.93}Al_{0.07}N$ QWs. *Phys. Rev. B*, 57, R9447–R9450, 1998.

98. Y.H. Cho, T.J. Schmidt, S. Bidnyk, G.H. Gainer, J.J. Song, S. Keller, U.K. Mishra, and S.P. DenBaars. Linear and nonlinear optical properties of In_xGa_{1-x}/GaN heterostructures. *Phys. Rev. B*, 61, 7571–7588, 2000.

99. N. Suzuki and N. Iizuka. Feasibility study on ultrafast nonlinear optical properties of 1.55-μm intersubband transition in AlGaN/GaN QWs. *Jpn. J. Appl. Phys., Part 2*, 36, L1006–L1008, 1997.

100. N. Suzuki and N. Iizuka. Effect of polarization field on intersubband transition in AlGaN/GaN QWs. *Jpn. J. Appl. Phys., Part 2*, 38, L363–L365, 1999.

101. C. Gmachl, H.M. Ng, and A.Y. Cho. Intersubband absorption in GaN/AlGaN multiple QWs in the wavelength range of $\lambda \sim 1.75$–4.2 μm. *Appl. Phys. Lett.*, 77, 334336, 2000.

102. H.M. Ng, C. Gmachl, T. Siegrist, S.N.G. Chu, and A.Y. Cho. Growth and characterization of GaN/AlGaN superlattices for near-infrared intersubband transitions. *Phys. Stat. Sol. (a)*, 188, 825–831, 2001.

103. J. Faist, F. Capasso, S. Sirtori, D.L. Sivco, A.L. Hutchinson, S.N.G. Chu, and A.Y. Cho. Measurement of the intersubband scattering rate in semiconductor QWs by excited state differential absorption spectroscopy. *Appl. Phys. Lett.*, 63, 1354–1356, 1993.

104. S. Lutgen, R.A. Kaindl, M. Woerner, T. Elsaesser, A. Hase, and H. Künzel. Nonlinear intersubband absorption of a hot quasi-two-dimensional electron plasma studied by femtosecond infrared spectroscopy. *Phys. Rev. B*, 54, R17343–R17346, 1996.

105. S. Lutgen, R.A. Kaindl, M. Woerner, T. Elsaesser, A. Hase, H. Künzel, M. Gulia, D. Meglio, and P. Lugli. Nonequilibrium dynamics in a quasi-two-dimensional electron plasma after ultrafast intersubband excitation. *Phys. Rev. Lett.*, 77, 3657–3660, 1996.

106. N. Iizuka, K. Kaneko, N. Suzuki, T. Asano, S. Noda, and O. Wada. Ultrafast intersubband relaxation (150 fs) in AlGaN/GaN multiple QWs. *Appl. Phys. Lett.*, 77, 648650, 2000.

107. C. Gmachl, S.V. Frolov, H.M. Ng, S.N.G. Chu, and A.Y. Cho. Sub-picosecond electron scattering time for $\lambda \sim 1.55$ μm intersubband transitions in GaN/AlGaN multiple quantum wells. *Electron. Lett.*, 37, 378380, 2001.

108. C. Gmachl, H.M. Ng, S.N.G. Chu, and A.Y. Cho. Intersubband absorption at $\lambda \sim 1.55$ μm in well- and modulation-doped GaN/AlGaN multiple QWs with superlattice barriers, *Appl. Phys. Lett.*, 77, 37223724, 2000.

109. C. Gmachl, H.M. Ng, and A.Y. Cho. Intersubband absorption in degenerately doped $GaN/Al_xGa_{1-x}N$ coupled double QWs, *Appl. Phys. Lett.*, 79, 1590–1592, 2001.

110. K. Kishino, A. Kikuchi, H. Kanazawa, and T. Tachibana. Intersubband transition in $(GaN)_m/(AlN)_n$ superlattices in the wavelength range from 1.08 to 1.61 μm. *Appl. Phys. Lett.*, 81, 1234–1236, 2002.

111. I. Waki, C. Kumtornkittikul, Y. Shimogaki, and Y. Nakano. Shortest intersubband transition wavelength (1.68 µm) achieved in AlN/GaN multiple QWs by metalorganic vapor phase epitaxy. *Appl. Phys. Lett.*, 82, 44654467, 2003.

112. A. Helman, M. Tchernycheva, A. Lusson, E. Warde, F.H. Julien, K. Moumanis, G. Fishman, E. Monroy, B. Daudin, D.L.S. Dang, E. Bellet-Amalric, and D. Jalabert. Intersubband spectroscopy of doped and undoped GaN/AlN QWs grown by molecular-beam epitaxy. *Appl. Phys. Lett.*, 83, 5196–5198, 2003.

113. K. Moumanis, A. Helman, F. Fossard, M. Tchernycheva, A. Lusson, F.H. Julien, B. Damilano, N. Grandjean, and J. Massies. Intraband absorptions in GaN/AlN quantum dots in the wavelength range of 1.27–2.4 µm. *Appl. Phys. Lett.*, 82, 868870, 2003.

114. Y. Hori, T. Andreev, X. Biquard, E. Monroy, D. Jalabert, A. Farchi, M. Tanaka, O. Oda, Le Si Dang, and B. Daudin, unpublished.

5 Ultrafast Non-Equilibrium Electron Dynamics in Metal Nanoparticles

Fabrice Vallée

CONTENTS

5.1   Introduction 101
5.2   Optical Properties of Metal Nanoparticles 104
    5.2.1   Metal Nanoparticle Optical Response 104
    5.2.2   Dielectric Function of a Metal Nanoparticle 106
    5.2.3   Surface Plasmon Resonance 109
    5.2.4   Nonlinear Optical Response: Sample Optical Property
           Changes 110
5.3   Optical Creation of Non-Equilibrium Electrons 111
    5.3.1   Coherent Electron-Light Coupling 111
    5.3.2   Non-Equilibrium Electron Gas Energy 115
5.4   Non-Equilibrium Electron Kinetics 116
    5.4.1   Energy Relaxation Kinetics of Non-Equilibrium Electrons 116
    5.4.2   Optical Probing of the Non-Equilibrium Electron Kinetics 118
    5.4.3   Time-Dependent Optical Response 125
5.5   Electron-Electron Energy Exchanges 125
5.6   Electron-Lattice Energy Exchanges 130
5.7   Non-Equilibrium Electron-Lattice Energy Exchanges 135
5.8   Conclusion 136
Acknowledgments 137
References 138

5.1   INTRODUCTION

The possibility of designing and mastering the physical and chemical properties of nanoobjects and nanostructured materials has been at the center of the large interest they have received in the academic and industrial domains. Confinement effects and the enhanced role of the interfaces, which roughly scales as the surface-over-volume ratio of the nanoobjects, are key parameters that can, to some extent, be modified and controlled to meet technological requirements or perform certain functions. Understanding and modeling the new physical properties of nanoobjects and nanomaterials has thus been an intense field of research during the past decades, motivated by both the fundamental interest and the importance for their design and the development of their technological applications.

The two main features that underlie the new properties of nanostructrured materials with delocalized electrons, metals, or semiconductors are the quantum confinement of the electron and phonon inside a nanoparticle and the size reduction below a certain characteristic length associated with the studied property, such as the wavelength for the optical properties. The high sensitivity of these properties to size reduction is of major interest for applications but also makes them very sensitive probes of the impact of confinement on the fundamental electronic or vibrational properties with which they are linked. The most conspicuous feature is the appearance of morphological resonances whose origins are drastically different in semiconductor and metal nanoparticles: they are associated with the quantum mechanical effect in the former, and essentially due to the classical dielectric confinement effect in the metal case. The latter effect is a consequence of the particle size reduction well below the optical wavelength and, in the simple quasi-static description, related a local field effect (Section 5.2).[1-9] Its large impact in metal nanoparticles is a direct consequence of their negative dielectric function over a large frequency range. It leads to a strong enhancement of the linear and nonlinear optical responses of metallic nanoobjects around a specific frequency, the surface plasmon resonance. Its characteristics depend on the structure of the nanoobjects (size, shape, composition, structure) and of their local environment (surrounding matrix). It is making their optical response (in particular, their color) controllable, provided that these parameters are controlled during the growth process, a property that has been exploited for many centuries and first investigated by Faraday[10] and modeled by Mie.[11]

The weaker role of quantum confinement in metal nanoparticles is a consequence of their large electron density as compared to the semiconductor case (Section 5.2). In contrast to semiconductors, one is always dealing with a quasi-continuum of electronic states. However, modification of the electronic and vibrational wave functions, and, consequently, the confined state energy and relaxation has a significant impact on the interaction processes inside a nanoobject (electron-electron, electron-lattice, etc.) and its coupling with the environment (energy or charge exchanges with its environment such as a liquid or solid matrix, adsorbed molecules, other nanoparticles, etc.). These are important effects in many physical and chemical properties of nanoobjects but are difficult to selectively address with conventional experimental techniques, especially for embedded nanoobjects. Femtosecond lasers and time-resolved nonlinear optical techniques are powerful tools for the analysis of these effects and, more generally, for understanding the fundamental interaction mechanisms of the elementary excitations in bulk and nanostructured materials. In the case of reduced dimensionality systems, they have been extensively applied to

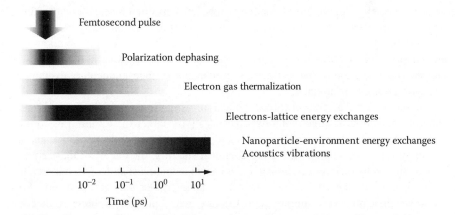

Femtosecond pulse

Polarization dephasing

Electron gas thermalization

Electrons-lattice energy exchanges

Nanoparticle-environment energy exchanges
Acoustics vibrations

10^{-2} 10^{-1} 10^{0} 10^{1}

Time (ps)

FIGURE 5.1 Time scale of the different non-equilibrium electron relaxation processes in a nanoparticle after excitation by a femtosecond pulse: coherent polarization decay, electron-electron energy exchanges (establishment of the electron temperature in few hundred femtoseconds), electron-lattice energy transfer (a picosecond scale), and damping of the nanoparticle energy to its environment (in few to few hundred picoseconds). Nanoparticle acoustic vibrations are observed on a picosecond to few tens of picoseconds scale.

the study of two-dimensional, one-dimensional, and zero-dimensional semiconductor systems,[12] and, more recently to metal nanoparticles and clusters.[13,14]

In the case of metallic materials, these techniques are based on selective electron excitation by an optical pump pulse. The electron kinetics are subsequently followed with a time-delayed probe pulse monitoring the time evolution of an optical property of the material that depends on the electron and/or lattice properties. This approach, discussed by Anisimov et al.,[15] was first applied with picosecond resolution[16] and further extended to the femtosecond time-domain, demonstrating the creation of a non-equilibrium electron distribution.[17] Different femtosecond techniques were developed to study the bulk metal electron kinetics, monitoring, for example, the transient optical response, transient harmonic generation, or induced photoelectron emission.[18]

Depending on the probing condition and pulse duration, the different steps of metal relaxation can be followed. They take place within a few femtoseconds to a few hundred picoseconds, as schematically shown in Figure 5.1. Coherent electron-light coupling takes place first, the induced polarization decaying on a sub-10-fs time scale.[19–22] It is much shorter than that of the electron-electron and electron-lattice energy redistributions and a strongly non-equilibrium electron distribution is thus created.[18,23–27] The injected energy is subsequently redistributed among the electrons by electron-electron (e-e) scattering, establishing a Fermi-Dirac distribution in few hundred femtoseconds,[18,23–28] transferred to the lattice by electron-phonon (e-ph) interaction on a slightly longer time scale (typically one picosecond),[23–25,29–33] and eventually damped to the environment in few to a few hundred picoseconds.[5,34,35] Material acoustic vibration (Lamb mode vibration, acoustic echoes, etc.) also takes place on a picosecond to a few hundred pisocecond time scale and can be observed in the time domain.[26,36] Extension of these techniques to metal nanoparticles and clusters, as well as metal-based nanomaterials, has permitted researchers to address new problems: electronic motion coherence losses or polarization decay,[19–22] ultrafast electron interactions (e-e and e-ph energy exchanges),[13,14,37–55] acoustic vibrations,[56–62] and energy exchanges with the matrix.[34,35] It opens up a new area in the physics of clusters: investigation of the interaction mechanisms of their elementary excitations and with their environment.

In the following, time-resolved optical investigation of the non-equilibrium electron relaxation in large metal nanoparticles (size larger than 2 nm, i.e., formed by more than about 200 atoms for a nanosphere) is discussed. Noble metals are model systems because of their good chemical stability, the excellent mastering of their synthesis by different chemical and physical techniques, and, consequently, the large diversity of the environment in which nanoparticles can be dispersed.[6] Furthermore, their relatively simple band structure makes possible a direct connection between the electron kinetics and the observed transient optical properties. This permits a simple modeling and interpretation of the time-domain optical data based on modeling of their optical nonlinearity (Section 5.4). It is particularly straightforward in the weak excitation regime, permitting direct extraction of the relevant electron relaxation characteristics. We focus here on the excitation step (Section 5.3) and electron energy redistribution processes by electron-electron (Section 5.5), electron-lattice coupling (Section 5.6), and their interplay (Section 5.7).

5.2   OPTICAL PROPERTIES OF METAL NANOPARTICLES

Time-resolved, pump-probe femtosecond experiments relying on the close connection between the electron properties and the optical response of the material (i.e., its dielectric function), the main features of the latter are summarized in the following for metal nanoparticles and nanomaterials.

5.2.1 METAL NANOPARTICLE OPTICAL RESPONSE

The optical response of a small metal sphere of diameter D embedded in a dielectric medium was described by Mie more than 100 years ago.[11] This general theory can be greatly simplified when one is dealing with objects sufficiently small as compared to the optical wavelength λ (i.e., typically $D \leq \lambda/10$). In this regime, the electromagnetic field can be considered as constant over the nanoparticle and the problem reduces to a static one (quasi-static approximation).[7] It is similar to the local field effect modeled by Lorenz-Lorentz and Clausius-Mossotti in solids and molecular systems, respectively.[3,63]

These approaches allow one to describe the absorption and scattering of individual nanospheres and some other shapes and structures as rods, disks, or multilayered materials.[6,7,11] For example, in the quasi-static limit, the scattering and extinction cross-sections of a nanosphere embedded in a transparent matrix of dielectric constant ε_d are given by:

$$S_{diff} = \frac{24\pi^3 V_{np}^2 \varepsilon_d^2}{\lambda^4} \left| \frac{\varepsilon - \varepsilon_d}{\varepsilon + 2\varepsilon_d} \right|^2 \;\; ; \;\; S_{ext} = \frac{18\pi V_{np} \varepsilon_d^{3/2}}{\lambda} \frac{\varepsilon_2}{\left| \varepsilon + 2\varepsilon_d \right|^2} \tag{5.1}$$

where $\varepsilon(\omega) = \varepsilon_1(\omega) + i\varepsilon_2(\omega)$ is the dielectric function of the metal in the nanoparticle and V_{np} its volume.

In optical experiments, one is generally dealing with a composite material formed by an ensemble of nanoparticles dispersed in a dielectric matrix (liquid or solid). If their density is sufficiently weak to neglect inter-particle interactions, the material optical properties are directly related to the individual particle ones (Equation 5.1) and metal volume fraction p. In the effective material approach, they can be described by introducing an effective dielectric constant (Maxwell Garnett expression)[1,2,5,6,8]:

$$\tilde{\varepsilon}(\omega) = \varepsilon_d + 3p\varepsilon_d \frac{\varepsilon(\omega) - \varepsilon_d}{\varepsilon(\omega) + 2\varepsilon_d} \tag{5.2}$$

where p is, typically, less than 1%. The absorption coefficient can then be written[1,6,8]:

$$\tilde{\alpha}(\omega) = \frac{9p\omega\varepsilon_d^{3/2}}{c} \frac{\varepsilon_2(\omega)}{\left[\varepsilon_1(\omega) + 2\varepsilon_d\right]^2 + \left[\varepsilon_2(\omega)\right]^2} = \frac{\omega}{\varepsilon_d^{1/2}c} p\left|f_{lf}(\omega)\right|^2 \varepsilon_2(\omega) \tag{5.3}$$

and is equal to NS_{ext}, where N is the number of nanoparticles per unit volume. Confinement introduces a factor $|f_{lf}(\omega)|^2$ that modifies the optical absorption spectrum (without local field enhancement effect, $\tilde{\alpha}$ would be given by $\left(\omega/\varepsilon_m^{1/2}c\right) p\varepsilon_2(\omega)$). This factor can be resonant if ε_1 is negative, a condition realized in metals when the conduction electron contribution to the dielectric function dominates (Section 5.2.2). The composite material absorption is then resonantly enhanced close to the frequency Ω_R, maximizing f_{lf}, (i.e., minimizing the denominator). It is the well-known condition for the surface plasmon resonance (SPR). It is concomitant with enhancement of the electric field by a factor f_{lf} inside the particle, and on a scale of about its size around it, as compared to the applied field.[5,6] This effect, referred to as the dielectric confinement effect, can be related to a resonance between the applied optical field and the collective electron oscillation.[1] It is also responsible for the enhancement of the nonlinear optical response whose physical origin is, as in bulk metal, directly connected to the electron distribution dependence of $\varepsilon(\omega)$ (Section 5.2.2).

If the imaginary part of ε_2 is small or weakly dispersed around Ω_R, the resonance condition simply reads:

$$\varepsilon_1(\Omega_R) + 2\varepsilon_d = 0 \tag{5.4}$$

The SPR effect is illustrated in Figure 5.2, which shows the absorption spectrum of silver and gold nanospheres. It must be noted that it is well marked only in noble and alkali metals. In other metals, the SPR overlaps with interband transitions, strongly broadening the resonance.[6]

The optical absorption of small nano-ellipsoids can be similarly calculated in the quasi-static approximation. For instance, for an ensemble of identical randomly

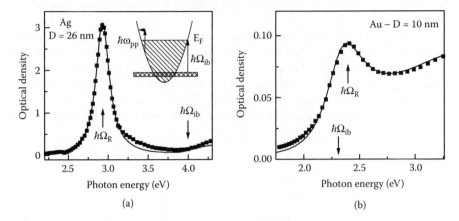

FIGURE 5.2 Measured absorption spectrum of: (a) $D = 26$ nm silver nanospheres embedded in a 50BaO-50P$_2$O$_5$ glass matrix and (b) $D = 10$ nm gold nanospheres in colloidal solution (squares). The absorption for $\hbar\omega \geq \hbar\Omega_{ib}$ with $\hbar\Omega_{ib} \approx 4$ eV in Ag and $\hbar\Omega_{ib} \approx 2.3$ eV in Au is due to the onset of the interband transitions. The lines are calculated using Equation 5.3 with the confinement modified metal dielectric function (Equation 5.13). The inset shows the schematic electron band structure of noble metals. $\hbar\Omega_{ib}$ is the interband transition threshold from the top of the d-bands to the Fermi surface. The arrows indicate intraband electron absorption of a $\hbar\omega_{pp}$ photon.

oriented prolate ellipsoids of eccentricity e, dispersed in a matrix of dielectric constant ε_d, the composite material absorption is given by[6,7]:

$$\tilde{\alpha}(\omega) = \frac{p\omega\varepsilon_d^{3/2}}{c} \sum_{i=x,y,z} \frac{\varepsilon_2(\omega)}{\left[L_i\varepsilon_1(\omega)+(1-L_i)\varepsilon_d\right]^2 + \left[L_i\varepsilon_2(\omega)\right]^2} \tag{5.5}$$

where x is the long axis direction of length l_x and y,z the short axis ones of length $l_y = l_z$ with ($e^2 = 1 - (l_y/l_x.)^2$). L_i are geometrical factors given by:

$$L_x = \frac{1-e^2}{e^2}\left(-1+\frac{1}{2e}\ln\left(\frac{1+e}{1-e}\right)\right) \tag{5.6}$$

and $L_y = L_z = (1 - L_x)/2$. The nanosphere SPR degeneracy is partly lifted and the system exhibits two SPR, associated with electron oscillations along the long and short axes of the ellipsoids.[6,7]

5.2.2 DIELECTRIC FUNCTION OF A METAL NANOPARTICLE

The dielectric constant ε entering the above expressions is that of the metal in a nanoparticle and includes the quantum confinement effects of the electrons. In bulk metals, ε can be separated into conduction and bound electron contributions. The

TABLE 5.1
Fermi Energy E_F, Conduction Electron Density n_e, Static Interband Dielectric Function $\varepsilon^{ib}(0)$, Interband Transition Threshold $\hbar\Omega_{ib}$, Density ρ, Lattice Heat Capacity C_L, Constant a Defining the Electron Heat Capacity $C_e = aT_e$, and Electron-Phonon Coupling Constant for Silver and Gold

	E_F (eV)	n_e (cm^{-3})	$\varepsilon^{ib}(0)$	$\hbar\Omega_{ib}$ (eV)	ρ (kg/m^3)	C_L (Jm^{-3}K^{-1})	a (Jm^{-3}K^{-2})	G (W/Km^3)
Ag	5.49	5.86×10^{22}	3.7	4	10.4×10^3	2.48×10^6	65	2.3×10^{16}
Au	5.53	5.90×10^{22}	6.7	2.3	19.7×10^3	2.55×10^6	65	2×10^{16}

former is related to intraband optical absorption and is well described by a Drude expression (free electron response),[64,65] so that ε can be written:

$$\varepsilon(\omega) = \varepsilon^{ib}(\omega) - \frac{\omega_p^2}{\omega\left[\omega + i/\tau_0(\omega)\right]} \tag{5.7}$$

where ω_p is the plasma frequency ($\omega_p^2 = n_e e^2/\varepsilon_0 m_e$, where n_e is the conduction electron density and m_e the electron mass). $\tau_0(\omega)$ is the electron optical relaxation time determined by electron-phonon and electron-electron scattering (neglecting electron-defect scattering) with simultaneous exchange of the energy $\hbar\omega$ of a photon.[66–68]

ε^{ib} is the bound electron contribution, related to transitions between two different bands (interband transitions). The dominant contribution comes from transitions from the full d-bands to the conduction band in noble metals (Figure 5.2-inset).[64] Due to the Pauli exclusion principle, it takes place only to empty conduction band states (from the top of the d-bands), imposing a photon energy threshold $\hbar\Omega_{ib}$ above which transitions are permitted (Table 5.1). Because of the weak dispersion of the d-bands, a narrow ensemble of conduction band states is involved for a fixed ω frequency. This is in contrast to the intraband transitions, for which a broad continuum of states contributes.

The quasi-free conduction electron motion is modified in a nanocrystal by the interface and the concomitant breakdown of translational invariance. A proper description of any electron-related effect should then use the confined electron wavefunctions and eigenenergies. However, conversely to the semiconductor nanoparticle case, quantum confinement plays a relatively weak role in metal nanoparticles and can be introduced as a correction for sizes down to 1 or 2 nm. This behavior is a consequence of the large electronic density of metals (5×10^{22} cm^{-3}) as compared to those photoexcited or injected in semiconductors (typically 10^{18} cm^{-3}). One is thus dealing with strongly degenerated electron distributions, and most of the electron-related properties involve electronic states close to the Fermi energy. In nanometric size metallic particles, they correspond to large quantum number states, with, for a realistic nanoparticle, weak energy separation.[69]

This can be illustrated using the simple model of electrons in a cubic box of size L^3, confined by an infinite potential.[69] It is well-known that the quantized energy of the electronic states is given by:

$$E_n = \frac{\hbar^2}{2m_e}\left(\frac{\pi}{L}\right)^2 n^2 = \Delta_E n^2 \text{ with } n^2 = n_x^2 + n_y^2 + n_z^2 \qquad (5.8)$$

where $n_{x,y,z}$ are integers. For large n, two consecutive states correspond to increase in n^2 by one (i.e., of n by $dn \approx 1/2n$) and are thus equally spaced by Δ_E. Their degeneracy g_n is the number of states in the shell volume of radius n and thickness dn in the (n_x, n_y, n_z) space region:

$$g_n = \pi n^2 dn = \frac{\pi n}{2} \qquad (5.9)$$

The last occupied states are defined by the Fermi energy E_F and correspond to a quantum number:

$$n_F = \frac{2L}{\lambda_F} \text{ with } E_F = \frac{\hbar^2}{2m_e}\left(\frac{2\pi}{\lambda_F}\right)^2 \qquad (5.10)$$

For a system with perfect symmetry the states around E_F are highly degenerated and equally spaced with a spacing inversely proportional to L^2.[70] Quantum confinement should thus play an important role even for few nanometer particles, with, for instance, for $\Delta_E \approx 90$ meV and a degeneracy $g_{n_F} \approx 12$ ($n_F \approx 8$) for $L = 2$ nm.

Such an effect has not been observed experimentally as a consequence of the nonperfect shape and surfaces of real systems. Imperfections lift the degeneracy of the states and, assuming that they are equally distributed, their spacing decreases with n[71]:

$$\Delta_{En} = \frac{2}{g_n}\Delta_E \text{ yielding } \Delta_{EF} = \frac{\pi\hbar^2}{m_e}\frac{\lambda_F}{L^3} = \frac{4E_F}{3N_e} \qquad (5.11)$$

where N_e is the number of electrons in the particle. The level spacing around E_F is thus strongly reduced, at least for not too small particles. The relevant parameter being n_F, the characteristic size is the Fermi wavelength λ_F (Equation 5.10), of the order of 5 Å in metals. For example, for $L = 2$ nm particles, one obtains $\Delta_{EF} \approx 15$ meV, which is smaller than the room-temperature thermal energy (≈ 25 meV) and the width of the electronic states (≈ 20 meV). As a good approximation, the states around E_F can thus be replaced by a quasi-continuum similar to the bulk metal one, down to nanometer-size particles.

Intraband optical absorption in a nanoparticle corresponds to transitions between confined electronic states with final states above E_F, leading to a structureless absorption spectrum. The associated dielectric constant has been shown to keep a Drude-type form (Equation 5.7),[1,6] with the a modified size-dependent scattering time:

$$\frac{1}{\tau^{nano}} = \frac{1}{\tau_0^{nano}} + 2g_S \frac{V_F}{D} \tag{5.12}$$

The first term on the right-hand side reflects bulklike electron scattering in the particle, with the electron interaction rates possibly altered by confinement. The second term is proportional to the inverse of the classical electron travel time in a particle (i.e., the Fermi velocity V_F divided by the diameter D of the assumed spherical nanoparticle). The frequency- and electron distribution-dependent proportionality factor g_S is on the order of 1.[6,69] This term is a manifestation of quantum confinement in the metal nanoparticle optical response. It is a consequence of the fact that in a particle, the wavefunctions of the confined electrons must be used instead of the Bloch ones in the bulk. The wavevector is no more a good quantum number, and photon absorption is then possible between confined electronic states without any collision.[1,69,72,73] Classically, it can be interpreted as being due to electron scattering off the surface, which as the other electron scattering processes induced breaking of wavevector conservation. This contribution is significant in sufficiently small nanoparticles (i.e., using $\tau_0 \approx 10$ fs in noble metals[66] and $g_S \approx 1$, the surface and bulk contributions are identical for $D \approx 30$ nm).

Investigations of the interband absorption spectrum of gold and silver nanoparticles have shown that the interband dielectric function is weakly modified for sizes down to 2 nm.[74] Above this size, the dielectric constant of metal nanoparticles can be written:

$$\varepsilon^{nano}(\omega) = \varepsilon^{ib}(\omega) - \frac{\omega_p^2}{\omega\left[\omega + i\left(1/\tau_0^{nano} + 2g_S V_F/D\right)\right]} \tag{5.13}$$

The linear absorption spectrum of metal nanoparticles embedded in a dielectric matrix can thus be reproduced using the measured bulk metal interband term and adding a Drude component and introducing a surface correction (from a practical point of view, ε^{ib} is extracted by subtracting the bulk Drude term from the measured ε).[66] This expression of ε is used in the following.

5.2.3 SURFACE PLASMON RESONANCE

In the quasi-static approximation, and provided that the resonance condition (Equation 5.4) is applicable, the SPR frequency is size independent (note that for small particles, quantum effects induce a frequency shift of the SPR[75]):

$$\Omega_R = \omega_p \Big/ \sqrt{\varepsilon_1^{ib}(\Omega_R) + 2\varepsilon_m} \qquad (5.14)$$

This approach yields a good reproduction of the measured absorption spectra in noble metals, provided that g_S is used as a parameter (Figure 5.2).[13] It is justified by the large variation of g_S, from 0.5 to 1.5, depending on the theoretical quantum model used to compute the confined metal dielectric function.[6] However, this parameter essentially influences the broadening of the SPR; its adjustment actually also compensates for other nonquantum effects not introduced here.

In particular, the above modeling assumes an ideal composite material with identical particles experiencing the same dielectric environment. This is not the case in real systems, and it has been shown that the SPR broadening strongly depends on their shape distribution and on their environment (i.e., porous matrix, vacuum, solvent, glass).[6,74] The dependence on the nature of the environment is not predicted by the quantum mechanical models, which include only internal particle effects. It has not been described theoretically and has been qualitatively attributed to modification of the electronic wave function by the surrounding material and called chemical damping.[74] Other effects are also important, such as the inhomogeneity of the matrix. Local fluctuations in the dielectric function ε_d of the surrounding environment on a nanometric scale induce changes in Ω_R from particle to particle and thus an inhomogeneous broadening of the SPR line as observed in porous matrices.[75,76] Ω_R is shape dependent, and shape distribution also leads to an inhomogeneous broadening of the SPR broadening.

5.2.4 NONLINEAR OPTICAL RESPONSE: SAMPLE OPTICAL PROPERTY CHANGES

In time-resolved pump-probe experiments, the time-dependent changes of the sample transmission, $\Delta T(\omega_{pr}, t_D) = T(\omega_{pr}, t_D) - T(\omega_{pr}, -\infty)$, or reflectivity, $\Delta R(\omega_{pr}, t_D)$, induced by a pump pulse at time t = 0, are monitored by a probe pulse of frequency ω_{pr} delayed by t_D. These reflect alterations of the material dielectric function, $\Delta\varepsilon(\omega_{pr})$, that is, the nonlinear optical response of the material. If they are sufficiently weak, a perturbational approach can be used:

$$\frac{\Delta T}{T}(\omega_{pr}, t_D) = \frac{\partial \ln T}{\partial \varepsilon_1} \Delta\varepsilon_1(\omega_{pr}, t_D) + \frac{\partial \ln T}{\partial \varepsilon_2} \Delta\varepsilon_2(\omega_{pr}, t_D)$$

$$\frac{\Delta R}{R}(\omega_{pr}, t_D) = \frac{\partial \ln R}{\partial \varepsilon_1} \Delta\varepsilon_1(\omega_{pr}, t_D) + \frac{\partial \ln R}{\partial \varepsilon_2} \Delta\varepsilon_2(\omega_{pr}, t_D) \qquad (5.15)$$

with $\varepsilon = \varepsilon^{bulk}$ in a metal film and $\varepsilon = \tilde{\varepsilon}$ in a composite material.

In metal film, $\Delta T(t_D)$ and $\Delta R(t_D)$ exhibit similar amplitudes, permitting experimental determination of $\Delta\varepsilon_1^{bulk}(t_D)$ and $\Delta\varepsilon_2^{bulk}(t_D)$. This cannot be done in dilute composite materials because of their small reflectivity and induced reflectivity changes (because $\tilde{\varepsilon}_1 \approx \varepsilon_d$). In these systems, $\Delta T/T$ is only determined by $\Delta\tilde{\varepsilon}_2$, or, equivalently, by the sample absorption change $\Delta\tilde{\alpha}$[13,48]:

$$\frac{\Delta T}{T}(\omega_{pr}, t_D) = -\Delta \tilde{\alpha}(\omega_{pr}, t_D)L = a_1 \Delta \varepsilon_1^{nano}(\omega_{pr}, t_D) + a_2 \Delta \varepsilon_2^{nano}(\omega_{pr}, t_D) \quad (5.16)$$

where L is the sample thickness. It is a linear combination of the changes of the real and imaginary parts of $\Delta \varepsilon^{nano}(\omega_{pr})$. The ω_{pr}-dependent coefficients a_1 and a_2 are related to the equilibrium $\varepsilon^{nano}(\omega_{pr})$ (Equation 5.13) and are entirely determined by the linear absorption properties.[13]

5.3 OPTICAL CREATION OF NON-EQUILIBRIUM ELECTRONS

In ultrafast optical investigations, a femtosecond pump pulse first excites a non-equilibrium electron distribution. In noble metals, if the pulse frequency ω_{pp} is smaller than the interband transition threshold Ω_{ib}, only intra-conduction band transitions are effective. Excitation thus takes place at constant electron density n_e in the bands.

When interband absorption takes place (i.e., for $\omega_{pp} \geq \Omega_{ib}$), the number of electrons in the conduction band increases. The created d-band holes recombine in a few tens of femtoseconds via an Auger process, leading to indirect conduction electron excitation.[77,78] The difference between the created non-equilibrium distributions must be taken into account only when processes taking place in the first tens of femtoseconds are studied, the subsequent electron kinetics being very similar for intra- and interband absorptions. For the sake of simplicity, only intraband excitation will be considered in detail here.

5.3.1 COHERENT ELECTRON-LIGHT COUPLING

Optical excitation of the conduction electrons first involves their coherent coupling with the pulses. In a bulk metal, the optical electric field couples with the electrons, inducing their forced oscillation at the optical frequency ω_{pp}. The electron motion generates a polarization (or equivalently, a current) that radiates an electromagnetic field, this simple picture leading to the Drude expression for the quasi-free electron dielectric function (Equation 5.7). Light absorption takes place with polarization decay due to electron scattering, leading to single-electron excitation (Landau damping), the relevant time being the electron scattering times τ_0 (Equation 5.7). It can be globally interpreted in the quasi-particle interaction model of the free electron absorption: one photon is absorbed by one conduction band electron with the assistance of a third quasi-particle (i.e., a phonon or an electron) to conserve energy and momentum.

In the case of metal nanoparticles dispersed in a dielectric matrix, a similar approach can be introduced, taking into account electron localization at a nanoscale. The optical field induces an electron density oscillation at the optical frequency ω_{pp} in the nanoparticles. This electron movement generates an oscillating dipole in each particle that radiates at the same frequency, modifying the applied electromagnetic field. Such coherent superposition of material polarization and electromagnetic field has been described in bulk dielectric material using mixed material-photon excitations, introducing the polariton concept.[79] Light absorption takes place with polariton damping, which usually reflects damping of its material part (i.e., decay of the induced material polarization). The same approach can be used here, the polarizable entities being the nanoparticles. Electron optical excitation in a nanoparticle can be understood in terms of electron polarization decay. As discussed by Kawabata and Kubo, it takes place with single-electron excitation, which is similar to Landau damping of the collective plasmon mode in a plasma.[1] As in bulk metal, decay is induced by electron scattering in each nanoparticle, that is, by electron-phonon, electron-electron, and, using a classical approach, the additional contribution of electron-surface scattering (Equation 5.2).

In both bulk and confined metallic materials, optical absorption eventually leads to the excitation of single electrons, each of them increasing its energy by $\hbar\omega_{pp}$. In this intraband process, conduction electrons with an energy E between $E_F - \hbar\omega_{pp}$ and E_F are excited above the Fermi energy with a final energy between E_F and $E_F + \hbar\omega_{pp}$ (Figure 5.3). Describing the electron distribution by a one-particle function

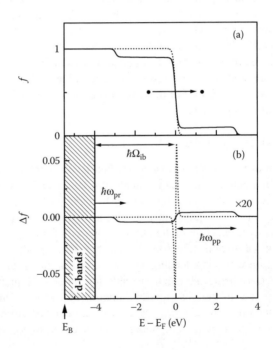

FIGURE 5.3 (a) Equilibrium (f_0, dashed line) and initial athermal (f, full line) electron occupation number for instantaneous intraband single electron excitation by a $\hbar\omega_{pp}$ pump pulse (arrow). (b) Corresponding electron distribution change $\Delta f = f - f_0$ in silver and after thermalization (the injected energy is the same in the two cases). The interband transition threshold $\hbar\Omega_{ib}$ and pump photon energy $\hbar\omega_{pp} = 3$ eV are shown. The arrow indicates probing at $\hbar\omega_{pr} = 1.5$ eV the freed conduction band states from the top of the d-bands. E_B is the energy of the conduction band bottom for a parabolic isotropic bandshape.

f and assuming an isotropic parabolic conduction band, the induced distribution change $\Delta f(E) = f(E) - f_0(E)$ in the constant transition matrix element approximation and weak excitation limit, is given by:

$$\Delta f^{exc}(E) = A\left\{\sqrt{E - \hbar\omega_{pp}}\, f_0(E - \hbar\omega_{pp})\left[1 - f_0(E)\right]\right.$$
$$\left. - \sqrt{E + \hbar\omega_{pp}}\, f_0(E)\left[1 - f_0(E + \hbar\omega_{pp})\right]\right\}$$

(5.17)

where f_0 is the electron distribution before optical excitation (Fermi-Dirac distribution at temperature T_0) and A is a constant defining the injected energy. Electron energy relaxation during the excitation process has been disregarded, and Δf^{exc} exhibits a steplike shape (Figure 5.3).

An important aspect is the time scale of the single-electron excitation (i.e., of that of the coherent polarization decay). For an optical pulse in resonance with the surface plasmon resonance, the induced polarization is usually described in terms of collective electron oscillation in a particle (i.e., of coherent superposition of single electron modes) and its decay discussed in terms of surface plasmon resonance dephasing. The polarization decay in large disk-shaped noble-metal particles has been monitored in the time domain using time-resolved second and third harmonic generation.[19–21] Sub-10-fs decay times have been deduced, consistent with the estimated scattering times τ_0 in noble metals. In the spectral domain, hole burning measurements have been recently performed to estimate the homogeneous surface plasmon resonance linewidth as a function of the size, shape, and environment of oblate silver particles. Dephasing time in the same time ranges are inferred by these spectral results.[22]

The Landau type of decay mechanism can be confirmed by directly monitoring the reduction of the occupation of electron states well below the Fermi energy (i.e., for E energy states with $E_F - E \gg k_B T_e$) that are fully occupied in thermal equilibrium (Figure 5.2-inset and Figure 5.3).[80] This depopulation induces absorption for probe pulses that can excite d-band electrons to these freed conduction band states, that is, satisfying:

$$\hbar\omega_{pr} > \hbar(\Omega_{ib} - \omega_{pp}) \gg k_B T_e$$

(5.18)

leading to a transient increase of the imaginary part of the dielectric function ε_2^{ib}.

In the case of a bulk metal, the temporal evolutions of the changes of the real and imaginary parts of the metal dielectric constant can be deduced from simultaneous measurements of the transmission and reflectivity changes in an optically thick film (Equation 5.15).[31] The results are shown in Figure 5.4 for a 23-nm-thick silver film using $\hbar\omega_{pr} = 1.5$ eV and $\hbar\omega_{pp} = 3$ eV. The measured $\Delta\varepsilon_2^{bulk}$ is influenced by both the interband and intraband contributions to the metal dielectric function. The latter is responsible for the long delay ($t_D > 50$ fs) signal. This part of the response is consistent with the one observed with $\hbar\omega_{pp} = \hbar\omega_{pr} \approx 1.5$ eV for which

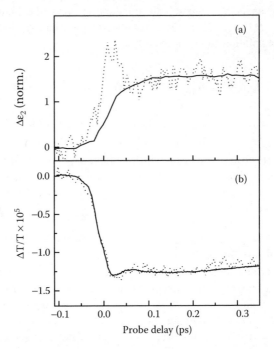

FIGURE 5.4 (a) Change of the imaginary part of the dielectric function $\Delta\varepsilon_2^{bulk}(\omega_{pr})$ measured in a 23-nm-thick silver film for $\hbar\omega_{pr} = 1.5$ eV and a near-infrared or blue pump pulse $\hbar\omega_{pp} = 1.5$ or 3 eV (full and dotted line, respectively). (b) Time dependence of the transmission change $-\Delta T/T$ for the same probing condition and resonant excitation of the surface plasmon resonance $\hbar\omega_{pp} \approx \hbar\Omega_R \approx 3$ eV in $D = 26$ nm Ag nanoparticles in a 50BaO-50P$_2$O$_5$ glass matrix (full line). The dashed line is the normalized $-\Delta T/T$ calculated from the measured $\Delta\varepsilon_1^{bulk}(\omega_{pr})$ and $\Delta\varepsilon_2^{bulk}(\omega_{pr})$ using Equation 5.16.

the condition of Equation 5.18 is not satisfied. It has been ascribed to a delayed modification of the electron scattering time τ_0.[31]

On a shorter time scale, $\Delta\varepsilon_2^{bulk}$ exhibits a fast rise and fall, a signature of the induced interband absorption. It involves final conduction band states around $E = E_F - \hbar(\Omega_{ib} - \omega_{pr}) \approx E_F - 2.5$ eV, in the energy region where the electron occupation number is reduced by single-electron excitation (Figure 5.3). These states relax in a few femtoseconds (about 3 fs for electrons 2.5 eV above E_F,[81] the "holes" below E_F having a similar dynamics as the corresponding electrons above E_F).[25] A transient $\Delta\varepsilon_2^{bulk}$ increase is thus observed, with a temporal shape essentially limited by the pump-probe cross-correlation.

A similar behavior is observed in clusters as shown in Figure 5.4 for $\hbar\omega_{pr} = 1.5$ eV and $\hbar\omega_{pp} = 3$ eV, corresponding to resonant excitation of the nanoparticle surface plasmon resonance. For this probe frequency, the imaginary part of the metal dielectric constant dominates the $\Delta T/T$ signal ($a_2/a_1 \approx 2.7$ in Equation 5.16). The observed time behavior is consistent with the one measured for $\Delta\varepsilon_2^{bulk}$. In particular, a short time delay peak is observed, followed by an increase over a 100-fs scale and picosecond decay due to electron energy transfer to the lattice. For this nanoparticle size ($D = 26$ nm), the electron dynamics is almost identical to the bulk metal one.[13,14] The measured $\Delta T/T$ can be quantitatively compared to the bulk response using Equation 5.16 and $\Delta \varepsilon^{bulk}$ measured for the same pump and probe frequencies. A very good agreement between the experimental and calculated $\Delta T/T$ responses is obtained (Figure 5.4), showing that the bulk and confined materials behave in very similar ways.

For optical pulses down to 15 to 20 fs, no deviation from "instantaneous" single-electron excitation has been observed, consistent with polarization decay with the electron scattering time τ_0. This is even faster in small nanoparticles due to electron-surface scattering (Equation 5.12). In the following, only single-electron excitation and energy relaxation effects are considered.

5.3.2 NON-EQUILIBRIUM ELECTRON GAS ENERGY

The electron gas is collectively described by its temperature T_e, and the electron gas kinetics by the T_e time evolution, when a Fermi-Dirac distribution f_{FD} is established. This is not the case for short times (i.e., a few hundred femtoseconds), for which the distribution is athermal and the non-Fermi-Dirac time-dependent f must be used. An important collective parameter here is the transient excess energy of the electron gas Δu_e, that is, its energy at time t minus the one before excitation:

$$\Delta u_e(t) = \frac{\sqrt{2}m_e^{3/2}}{\pi^2 \hbar^3} \int E^{3/2} \Delta f(E,t) dE = a\left[T_e^2(t) - T_0^2 \right] \tag{5.19}$$

the second equality being of course valid when the electron temperature is established. The temperature dependence of the electron heat capacity has been taken into account, $C_e(T_e) = aT_e$, a being a constant ($a = \pi^2 n_e k_B / 2T_F$, where T_F is the Fermi temperature, for a free electron gas) (Table 5.1).[64]

It is convenient to characterize the total energy injected by the pump pulse by defining a maximum equivalent electron temperature rise, ΔT_e^{me}, as the temperature rise of a thermalized electron gas for the same energy increase:

$$\Delta T_e^{me} = \left[T_0^2 + 2N_{pp} \hbar \omega_{pp} / a \right]^{1/2} - T_0 \tag{5.20}$$

where T_0 is the initial equilibrium temperature of the system and N_{pp} is the number of absorbed pump photons per unit volume.

A consequence of electron confinement in a nanoparticle is localization of the injected energy, excitation being much faster than energy transfer to the surrounding matrix. The minimum energy injection in a nanoparticle corresponds to absorption of one photon. There is thus a minimum ΔT_e^{me} that can be induced in each excited particle, which is set by the pump photon energy. This can be fairly large for small sizes (Figure 5.5), an effect that must be taken into account in interpreting the experimental data.[13]

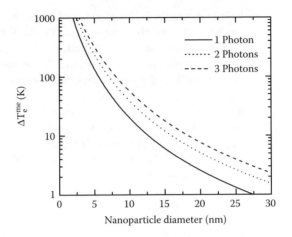

FIGURE 5.5 Maximum equivalent electron temperature rise ΔT_e^{me} induced by absorption of one, two, or three near-infrared photons $\hbar\omega_{pp} = 1.3$ eV in a silver nanosphere as a function of its diameter (using Equation 5.20 with $a = 65$ Jm^{-3}K^{-2}).

The lattice heat capacity C_L being much larger than the electronic one (Table 5.1), a large transient electron perturbation can be achieved with only a small lattice heating when the system eventually reaches quasi-equilibrium. The final temperature rise $T_L - T_0$ of the thermalized nanoparticle is of the order of $(C_e/C_L)\Delta T_e^{me} \approx \Delta T_e^{me}/100$.

5.4   NON-EQUILIBRIUM ELECTRON KINETICS

The excitation step leads to the creation of an incoherent non-equilibrium electron distribution whose time evolution can be described in terms of energy redistribution kinetics. Electron-electron (e-e) scattering redistributes the injected energy among the electrons, eventually leading to a Fermi-Dirac distribution (i.e., the establishment of an electron temperature). Electron-phonon (e-ph) coupling transfers the electron energy to the lattice, thermalizing the nanoparticle (neglecting exchanges with the environment that usually take place on a longer time scale). Modeling of the energy redistribution kinetics and its optical probing is described in the following using a bulk metal approach.

5.4.1   Energy Relaxation Kinetics of Non-Equilibrium Electrons

In bulk noble metals, the time evolution of the conduction electron energy distribution function f can be described by the Boltzmann equation[23,25,82,83]:

$$\frac{df(E,t)}{dt} = \frac{\partial f(E,t)}{\partial t}\bigg|_{e-e} + \frac{\partial f(E,t)}{\partial t}\bigg|_{e-ph} + \frac{\partial f(E,t)}{\partial t}\bigg|_{exc} \tag{5.21}$$

where an isotropic conduction band has been assumed (leading to $f(\mathbf{k}) = f(E(k))$). The first term describes e-e scattering, which, taking energy and momentum conservation into account, reads[23,25,84]:

$$\left.\frac{\partial f(E(\mathbf{k}))}{\partial t}\right|_{e-e} = \frac{2\pi}{\hbar}\sum_{k_1,k_2,k_3}\left|M_{ee}\left(\left|\mathbf{k}-\mathbf{k}_2\right|\right)\right|^2 F(E,E_1,E_2,E_3)\delta_k\delta_E \qquad (5.22)$$

where δk and δE stand for electron momentum and energy conservations: $\mathbf{k} + \mathbf{k}_1 - \mathbf{k}_2 - \mathbf{k}_3 = 0$ and $E(k) + E_1(k_1) - E_2(k_2) - E_3(k_3) = 0$. F includes the Pauli exclusion effect for electron scattering in and out of the E state:

$$F = -f(E)f(E_1)\left[1-f(E_2)\right]\left[1-f(E_3)\right]+\left[1-f(E)\right]\left[1-f(E_1)\right]f(E_2)f(E_3) \quad (5.23)$$

Assuming statically screened Coulomb electron interaction, the e-e scattering matrix element is:

$$M_{ee}(q) = \frac{e^2}{\varepsilon_0\varepsilon^{ib}(0)}\frac{1}{q^2+q_S^2} \qquad (5.24)$$

The static description of screening overestimating reduction of the e-e scattering amplitude, a phenomenological screening reduction has been introduced using a screening wavevector

$$q_S = \beta q_{TF} \approx \frac{\beta e\sqrt{m_e}}{\pi\hbar\sqrt{\varepsilon_0\varepsilon^{ib}(0)}}\left(3\pi^2 n_e\right)^{1/6} \qquad (5.25)$$

instead of the Thomas Fermi one q_{TF} (with $\beta = 0.73$ in bulk silver and gold).[23,25,85] Assuming a Fermi-Dirac distribution at zero temperature, this term yields the usual expression for the electron lifetime in the vicinity of the Fermi surface due to inelastic e-e collisions[28,84]:

$$\frac{1}{\tau_{e-e}(E)} = \frac{m_e e^4(E-E_F)^2}{64\pi^3\hbar^3\varepsilon_0^2\varepsilon_{ib}^2(0)E_S^{3/2}\sqrt{E_F}}\left[\frac{2\sqrt{E_F E_S}}{4E_F+E_S}+\arctan\left(\sqrt{\frac{4E_F}{E_S}}\right)\right] \qquad (5.26)$$

with $E_S = \hbar^2 q_S^2/2m_e$. It exhibits the characteristic $(E-E_F)^{-2}$ variation as a consequence of phase space filling (Pauli exclusion effect).

The second term on the right-hand side is the e-ph scattering rate[25,82]:

$$\left.\frac{\partial f(E(\mathbf{k}))}{\partial t}\right|_{e-ph} = \frac{2\pi}{\hbar}\sum_{q}\left|M_{eph}\right|^2 F^-(\mathbf{k},\mathbf{q})\cdot\delta(E(\mathbf{k})-E(\mathbf{k}-\mathbf{q})-\hbar\omega_q)+$$

$$+\frac{2\pi}{\hbar}\sum_{q}\left|M_{eph}\right|^2 F^+(\mathbf{k},\mathbf{q})\cdot\delta(E(\mathbf{k})-E(\mathbf{k}+\mathbf{q})+\hbar\omega_q) \qquad (5.27)$$

where $\hbar\omega_q$ is the energy of the q wavevector phonon and:

$$F^-(k,q) = -f(E(k))[1 - f(E(|k - q|))][1 + N(\hbar\omega_q)] +$$
$$+ [1 - f(E(k))]f(E(|k - q|))N(\hbar\omega_q) \tag{5.28}$$

and $F^+ (k,q) = -F^- (k + q,q)$. $N(\hbar\omega_q)$ is the occupation number of the q phonon. Assuming deformation potential coupling, the e-ph interaction matrix element is:

$$\left|M_{eph}(q)\right|^2 = \frac{\hbar^2 \Xi^2}{2\rho V} \frac{q^2}{\hbar\omega_q} \tag{5.29}$$

where Ξ is the deformation potential and ρ the material density. Although it is a rough approximation in metal, it has been shown that the exact nature of the e-ph coupling does not influence the computed electron dynamics. This is a consequence of the fact that the lattice temperature T_L is larger than the Debye temperature Θ_D (Table 5.1); electron distribution changes on the energy scale of a phonon have a minor influence on the overall dynamics. In the following, the Debye model for the phonon dispersion will be used: $\omega_q = v_s q$, v_s being the material sound velocity (note that, as above the interaction Hamiltonian, the exact phonon dispersion has no impact on the computed data for T_L much larger than Θ_D).

The last term in Equation 5.21 describes incoherent single-electron excitation. It is identical to Δf^{exc} (Equation 5.17), replacing f_0 by f and A by $BI_{pp}(t)$ to take into account the finite duration of the pump pulses (B is a constant and $I_{pp}(t)$ is the time-dependent pump-pulse intensity). The computed time-dependent electron distribution function is shown in Figure 5.6 in the case of silver for a weak excitation, $\Delta T_e^{me} = 100 \text{K}$. Fast relaxation of the high-energy electron leads to a buildup of the induced distribution change amplitude $\Delta f(E,t) = f(E,t) - f_0(E)$ around E_F that takes place concurrently with excitation by the pump pulse. This fast energy redistribution is at the origin of the $\Delta f(E)$ shape distortion as compared to instantaneous excitation (Figure 5.3).

Δf initially extends over a very broad energy range and subsequently strongly narrows as the electron gas internally thermalizes (Figure 5.6). It has been shown that these kinetics are independent of the injected energy in the weak perturbation regime (i.e., for $\Delta T_e^{me} \leq 200 - 300 \text{K}$) and strongly depend on the excitation amplitude for larger energy injection (Section 5.4.3).[25]

5.4.2 OPTICAL PROBING OF THE NON-EQUILIBRIUM ELECTRON KINETICS

Information on the electron kinetics is obtained in optical pump-probe experiments, provided that the dielectric function change $\Delta\varepsilon$ is connected to Δf. In noble metals, the interband contribution $\Delta\varepsilon^{ib}$ usually dominates the observed response. This change is a consequence of the interband absorption spectrum change around $\hbar\Omega_{ib}$ (except

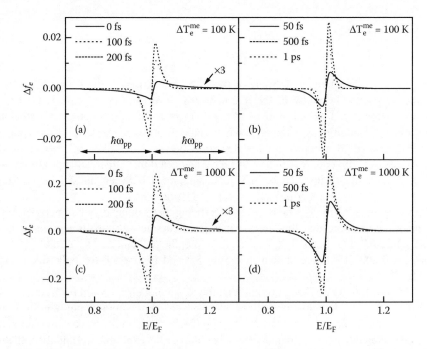

FIGURE 5.6 Distribution changes $\Delta f = f - f_0$ computed in bulk Ag with the Boltzmann Equation 5.21 for different time delay after intraband excitation with a $\hbar\omega_{pp} = 0.24\,E_F$ 25 fs pump pulse,[25] in the weak and strong excitation regime $\Delta T_e^{me} = 100K$ and 1000K, respectively ($t = 0$ is defined as the pump pulse intensity maximum).

for very non-equilibrium situations) due to electron distribution smearing induced by the pump pulse (i.e., the change of the occupation of the electronic states close to E_F).[23–25] To quantitatively describe this effect for interpreting cw thermomodulation measurements, Rosei and co-workers have modeled the bulk noble metal band structure around the Brillouin zone points contributing to the absorption.[86,87]

In Ag, absorption takes place around the L point of the Brillouin zone, with a dominant contribution due to transitions from the d-bands to the conduction band (d → c) and a weaker one due to transition from the conduction band to a higher energy empty s-band (c → s). Using this model, in the constant transition matrix element approximation, $\Delta\varepsilon_2^{ib}$ is related to Δf by[86]:

$$\Delta\varepsilon_2^{ib} \propto -\frac{1}{\omega^2}\left[\int_{E_{dc}^m}^{E_{dc}^M} D_{d\to c}(E,\omega)\Delta f(E)dE - K_{sd}\int_{E_{cs}^m}^{E_{cs}^M} D_{c\to s}(E,\omega)\Delta f(E)dE\right] \quad (5.30)$$

where $D(E,\omega)$ is the energy-dependent joint density of state for the c → s or d → c transitions and K_{sd} is their relative amplitude. In Au, only d-band to conduction band transitions are taking place close to Ω_{ib}, with a dominant contribution at the L point and a weaker one around the X point, at lower energy[87]:

$$\Delta\varepsilon_2^{ib} \propto -\frac{1}{\omega^2}\left[\int_{E_{dc,L}^m}^{E_{dc,L}^M} D_{d\to c}^L(E,\omega)\Delta f(E)dE - K_{XL}\int_{E_{dc,X}^m}^{E_{dc,X}^M} D_{d\to c}^X(E,\omega)\Delta f(E)dE\right]$$ (5.31)

where K_{XL} is the relative amplitude of the d → c transitions at the X and L points. In these models, the width of the interband transitions has been neglected.[39] Estimation of the actual influence of its change on $\Delta\varepsilon^{ib}$ is difficult because its inclusion introduces an additional parameter, when comparing experimental and theoretical results, that could compensate for deviation between the real and model band structures. As a first approximation, we have neglected this effect, an approximation phenomenologically justified by the good agreement between the calculated and measured transient optical property spectra (Section 5.5).

These models link $\Delta\varepsilon_2^{ib}(\omega)$ to Δf, $\Delta\varepsilon_1^{ib}(\omega)$ being then obtained by Kramers-Kronig transformation (note that this approach is justified since $\Delta\varepsilon_2^{ib}(\omega)$ is non-zero only over a limited range around Ω_{ib}). As an example, the $\Delta\varepsilon_1^{ib}$ and $\Delta\varepsilon_2^{ib}$ spectra computed for $t_D = 0$ and 500 fs are shown in Figure 5.7 and Figure 5.8 for bulk silver and gold.[23,25]

One must, however, keep in mind that the Rosei models are valid for frequencies close to Ω_{ib} and a quasi-thermalized electron gas. For strongly non-equilibrium electrons and frequency away from Ω_{ib}, significant deviations can be observed and must be taken into account to quantitatively compare the experimental and theoretical results.[25] In this regime, the simple model of a parabolic conduction band and undispersed d-bands can be used,[25] where, in contrast to the Rosei's description, a frequency ω_{pr} probes a single energy class of electron states defined by: $E = \hbar\omega_{pr} + (E_F - \hbar\Omega_{ib})$.

In nanoparticles, the electron kinetics can *a priori* not be described by the above e-e and e-ph scattering rates, which have been computed using the bulk wavefunctions and taking into account momentum conservation. However, as a first approximation to describe the main features of the time-resolved optical data, we will assume that electron kinetics are not deeply modified for not too small particles. It turns out that this small solid approach is correct for sizes down to 10 nm, a large increase of the electron-scattering efficiency as compared to the bulk ones being observed for the smaller ones (Section 5.5 and 5.6). Keeping in mind this limitation and taking into account that the interband transitions are weakly modified for sizes down to about 2 nm, the interband contribution $(\Delta T/T)_{ib}$ to $\Delta T/T$ is obtained using the computed bulk $\Delta\varepsilon^{ib}$ and Equation 5.16. The difference with the bulk response only lies in the coefficients a_1 and a_2 linking $\Delta T/T$ to $\Delta\varepsilon_1$ and $\Delta\varepsilon_2$.

In silver nanospheres, the computed $\Delta T/T$ exhibits two distinct spectral features around Ω_{ib} and Ω_R with very different time behaviors (Figure 5.7). The former is similar to that observed in bulk silver and corresponds to resonant probing of the interband transitions.[25] In this region, a_1 and a_2 are small and almost undispersed. The $(\Delta T/T)_{ib}$ spectral shape thus reflects the strong dispersion of the real and imaginary parts of $\Delta\varepsilon^{ib}$. Its time behavior is directly related to the electron distribution around the Fermi energy and thus to the internal electron thermalization dynamics. In particular, it exhibits a delayed rise, a signature of the noninstantaneous establishment of an electron temperature (Section 5.5).[25,37]

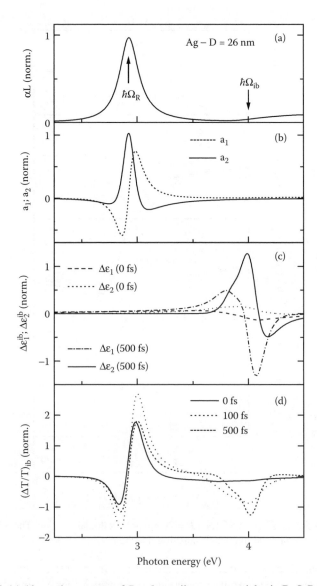

FIGURE 5.7 (a) Absorption spectra of $D = 6$ nm silver nanoparticles in BaO-P_2O_5. The SPR frequency $\hbar\Omega_R = 2.93$ eV and interband transition threshold $\hbar\Omega_{ib} \approx 4$ eV are indicated. (b) Dispersion of the coefficients a_1 and a_2 linking $\Delta T/T$ to $\Delta\varepsilon_1$ and $\Delta\varepsilon_2$ (Equation 5.16) for the same sample. (c) Interband contribution to $\Delta\varepsilon_1$ and $\Delta\varepsilon_2$ calculated for $t_D = 0$ fs and 500 fs for excitation with a 25-fs near-infrared pulse and $\Delta T_e^{me} = 100$K. (d) Corresponding induced transmission change due to the interband contribution $(\Delta T/T)_{ib}$ for $t_D = 0$, 100, and 500 fs.

In contrast, the structure around Ω_R is a consequence of the well-known resonant enhancement of the nonlinear optical response around the SPR due to the dielectric confinement effect.[5,72,88] It is specific to confined systems and reflects in a large

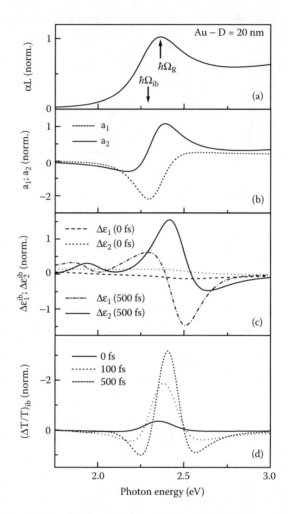

FIGURE 5.8 Same as Figure 5.7 for a $D = 20$ nm gold colloidal solution. The SPR frequency is $\hbar\Omega_R = 2.38$ eV and the interband transition threshold $\hbar\Omega_{ib} \approx 2.3$ eV.

amplitude and dispersion of a_1 and a_2 around Ω_R. In this region, $\Delta\varepsilon^{ib}$ is small and almost undispersed because it corresponds to frequencies away from the interband transitions threshold. Actually, Ω_R being much smaller than Ω_{ib}, $\Delta\varepsilon_2^{ib}(\omega \approx \Omega_R)$ is non-zero only for very short times t_D, when the conduction electrons are strongly out of equilibrium (typically for $t_D \leq 50$ fs). The short time scale $\Delta\varepsilon_2^{ib}$ contribution shows up in a small red shift of the computed $(\Delta T/T)_{ib}$ spectral shape for $t_D = 0$ fs, as compared to the ones for $t_D = 100$ and 500 fs. For these longer delays, only $\Delta\varepsilon_1^{ib}$ is non-zero around Ω_R (optical Kerr effect) and the $(\Delta T/T)_{ib}$ spectral shape reflects that of a_1. Its time behavior essentially follows that of $\Delta\varepsilon_1^{ib}(\Omega_R)$, which has been shown to be almost identical to the time evolution of the electron gas excess energy.[24,25,31] Consequently, in this spectral range, $(\Delta T/T)_{ib}$ "instantaneously" rises with energy injection in the electrons and decays with the electron energy losses to the lattice (note that $(\Delta T/T)_{ib}$ is smaller for $t_D = 0$ fs than 100 fs because only part of the energy is injected into the electron gas by a finite-duration pulse for 0 delay).

Knowing the dielectric function change of a metal, the transient $(\Delta T/T)_{ib}$ spectrum can be computed for different composite nanomaterials made of nanoparticle of different shapes dispersed in different environments using the relevant a_1 and a_2 coefficients, Equation 5.16 (at least for not too small objects so that the bulk electron kinetics can be used). It is illustrated in Figure 5.9, in the case of an ensemble of randomly oriented silver ellipsoids (prolate shape with small-over-large length ratio $l_x/l_y = 0.5$; i.e., eccentricity $e = 0.87$) dispersed in fused silica. The a_1 and a_2 coefficients are calculated using Equations 5.5 and 5.6. As compared to the nanosphere case, an additional low-frequency structure shows up in the computed $(\Delta T/T)_{ib}$ spectrum. It is a consequence of the splitting of the SPR and is associated with electron oscillation along the long axis of the ellipsoids. Concomitantly, the high-frequency feature is blue shifted, following the second SPR associated to the short axis (Figure 5.9). Note that the alteration of the observed spectrum is only a consequence of the way the electron properties are probed, the underlying fundamental physical processes remaining unchanged.

The clear frequency separation of the interband- and confinement-induced resonances in the nonlinear optical response only takes place in silver. In other metals, these features overlap, making their time-dependent optical response more complex.[89] In gold nanospheres, the SPR is close to Ω_{ib}, and these two features overlap, leading to more complicated spectral and temporal behaviors of $(\Delta T/T)_{ib}$ (Figure 5.8). The spectral shape is actually determined simultaneously by the SPR nonlinear enhancement effects (a_1 and a_2 dispersions) and the interband transition feature ($\Delta\varepsilon_1^{ib}$ and $\Delta\varepsilon_2^{ib}$ dispersions). In particular, this leads to a small drift with time of the frequency where $(\Delta T/T)_{ib}$ is maximum, mostly due to the evolution of the $\Delta\varepsilon$ spectral shape during electron thermalization.[23,89] The time evolution of $(\Delta T/T)_{ib}$ is largely influenced by the internal electron thermalization, showing a delayed rise and frequency-dependent decay (Section 5.4.3). Similar effects are expected in copper, which exhibits the same degeneracy between the interband transitions and SPR.[6,38,39]

It is interesting to note that in the case of silver, if the real and imaginary parts of $\Delta\varepsilon$ are weakly dispersed, the characteristic spectral shape of their contributions to $\Delta T/T$ imposed by the a_1 and a_2 dispersion closely corresponds to that obtained for a SPR spectral shift and broadening, respectively. This is simply related to the fact that in silver, the SPR exhibits a quasi-Lorentzian shape with a frequency determined by $\varepsilon_1(\Omega_R)$ and a width by $\varepsilon_2(\Omega_R)$.[13,48] The full approach used here can thus be reduced to a simple analysis in terms of SPR frequency shift and induced broadening, related to $\Delta\varepsilon_1(\Omega_R)$ and $\Delta\varepsilon_2(\Omega_R)$.[13,48] This is not the case in gold, where the linear absorption spectrum around the SPR cannot be simply described by a quasi-Lorentzian shape. In particular, although the lineshape of the computed $(\Delta T/T)_{ib}$ is similar to the one obtained for an induced broadening effect, it cannot be interpreted in terms of SPR broadening (the SPR width cannot be correctly defined in the case of gold or copper), nor can it be related to an electron scattering rate increase. Its physical origin is much more complex and is related to both $\Delta\varepsilon_1$ and $\Delta\varepsilon_2$, with contributions weighted by the enhancement of the nonlinear response around the SPR (Figure 5.8).

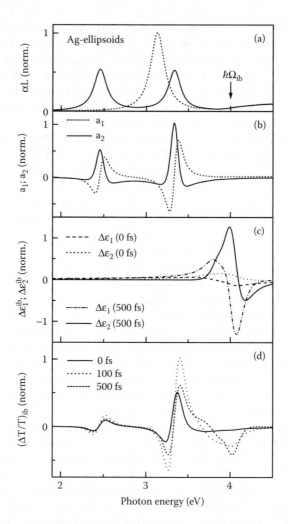

FIGURE 5.9 Same as Figure 5.7 for randomly oriented silver nanoellipsoids with prolate shape and a short-over-long axis ratio $l_y/l_x = 0.5$, embedded in fused silica. A size effect on the metal dielectric function for a 15-nm particle has been assumed (Equation 5.13). Also shown is the absorption spectrum of nanospheres (dashed line in Figure 5.9a).

The intraband term ε is also modified by electron excitation. Its contribution to $\Delta\varepsilon$ is, however, weaker than the interband one, except when probing away from the interband transitions (Figure 5.4). Its physical origin takes place in the dependence of the probabilities of all the electron scattering processes on the electron distribution (or temperature T_e when it is established) and lattice temperature T_L. In nanoparticles, we can write:

$$\Delta\gamma = \Delta\gamma_{e-ph}(T_e, T_L) + \Delta\gamma_{e-e}(T_e, T_L) + \Delta\gamma_S(T_e) \tag{5.32}$$

where $\gamma_S = 2g_S(T_e)V_F/D$. The electron-electron and electron-surface terms $\Delta\gamma_{e-e}$ and $\Delta\gamma_S$ rise and decay with the electron temperature, while $\Delta\gamma_{e-ph}$ rises with the lattice temperature (i.e., T_e decay). Note that the T_e dependencies of the electron optical scattering rates are, however, much weaker than for the DC rates $\gamma(\omega = 0)$. The surface term has been shown to yield an important contribution in small metal nanoparticles. It is actually the manifestation of quantum confinement in their non-linear optical response.[13,48] In the equilibrium regime ($T_e = T_L$), the purely electronic mechanisms are almost negligible (because $\Delta T_L \ll \Delta T_e^{me}$) and $\Delta\gamma$ is essentially determined by $\Delta\gamma_{e-ph}$ (i.e., by ΔT_L).

5.4.3 TIME-DEPENDENT OPTICAL RESPONSE

The optical property changes $\Delta T/T$ and $\Delta R/R$ accessed in time-domain experiments globally reflect the time behavior of the non-equilibrium conduction electron distribution, that is, on a few picosecond scale, mostly two processes: (1) internal electron thermalization and (2) electron-lattice energy exchanges. Their complex frequency dependencies are a consequence of the way these fundamental processes are probed. For some probe frequency, it also leads to complex time behavior of the signals, not associated with electron relaxation kinetics, and with interplays between the e-e and e-lattice processes. Extraction of characteristic electron interaction times must thus be done keeping in mind these effects, illustrated in Figure 5.10 in the case of gold for both weak and strong excitation regimes. Interpretation is frequently more complex in the strong excitation regime, for which the observed $(\Delta T/T)_{ib}$ spectral and temporal shapes are pump intensity dependent (Figure 5.10). They reflect the excitation-dependent electron kinetics, with acceleration of the establishment of an electron temperature and a large slowing-down of the electron-lattice energy exchanges with a ΔT_e^{me} increase.[25,40,44,45,56]

5.5 ELECTRON-ELECTRON ENERGY EXCHANGES

Electron-electron scattering and the correlated internal thermalization dynamics of the conduction electrons have been investigated in bulk gold and silver, taking advantage of the sensitivity of the interband absorption around Ω_{ib} on the electron occupation number change around E_F.[23,25] Electron internal thermalization is concomitant with a buildup of Δf close to E_F (Figure 5.6) and thus of $(\Delta T/T)_{ib}$ around Ω_{ib} (Figures 5.7 and 5.8). In this frequency range, $\Delta T/T$ is very sensitive to the thermal or athermal character of the distribution,[23,25] and to its evolution to a Fermi-Dirac distribution (i.e., establishment of the electron temperature, T_e).

The situation is similar in large clusters ($D \geq 2$ nm), their absorption around Ω_{ib} being almost unaffected by confinement. This is backed up by the very good agreement between the spectral shape of the measured $\Delta T/T$ and computed $(\Delta T/T)_{ib}$, which dominates the signal in this frequency range (Figure 5.11). In particular, in both silver and gold, they exhibit the same narrowing with time and concomitant large increase of the $\Delta T/T$ amplitude around $\omega_{pr} = \Omega_{ib}$, accompanied by sign changes with time on its blue and red wings, As in films for cw and transient measurements,[23,25,87,90]

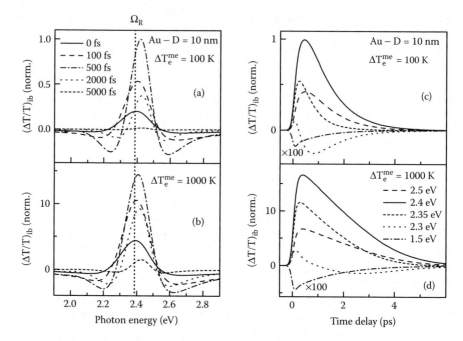

FIGURE 5.10 Computed-induced transmission change due to the interband contribution $(\Delta T/T)_{ib}$ around the surface plasmon frequency resonance $\hbar\Omega_R = 2.38$ eV in gold nanospheres in colloidal solution for different delays t_D after excitation by a 140-fs near-infrared pump pulse and $\Delta T_e^{me} = 100$K (a) and 1000K (b). (c) and (d): Corresponding time dependence of $(\Delta T/T)_{ib}$ for different probe photon energies.

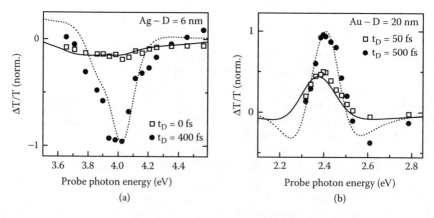

FIGURE 5.11 (a) Spectrum of the transmission change $\Delta T/T$ measured around the interband transition threshold $\hbar\Omega_{ib} \approx 4$ eV for pump-probe delays of 0 fs and 400 fs in $D = 6$ nm Ag nanospheres in BaO-P_2O_5. (b) Spectrum of $\Delta T/T$ for $D = 20$ nm gold colloid measured around the interband transition threshold $\hbar\Omega_{ib} \approx 2.3$ eV and SPR frequency $\hbar\Omega_R = 2.38$ eV for pump-probe delays of 50 fs and 500 fs. The full and dashed lines are the calculated $(\Delta T/T)_{ib}$ from the transient electron distribution (Section 5.4.2).

slightly broader experimental than calculated shapes are observed that can be ascribed to the approximations made in the band structure modeling and/or neglect of the width of the electronic states.[39]

The observed features, similar to those reported in metal films,[23,25] are signatures of the existence of a non-Fermi electron distribution in metal nanoparticles on a few hundred femtosecond time scale (i.e., a delayed electron gas internal thermalization). The data have been checked to be independent of the pump intensity and wavelength, showing that the measurements are performed in the weak excitation regime (both tests are necessary for the smallest nanoparticles, for which excitation takes place with absorption of a single photon, Figure 5.5).[89] It is thus possible to follow the internal electron thermalization dynamics and to compare it for different particle sizes, that is, to analyze the impact of confinement on the electron-electron interactions.[37,89]

This can be done by monitoring the spectral shape narrowing of $\Delta T/T$ around Ω_{ib} or, equivalently, the rise of its maximum amplitude around 4 eV and 2.3 eV in silver and gold, respectively. Characteristic internal thermalization time τ_{th} has been defined by fitting the measured signal around these probe photon energies, using a response function of the form[23,25,37,89]:

$$u(t) = H(t)\left[1 - \exp(-t/\tau_{th})\right]\exp(-t/\tau_{e-ph}) \qquad (5.33)$$

where $H(t)$ is the Heaviside function (Figure 5.12). The exponential decay with time τ_{e-ph} stands for electron energy transfer to the lattice. For D $\geq$ 10 nm, τ_{th} is close to its film value ($\tau_{th} \approx 350$ fs and 500 fs in Ag and Au, respectively),[23,25] and strongly decreases for smaller sizes. This effect is independent of the cluster environment and excitation conditions (Figure 5.13), demonstrating an intrinsic confinement-induced fastening of the electron-electron energy exchanges.[37,89] Similar measurements were performed in gold colloids with $D = 9$ to 48 nm by Link and El-Sayed.[14] A delayed signal rise has also been observed around Ω_{ib}, yielding τ_{th} values almost identical to the bulk one. This absence of variation in this size range is consistent with our results.

It has been shown in bulk metal that τ_{th} is essentially determined by e-e collisions around the Fermi surface. Their probabilities are strongly reduced by the Pauli exclusion principle effects, making them the slowest scattering processes involved in the internal thermalization.[23,25] The experimental results have been found to be in quantitative agreement with the computed ones describing e-e scattering by a statically screened Coulomb potential, with phenomenologic screening reduction (Section 5.4). For a weak perturbation, the dependence of τ_{th} on n_e and core electron screening amplitude (related to $\varepsilon_{sc} = \varepsilon^{ih}(0)$) is identical to that of the scattering time of an electron out of its state τ_{e-e}[25,85]:

$$\tau_{e-e} \propto n_e^{5/6}\varepsilon_{sc}^{1/2}\left[\frac{2\sqrt{E_F E_S}}{4E_F + E_S} + \arctan\sqrt{\frac{4E_F}{E_S}}\right] \propto n_e^{5/6}\varepsilon_{sc}^{1/2} \qquad (5.34)$$

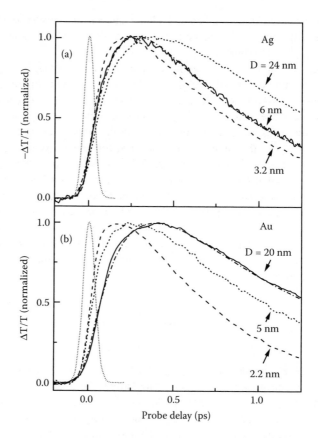

FIGURE 5.12 (a) Time behavior of the transmission change $-\Delta T/T$ measured for $\hbar\omega_{pr} =$ 3.95 eV and $\hbar\omega_{pp} = 1.32$ eV in spherical Ag nanoparticles of diameter $D = 24$ nm and 6 nm in BaO-P$_2$O$_5$ and $D = 3.2$ nm in Al$_2$O$_3$. (b) Time-dependent $\Delta T/T$ measured in $D = 20$ and 5 nm gold colloidal solutions and D = 2.2 nm nanoparticles in a Al$_2$O$_3$ matrix. The dash-dotted lines in (a) and (b) are fits to the Ag-$D = 6$ nm and Au-$D = 20$ nm cases using Equation 5.33. The dotted lines are the pump-probe cross-correlations.

This dependence quantitatively explains the faster electron scattering measured in bulk silver than in gold (350 fs as compared to 500 fs), d-band electron screening being larger in the latter (Table 5.1).[25]

The τ_{th} size dependence has been ascribed to intrinsic confinement-induced fastening of the electron-electron energy exchanges, due to less efficient screening of the e-e Coulomb interactions close to a surface.[37,89] It has been shown that the wave-functions of the conduction electrons extend beyond the particle radius defined by the ionic lattice (electron spillout).[91] This leads to an effective reduction in their density n_e and thus in ω_p. It manifests itself by a red shift of the SPR with size reduction in alkali metal clusters.[92] Conversely, the d-electron polarizability is reduced close to the surface, decreasing their contribution to ε^{ib}.[75,93] Both effects reduce the efficiency of the screening of the e-e Coulomb interactions. Using this model and the modeled spatial variations of ε_{sc} and n_e in a cluster, a good reproduction of the

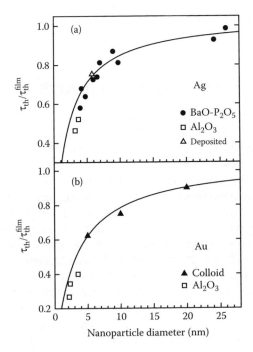

FIGURE 5.13 Size dependence of the characteristic time τ_{th} for the electron internal thermalization (establishment of the electron temperature) for Ag (a) and Au (b) nanoparticles embedded in a BaO-P_2O_5 (full dots) or Al_2O_3 (open squares) matrix, deposited on a substrate (open triangle) or in colloidal solution (full triangles). All data are normalized to the measured metal film values τ_{th}^{film}. The lines are the computed τ_{th} taking into account both the spillout and d-electron localization effects.[37,89]

experimental τ_{th} size dependence was obtained (Figure 5.13). It actually reflects enhancement of the e-e scattering close to the metal surface.[37,89]

The above model, based on a simple extension of the bulk calculations, overlooks specific features of the confined materials. In particular, the e-e scattering rate has been derived using the bulk electron wavefunctions and includes momentum conservation. It is relaxed in confined systems, due to, classically, electron scattering off the surfaces leading to opening of additional e-e scattering channels (i.e., further size-dependent modifications of the e-e interactions) (this effect is comparable to the increase of the electron optical scattering rate due to surface effects, Equation 5.12). For large excitation, other effects can also influence the observed kinetics, such as resonant dynamic screening modification.[93,94] Both in the weak and strong excitation regimes, a full electron kinetic description requires nonlocal calculations of the e-e scattering rate and many body effects in a nanoparticle using the confined electron wavefunctions and modified interaction potential.

These first measurements open up many possibilities for the investigation of the quantum confinement effect on electron kinetics and interactions. In particular, their extension to other metals or types of confinement (one- or two-dimensional systems) and the use of other techniques would be particularly interesting. This is, in particular, the case of time-resolved photoemission, which has been largely used to study electron kinetics in few atom clusters,[53-55] but only scarcely in few hundred atom nanoparticles.[51]

5.6 ELECTRON-LATTICE ENERGY EXCHANGES

The transient signals measured in metallic systems decay in a few picoseconds with the electron gas excess energy and thus contain information on the energy exchanges of the electrons with their environment. In the case of films, this is essentially limited to the metal lattice permitting direct measurement of the electron-phonon interaction time (i.e., the energy transfer time as defined using the two-temperature model, Equation 5.35). For nanoparticles, the electron energy can also be damped to the surrounding solvent or matrix, either directly or via the metal lattice, possibly modifying the observed relaxation. This coupling is frequently assumed to be sufficiently slow to be neglected on the scale of the metal electron-lattice energy exchange. However, it strongly increases with size reduction and may play a role in the observed electron cooling, especially for large electron excitation.[40,45,96]

As long as only energy exchanges inside a nanoparticle are considered, they can be described using the bulk metal two-temperature model.[15,97,98] Note that this model can be generalized to include energy transfer to the environment, adding a space-dependent temperature of the matrix, an extension not considered here.[96] The two-temperature model is based on the assumption that both the electron and the lattice are thermalized at different temperatures T_e and T_L and can only be used after internal electron thermalization (i.e., a few hundred femtoseconds to one picosecond) (note that the assumption that the lattice temperature is always maintained by phonon anharmonic interactions can certainly be questioned). The electron gas is then collectively described by its temperature, and its cooling dynamics can be simply modeled using the rate equation system[15,97,98]:

$$C_e(T_e)\frac{\partial T_e}{\partial t} = -G(T_e - T_L)$$

$$C_L\frac{\partial T_L}{\partial t} = G(T_e - T_L)$$

(5.35)

where G is the effective electron-phonon coupling constant. It can be deduced from the Boltzmann equation model (Section 5.4.1), computing the electron energy loss rate to the lattice in the thermalized regime (with $T_e, T_L > \Theta_D$):

$$C_e\frac{\partial T_e}{\partial t} = \frac{\partial \Delta u_e}{\partial t} = \frac{m_e^{3/2}\sqrt{2E}}{\pi^2\hbar^3}\int E\frac{\partial f(E)}{\partial t}\bigg|_{e-ph} dE = -G(T_e - T_L)$$

(5.36)

with

$$G = \Xi^2\frac{k_B m_e^2 q_D^2}{16\rho\pi^2\hbar^3}$$

(5.37)

where q_D is the Debye wave vector.[64]

For a weak excitation, $\Delta T_e^{me} \ll T_0$, $C_e(T_e)$ can be identified with $C_e(T_0)$, leading to an exponential decay of the electron temperature rise ΔT_e:

$$\Delta T_e = T_e(t) - T_L(t) = (T_{exc} - T_0) \exp\left[-(t - t_{th})/\tau_{e-ph}\right] \tag{5.38}$$

with the time constant[23,25]:

$$\tau_{e-ph} = \frac{C_e(T_0)C_L}{G(C_e(T_0) + C_L)} \approx C_e(T_0)/G = aT_0/G \tag{5.39}$$

T_{exc} is the temperature at time t_{th}, after which T_e can be defined. For the sake of simplicity, $t_{th} = 0$ will be used in the following; that is, instantaneous internal electron thermalization is assumed and $T_{exc} = T_0 + \Delta T_e^{me}$. The electron excess energy is proportional to ΔT_e: $\Delta u_e \approx 2aT_0\Delta T_e$ (Equation 5.19), and thus exhibits the same dynamics.

In contrast, for strong excitation, the T_e dependence of C_e must be taken into account. The above rate equation system can be solved analytically:

$$T_{eq} \ln\left[\frac{T_e - T_{eq}}{T_{exc} - T_{eq}}\right] + \tilde{T} \ln\left[\frac{T_e - \tilde{T}}{T_{exc} - \tilde{T}}\right] = -G\frac{\tilde{T} + T_{eq}}{2C_L}t \tag{5.40}$$

with

$$T_{eq} = -\frac{C_L}{a} + \left[\left(\frac{C_L}{a}\right)^2 + T_{exc}^2 + 2\frac{C_L}{a}T_0\right]^{1/2} \approx T_0 + \frac{a}{2C_L}T_{exc}^2 \tag{5.41}$$

$$\tilde{T} = T_{eq} + 2\frac{C_L}{a} \approx 2\frac{C_L}{a}$$

T_{eq} is the final temperature of the thermalized electron-lattice system. For a reasonable metal heating (i.e., $T_e, T_L \ll C_L/a \approx 100T_0$), a simplified expression can be used:

$$\frac{T_{exc} - T_e}{T_{eq}} - \ln\left[\frac{T_e - T_{eq}}{T_{exc} - T_{eq}}\right] = \frac{G}{aT_{eq}}t \tag{5.42}$$

The electron temperature T_e decay nonexponentially, with a large perturbation-dependent slowing down of its short time delay dynamics, exponential decay with the time constant $\tau_{e-ph}T_L/T_0 \approx \tau_{e-ph}$, being eventually recovered as the electron gas cools-down (Figure 5.14), in agreement with the experimental data.[25,99,100] In this nonlinear regime, quantitative extraction of the intrinsic electron-lattice interaction

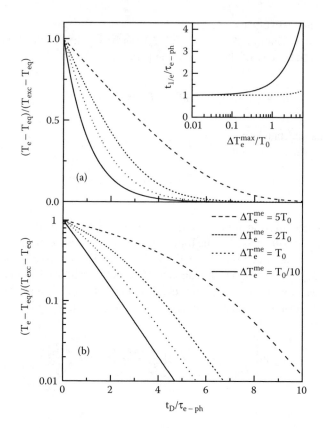

FIGURE 5.14 (a) Time dependence of the electron temperature T_e computed with the two-temperature model (Equation 5.35) for different excitation amplitudes $\Delta T_e^{me} = T_{exc} - T_0$ normalized to the initial temperature T_0 (the electron temperature is assumed to be established instantaneously). T_{eq} is the final quasi-equilibrium temperature of the electron-lattice system. The inset shows the excitation amplitude dependence of the time for the signal to decay of $1/e$ (full line). The dotted line is the rise of the long delay decay rate due to lattice heating (i.e., $\tau_{e\text{-}ph}T_{eq}/T_0$). (b) same as (a) but on a logarithmic scale. The times are normalized to the electron-lattice energy transfer time $\tau_{e\text{-}ph}$ (Equation 5.39).

parameter ($\tau_{e\text{-}ph}$ or G) requires precise knowledge of the injected energy and of the origin of the measured signal, i.e., if it is related to ΔT_e or $\Delta u_{e,}$ their connection being nonlinear (Equation 5.19).

The energy losses taking place on only a slightly longer time scale than internal thermalization, both effects can influence the measured temporal shapes. In noble metals, when probing away from the interband transition, the amplitude of the optical property change is almost independent of the exact electron distribution and proportional to Δu_{e}, permitting a direct determination of $\tau_{e\text{-}ph}$ in the weak excitation regime.[24,25,46,47] It is illustrated in Figure 5.15a, showing the measured $\Delta T/T$ time dependence for probing in and off resonance with the interband transitions. Note that after internal electron gas thermalization, both signals decay with the same time constant $\tau_{e\text{-}ph}$.

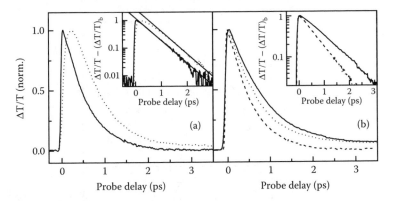

FIGURE 5.15 (a) Time behavior of the normalized transmission change $\Delta T/T$ measured in $D = 26$ nm Ag nanospheres off and on resonance with the interband transitions $\hbar\omega_{pr} \approx 1.3$ eV (full line) and 3.9 eV (dotted line), respectively, for $\hbar\omega_{pp} \approx 2.6$ eV in the weak excitation regime ($\Delta T_e^{me} = 100K$). The inset shows the same data on a logarithmic scale after subtraction of the long delay residual signal ($\Delta T/T)_b$. (b) $\Delta T/T$ time-dependence for $\hbar\omega_{pp} = 3$ eV and $\hbar\omega_{pr} = 1.5$ eV in $D = 26$ nm and $D = 6$ nm Ag nanospheres in BaO-P$_2$O$_5$ and $D = 3$ nm in Al$_2$O$_3$ (full, dotted, and dashed line, respectively). The inset shows the $D = 26$ and 3 nm data on a logarithmic scale after substraction of ($\Delta T/T)_b$.

In silver, the nonresonant conditions are realized for a near-infrared or blue probe pulse after few tens of femtoseconds, and for a near-infrared pulse in gold. As expected, the measured transmission changes $\Delta T/T$ decay exponentially with a decay time decreasing for small sizes (Figure 5.15b), indicating an increase in the electron-lattice coupling in small clusters.[46] Measurements performed in silver and gold nanospheres of diameters between 2 and 30 nm embedded in different matrices have shown that τ_{e-ph} is comparable to its bulk value for large particles ($D \geq 10$ nm, typically) but significantly decreases for smaller ones (Figure 5.16).[46,47] In contrast, using extrapolation of the large perturbation measurements, no size dependence of τ_{e-ph} (or G) has been estimated in the case of gold colloidal nanoparticles with size between 2.4 and 100 nm by Hartland et al.[44,57] A similar conclusion was drawn by Link and El-Sayed[14] for radii ranging from 2 to 50 nm. A nonmonotonic size dependence has been observed by Zhang et al.[101] in gold, with possible modification of the decay time due to strong excitation.

A decrease in the $\Delta T/T$ decay time with D has also been reported in tin[41] and gallium[43] nanoparticles for sizes ranging from 4 to 12 nm and 10 to 18 nm, respectively. The size dependence is, however, much larger than for silver, τ_{e-ph} being almost proportional to D over the investigated range. This variation has been interpreted in terms of electron coupling with the surface acoustic modes of the particles (i.e., the quantized vibration modes replacing the bulk ones) and quenching of the electron-bulk phonon interactions using the Belotskii and Tomchuk model.[102,103] However, this model, only applicable to small sizes, predicts a much slower electron-lattice energy transfer than in the bulk,[103] in contrast to the observations in noble metals.

The origin of the different behaviors observed by the different groups is not clear. As most of the experiments were performed in the strong excitation regime,

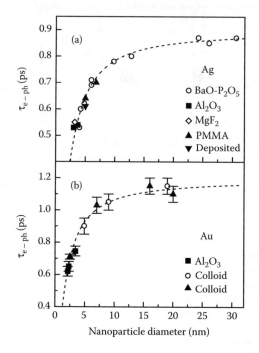

FIGURE 5.16 (a) Size dependence of the measured electron-lattice energy exchange time $\tau_{e\text{-}ph}$ for silver nanospheres in different matrices or deposited on a glass substrate. (b) Same as (a) for gold nanospheres in Al_2O_3 or in colloidal solutions prepared using a radiolysis technique (open circles) or commercially available (triangles).[46] The dashed lines are guides to the eye.

these discrepancies can be partly ascribed to the perturbation-dependent slowing down of the electron gas cooling, with a possible influence of the internal electron thermalization dynamics. This is confirmed by the results of systematic pump-power dependent measurements performed in the case of gold,[14,44,56] copper,[38] and silver[104,105] nanoparticles that clearly show a large increase of the observed decay time with the pump power. The intrinsic low perturbation values, $\tau_{e\text{-}ph}$, and thus the e-ph coupling constants $G = C_e\left(T_0\right)/\tau_{e-ph}$, inferred by these systematic studies are consistent with the ones reported in films.[23-25,29,106]

An important point here could also be the influence of the environment,[35,40,45,107] and of the interface layer and quality (as has been observed for the SPR width).[6] This can be particularly important for small sizes for which the particles are strongly coupled with their environment, the ratio of the surface to volume atom numbers strongly increasing as D decreases. Actually, the decay times measured in 18-nm gold particles have been shown to depend on the solvent (water or cyclohexane).[107] Similarly, relaxation has been found to be faster in Ag particles embedded in aluminate than in silica glass.[40] These matrix effects are much less important in the weak excitation regime (Figure 5.16). The reason for these different behaviors is not clear, but an important effect could be local heating of the matrix around the nanoparticles with thus a strong influence of the thermal conductivity of the matrix material.[35]

The τ_{e-ph} size dependence measured in the weak excitation regime in silver and gold[46] is similar to that of τ_{th} (Section 5.6). It suggests a confinement-induced acceleration of these exchanges due to an increase of electron-lattice mode coupling as in semiconductor nanoparticles.[108] Modeling, taking into account both the electron and phonon confinement effects, is necessary to quantitatively reproduce the experimental data. Additional experimental and theoretical investigations are clearly needed to understand electron-lattice coupling in confined metallic systems.

5.7 NON-EQUILIBRIUM ELECTRON-LATTICE ENERGY EXCHANGES

On a short time scale, the electron distribution is out of equilibrium and the two-temperature mode (Equation 5.35) is not applicable. In this regime, electron-phonon interaction must be described in a less global way, taking into account the nonthermal character of the distribution.[31,109] Experimentally, for both nanoparticles and films, the short time scale $\Delta T/T$ decay clearly shows that the electron gas energy loss rate to the lattice is initially slower and increases over a time scale of a few hundred femtoseconds to reach its long delay value (Figure 5.17). This behavior is a direct consequence of the existence of a long-living athermal electron gas. During the excitation process, a small number of electrons gain a large excess energy as compared to $k_B T_0$. As a crude approximation, separating the electron gas into unperturbed and non-equilibrium electrons, only the latter ones can lose energy by phonon emission. Their number increases with time as e-e scattering redistributes energy among the carriers; this leads to an overall increase in the energy loss rate to the lattice during the early stages of the internal electron gas thermalization.[24,31] As internal thermalization is approached, the above separation is no longer valid and a constant energy loss rate is eventually reached, corresponding to collective electron gas-lattice interaction as described by the two-temperature model. This evolution from a quasi-individual to a collective electron behavior is responsible for the observed short-time non-exponential decay of the excess energy. In small metal nanoparticles, the electron gas internal thermalization is faster (Section 5.5). The transient non-equilibrium regime for e-ph energy exchanges is thus observed over a shorter time scale and is barely observable for the smallest nanoparticles (Figure 5.17).

The above description of the interplay between electron thermalization and energy losses can be made more quantitative by computing the electron gas relaxation dynamics using the electron Boltzmann equation (Section 5.4.1). The time evolution of Δu_e calculated for the bulk silver conditions is shown in Figure 5.17 and exhibits the same behavior as the experimental ones for the film or large nanoparticle cases. Similar calculations performed assuming instantaneous internal thermalization of the electron gas only show an exponential decay of Δu_e in agreement with the two-temperature model.[31]

For nanoparticles, as a first approximation, the simulations were repeated, modifying the e-e and e-ph scattering rate to match their increase with cluster size

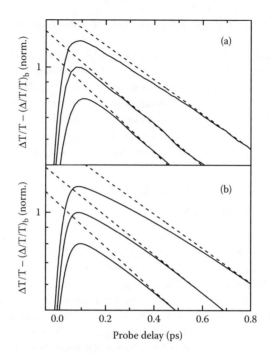

FIGURE 5.17 (a) Measured subpicosecond temporal evolution of the transmission change $\Delta T/T - (\Delta T/T)_b$ in Ag nanoparticles with average diameter $D = 26$ nm (top), 6 nm (middle), and 4 nm (bottom) on a logarithmic scale. (b) Time dependence of the electron excess energy Δu_e (Equation 5.19) computed for the same samples using the electron distribution computed with the Boltzmann equation (Equation 5.21).[109] The dashed lines correspond to exponential relaxation with the measured long decay times (Figure 5.15b).

reduction. This is partly justified by the similar quasi-continuum density of states of the bulk and confined systems for sizes down to a few nanometers (Section 5.2.2). The results are in very good agreement with the experimental ones showing the same transient evolution of the signal decay to an exponential behavior on a reduced time scale with size reduction (Figure 5.17). This confirms the strong interplay between e-e and e-ph scattering observed in bulk metals and yields further evidence of the acceleration of the electron energy exchanges in small clusters.[109]

5.8 CONCLUSION

Time-resolved experiments have emerged as powerful tools for the selective investigation of electron kinetics in metals. With their recent extensions and the improvement in femtosecond lasers, they now offer the unique possibility to directly address these problems in metal clusters. The first results, mostly obtained in noble metal clusters, are very promising and have already yielded new insight into the impact of confinement on the electron kinetics and the first steps of its evolution from a solid to a molecular type of behavior.

Results have now been obtained on electron-electron and electron-phonon scattering in noble metal particles over very large size ranges and for different types of environments. Electron-electron energy exchanges have been shown to be almost unaffected by confinement for sizes larger than 10 nm and to strongly increase for smaller sizes. Extension of these first measurements to other metals or types of confinement (one- or two-dimensional systems) and the use of other techniques, such as time-resolved photoemission,[49,53–55] for nanometric-size particles would be particularly interesting here. The size dependence of electron-lattice energy exchanges is still controversial in the strong excitation regime, probably because of the influence of the surrounding media and/or difficulty of comparing experiments performed in very different conditions. Further systematic investigations are clearly necessary, in particular as a function of the environment. In the weak excitation regime, direct measurement of the electron-lattice energy exchange time has permitted researchers to show a confinement-induced acceleration of the energy transfer for sizes smaller than 10 nm. A very interesting extension of these types of measurement would be time-resolved investigation of spin dynamics in clusters, similar to what has been done in bulk metals.[110–113] On the theoretical side, the electron dynamics modeling is based on a small solid approach; that is, modifications are introduced to the bulk material response. This is justified for not too small clusters, formed by more than a few hundred atoms, for which quantum mechanical confinement only introduces correction to their properties, but only constitutes a first approximation. Improved modeling of the electron scattering based on a quantum mechanical approach is necessary for understanding electron interactions in a confined system.

Up to now, all the experimental studies were performed on an ensemble of nanoparticles (a few tens of thousand to one million, typically) embedded in a matrix or deposited on a substrate. The responses are thus average over their size distribution and environment fluctuations. This is imposed by experimental constraints: the limited spatial resolution and measurement sensitivity. Improvement of the experimental techniques should permit researchers to address nanoparticle dynamics under different conditions, such as the ones of free clusters or of individual nanoparticles.[114,115] In particular, the development of new far-field techniques that permit one to directly monitor the optical absorption and nonlinear optical response of single metal (or semiconductor) nanoparticles is very promising.[115] These should allow for a closer comparison of the experimental and theoretical results, bringing new insight into understanding the physics of these systems.

On the other hand, most of the measurements were performed in relatively dilute material for which cluster coupling can be neglected. High density and, in particular, two- and three-dimensional self-organized nanoparticle systems can now be grown and are known to exhibit collective properties.[116–120] Study of the impact of this coupling on their dynamics; energy and charge exchange processes; and optical nonlinearities is a very interesting new area.

Acknowledgments

The authors wish to thank N. Del Fatti, P. Langot, D. Christofilos, C. Voisin, A. Arbouet, C. Guillon, and C. Flytzanis for their important contributions to these studies and for very helpful discussions. We are also indebted to A. Nakamura, Y. Hamanaka, and S. Omi for helpful discussions and for providing some of the silver nanoparticle samples. We also thank B. Prével, M. Gaudry, E. Cottancin, J. Lermé, M. Pellarin, M. Broyer, M. Maillard, and M. P. Pileni for their help in the theoretical and experimental parts of this work and for providing us with very good quality samples. We also acknowledge financial support by the Conseil Régional d'Aquitaine.

References

1. Kawabata, A. and Kubo, R., *J. Phys. Soc. Jap.*, 21, 1765, 1966; Kubo, R., Kawabata, A., and Kobayashi, S., *Ann. Rev. Mater. Sci.,* 14, 49, 1984.
2. Genzel, L., Martin, T.P., and Kreibig, U., *Z. Phys. B*, 21, 339, 1975.
3. Jackson, J.D. *Classical Electrodynamics*, John Wiley, New-York, 1962.
4. Van der Hulst, H.C., *Light Scattering by Small Particles*, Dover Publ. Inc., New York, 1981.
5. Flytzanis, C., Hache, F., Klein, M.C., Ricard, D., and Roussignol, P., in *Progress in Optics Vol. XXIX,* Wold, E., Ed., North Holland, Amsterdam, 1991, 321.
6. Kreibig, U. and Vollmer, M., *Optical Properties of Metal Clusters*, Springer, Berlin, 1995.
7. Bohren, C.F. and Huffman, D.R., *Absorption and Scattering of Light by Small Particles*, John Wiley & Sons, New York, 1998.
8. Vallée, F., Del Fatti, N., and Flytzanis, C., in *Nanostructured Materials*, Shalaev, V.M. and Moskovits, M., Eds., American Chemical Society, Washington, D.C., 1997, 70.
9. Kelly, K.L., Coronado, E., Lin Zhao, L., and Schatz, G.C., *J. Phys. Chem. B*, 107, 668, 2003.
10. Faraday, M., *Philos. Trans. R. Soc. (London),* 147, 145, 1857.
11. Mie, G., *Am. Phys. (Leipzig)*, 25, 377, 1908.
12. See, for example, Shah, J., *Ultrafast Spectroscopy of Semiconductors and Semiconductor Nanostructures*, Springer Verlag, Berlin, 1996.
13. For a review, see Voisin, C., Del Fatti, N., Christofilos, D., and Vallée, F., *J. Phys. Chem B*, 105, 2264, 2001.
14. Link, S. and El-Sayed, M.A., *J. Phys. Chem. B*, 103, 8410, 1999.
15. Anisimov, S.I., Kapeliovitch, B.L., and Perelman, T.L., *Sov. Phys. JETP*, 39, 375, 1974.
16. Eesley, G.L., *Phys. Rev. Lett.,* 51, 2140, 1983; *Phys. Rev. B*, 33, 2144, 1986.
17. Fujimoto, J.G., Liu, J.M., Ippen, E.P., and Bloembergen, N., *Phys. Rev. Lett.*, 53, 1837, 1984.
18. Vallée, F., *C. R. Acad. Sci.*, 2, 1469, 2001.
19. Lamprecht, B., Leitner, A., and Aussenegg, F.R., *Appl. Phys. B*, 68, 419, 1999.
20. Lamprecht, B., Krenn, J. R., Leitner, A., and Aussenegg, F.R., *Phys. Rev. Lett.*, 83, 4421, 1999.
21. Vartanyan, T., Simon, M., and Träger, F., *Appl. Phys. B*, 68, 425, 1999.
22. Bosbach, J., Hendrich, C., Stietz, F., Vartanyan, T., and Träger, F., *Phys. Rev. Lett.*, 89, 257404, 2002.
23. Sun, C.K., Vallée, F., Acioli L.H., Ippen, E.P., and Fujimoto, J.G., *Phys. Rev. B*, 50, 15337, 1994.
24. Groeneveld, R., Sprik, R., and Lagendijk, A., *Phys. Rev. B*, 51, 11433, 1995.

25. Del Fatti, N., Voisin, C., Achermann, M., Tzortzakis, S., Christofilos, D., and Vallée, F., *Phys. Rev. B,* 61, 16956, 2000.
26. Tas G. and Maris, H.J., *Phys. Rev. B*, 49, 15046, 1994.
27. Fann, W.S., Storz, R., Tom, H.W.K., and Bokor, J., *Phys. Rev. Lett.,* 68, 2834, 1992; *Phys. Rev. B,* 46, 13592, 1992.
28. For a review on time-resolved photoelectron emission in metals, see Petek, H. and Ogawa, S., *Progr. Surf. Science*, 56, 239, 1997.
29. Brorson, S.D., Kazeroonian, A., Modera, J.S., Face, D.W., Cheng, T.K., Ippen, E.P., Dresselhaus M.S., and Dresselhaus, G., *Phys. Rev. Lett.*, 64, 2172, 1990.
30. Elsayed-Ali, H.E., Juhasz, T., Smith G.O., and Bron W.E., *Phys. Rev. B*, 43, 4488, 1991.
31. Del Fatti, N., Bouffanais, R., Vallée, F., and Flytzanis, C., *Phys. Rev. Lett.,* 81, 922, 1998.
32. Juhasz, T., Elsayed-Ali, H.E., Hu, X.H., and Bron, W.E., *Phys. Rev. B*, 45, 13819, 1992.
33. Hohlfeld, J., Wellershoff, S.S., Güdde, J., Conrad, U., Jähnke, V., and Matthias, E., *Chem. Phys,* 251, 237, 2000.
34. Bloemer, M.J., Haus, J.W., and Ashley, P.R., *J. Opt. Soc. Am. B*, 7, 790, 1990.
35. Haglund, R.F., Lupke, G., Osborne, D.H., Chen, H., Magruder, R.H., and Zuhr, R.A., in *Ultrafast Phenomena XI*, Elsaesser, T., Fujimoto, J.G., Wiersma, D.A., and Zinth, W., Eds., Springer-Verlag, Berlin, 1998, 356.
36. Del Fatti, N., Voisin, C., Christofilos, D., Vallée F., and Flytzanis, C., *J. Phys. Chem. A*, 104, 4321, 2000.
37. Voisin, C., Christofilos, D., Del Fatti, N., Vallée, F., Prével, B., Cottancin, E., Lermé, J., Pellarin, M., and Broyer, M., *Phys. Rev. Lett.,* 85, 2200, 2000.
38. Tokizaki, T., Nakamura, A., Kavelo, S., Uchida, K., Omi, S., Tanji, H., and Asahara, Y., *Appl. Phys. Lett.*, 65, 941, 1994.
39. Bigot, J.-Y., Merle, J.-C., Cregut, O., and Daunois, A., *Phys. Rev. Lett.,* 75, 4702, 1995.
40. Halté, V., Bigot, J.-Y., Palpant, B., Broyer, M., Prével, B., and Pérez, A., *Appl. Phys. Lett.,* 75, 3799, 1999.
41. Stella, A., Nisoli, M., De Silvestri, S., Svelto, O., Lanzani, G., Cheyssac, P., and Kofman, R., *Phys. Rev. B*, 53, 15497, 1996.
42. Perner, M., Bost, P., Lemmer, U., von Plessen, G., Feldmann, J., Becker, U., Mennig M., Schmitt, M., and Schmidt, H., *Phys. Rev. Lett.,* 78, 2192, 1997.
43. Nisoli, M., Stagira, S., De Silvestri, S., Stella, A., Tognini, P., Cheyssac P., and Kofman, R., *Phys. Rev. Lett.,* 78, 3575, 1997.
44. Hodak, J.H., Henglein, A., and Hartland, G.V., *J. Chem. Phys.*, 112, 5942, 2000.
45. Hamanaka, Y., Kuwabata, J., Tanahashi, I., Omi, S., and Nakamura, A., *Phys. Rev. B*, 63, 104302, 2001.
46. Arbouet, A., Voisin, C., Christofilos, D., Langot, P., Del Fatti, N., Vallée, F., Lermé, J., Celep, G., Cottancin, E., Gaudry, M., Pellarin, M., Broyer, M., Maillard, M., Pileni, M.P., and Treguer, M., *Phys. Rev. Lett.,* 90, 177401, 2003.
47. Del Fatti, N., Flytzanis, C., and Vallée, F., *Appl. Phys. B*, 68, 433, 1999.
48. Del Fatti, N., Vallée, F., Flytzanis, C., Hamanaka, Y., and Nakamura, A., *Chem. Phys.,* 251, 215, 2000.
49. Bauer, C., Abid, J.-P., Fermin, D., and Girault, H.H., *J. Chem. Phys.,* 120, 9302, 2004.
50. Fierz, M., Siegmann, K., Scharte, M., and Aeschlimann, M., *Appl. Phys. B,* 68, 415, 1999.
51. Lehmann, J., Merschdorf, M., Pfeiffer, W., Voll, S., and Gerber, G., *Phys. Rev. Lett.*, 85, 2921, 2000.

52. Merschdorf, M., Kennerknecht, C., Willig, K., and Pfeiffer, W., *New J. Phys.*, 4, 95.1, 2002.
53. Pontius, N., Lüttgens, G., Bechthold, P.S., Neeb, M., and Eberhardt, W., *J. Chem. Phys.*, 115, 10479, 2001.
54. Pontius, N., Neeb, M., Eberhardt, W., Lüttgens, G., and Bechthold, P.S., *Phys. Rev. B*, 67, 035425, 2003.
55. Niemietz, M., Gerhardt, P., Ganteför, G., and Dok Kim, Y., *Chem. Phys. Lett.*, 380, 99, 2003.
56. Hodak, J.H., Martini, I., and Hartland, G.V., *J. Phys. Chem. B*, 102, 6958, 1998.
57. Hodak, J.H., Henglein, A., and Hartland, G.V., *J. Chem. Phys.*, 111, 8613, 1999.
58. Voisin, C., Del Fatti, N., Christofilos D., and Vallée, F., *Appl. Surf. Science*, 164, 131, 2000.
59. Nisoli, M., De Silvestri, S., Cavalleri, A., Malvezzi, A.M., Stella, A., Lanzani, G., Cheyssac, P., and Kofman, R., *Phys. Rev. B*, 55, R13424, 1997.
60. Del Fatti, N., Voisin, C., Chevy, F., Vallée, F., and Flytzanis, C., *J. Chem. Phys.*, 110, 11484, 1999.
61. Perner, M., Grésillon, S., März, J., Von Plessen, G., Feldmann, J., Porstendorfer, J., Berg, K.-J., and Berg, G., *Phys. Rev. Lett.*, 85, 792, 2000.
62. Hartland, G.V., Hu, M., Wilson, O., Mulvaney, P., and Sader, J.E., *J. Phys. Chem.*, 106, 743, 2002.
63. Böttcher, C.J., *Theory of Electric Polarization*, Elsevier, Amsterdam, 1973.
64. Ashcroft, N.W. and Mermin, N.D., *Solid State Physics*, Saunders College, Philadelphia, 1976.
65. Johnson, P.B. and Christy, R.W., *Phys. Rev. B*, 6, 4370 1972.
66. Smith, J.B. and Ehrenreich, H., *Phys. Rev. B*, 25, 923, 1982.
67. Gurzhi, R.N., *Sov. Phys. JETP*, 35, 673, 1959.
68. Tsai, C.-Yi, Tsai, C.-Yao, Chen, C.-H., Sung, T.-L., Wu, T.-Y., and Shih, F.-P., *IEEE J. Quant. Electr.*, 34, 552, 1998.
69. Perenboom, J.A.A, Wyder, P., and Meier, F., *Phys. Rep.*, 78, 173, 1981.
70. Fröhlich, H., *Physica*, 4, 406, 1937.
71. Kubo, R., *J. Phys. Soc. Jap.*, 17, 975, 1962.
72. Ricard, D., in *Nonlinear Optical Materials: Principles and Applications*, Degiorgio, V. and Flytzanis, C., Eds., IOS Press, Amsterdam, 1995, 289.
73. Hache, F., Ricard, D., and Flytzanis, C., *J. Opt. Soc. Am. B*, 3, 1647, 1986.
74. Hovel, H., Fritz, S., Hilger, A., Kreibig, U., and Vollmer, M., *Phys. Rev. B*, 48, 18178, 1993.
75. Lermé, J., Palpant, B., Prével, B., Pellarin, M., Treilleux, M., Vialle, J.L., Perez, A., and Broyer, M., *Phys. Rev. Lett.*, 80, 5105, 1998.
76. Lermé, J., Palpant, B., Prével, B., Cottancin, E., Pellarin, M., Treilleux, M., Vialle, J.L., Perez, A., and Broyer, M., *Eur. Phys. J. D*, 4, 95, 1998.
77. Knoesel, A., Hotzel, A., and Wolf, M., *Phys. Rev. B*, 57, 12812, 1998.
78. Matzdorf, R., Gerlach, A., Theilmann, F., Meister, G., and Goldmann, A., *Appl. Phys. B*, 68, 393, 1999.
79. Mills, D.L. and Burstein, E., *Rep. Prog. Phys.*, 37, 817, 1974.
80. Voisin, C., Christofilos, D., Del Fatti, N., and Vallée, F., *Eur. Phys. J. D*, 16, 139, 2001.
81. Aeschlimann, M., Bauer, M., Pawlik, S., Weber, W., Burgermeister, R., Oberli, D., and Siegmann, H.C., *Phys. Rev. Lett.*, 79, 5158, 1997.
82. Ziman, J.M., *Principles of the Theory of Solids*, Cambridge University Press, Cambridge, 1969.

83. Knorren, R., Bennemann, K.H., Burgermeister, R., and Aeschlimann, M., *Phys. Rev. B*, 61, 9427, 2000.
84. Pines, D. and Nozières, P., *The Theory of Quantum Liquids*, Benjamin, New York, 1966.
85. Juaristi, J.I., Alducin, M., and Nagy, I., *J. Phys.: Condens. Matter*, 15, 7859, 2003.
86. Rosei, R., *Phys. Rev. B*, 10, 474, 1974.
87. Rosei, R., Antonangeli, F., and Grassano, U.M., *Surf. Sci.*, 37, 689, 1973.
88. Haglund, R.F., in *Handbook of Optical Properties*, Vol. 2, Hummel, R.E. and Wissmann, P., Eds., CRC Press, New York, 1997, 191.
89. Voisin, C., Christofilos, D., Loukakos, P.A., Del Fatti, N., Vallée, F., Lermé, J., Gaudry, M., Cottancin, E., Pellarin, M., and Broyer, M., *Phys. Rev. B*, 69, 195416, 2004.
90. Rosei, R., Culp, C.H., and Weaver, J.H., *Phys. Rev. B*, 10, 484, 1974.
91. Ekardt, W., *Phys. Rev. B*, 29, 1558, 1984.
92. Bréchignac, C., Cahuzac, P., Leygnier, J., and Sarfati, A., *Phys. Rev. Lett.*, 70, 2036, 1993.
93. Liebsch, A., *Phys. Rev. B*, 48, 11317, 1993.
94. Shahbazyan, T.V. and Perakis, I.E., *Phys. Rev. B*, 60, 9090, 1999; *Chem. Phys.*, 251, 37, 2000.
95. Shahbazyan, T.V., Perakis, I.E., and Bigot, J.-Y., *Phys. Rev. Lett.*, 81, 3120, 1998.
96. Grua, P., Morreeuw, J.P., Bercegol, H., Jonusauskas, G., and Vallée, F., *Phys. Rev. B*, 68, 035424, 2003.
97. Kaganov, M.I., Lifshitz I.M., and Tanatarov, L.V., *Sov. Phys. JETP*, 4, 173, 1957.
98. Allen, P.B., *Phys. Rev. Lett.*, 59, 1460, 1987.
99. Schoenlein, R.W., Lin, W.Z., Fujimoto J.G., and Eesley, G.L., *Phys. Rev. Lett.*, 58, 1680, 1987.
100. Sun, C.K., Vallée, F., Acioli, L.H., Ippen, E.P., and Fujimoto, J.G., *Phys. Rev. B*, 48, 12365, 1993.
101. Smith, B.A., Zhang, J.Z., Giebel, U., and Schmid, G., *Chem. Phys. Lett.*, 270, 139, 1997.
102. Belotskii, E.D. and Tomchuk, P.M., *Surf. Science*, 239, 143, 1990.
103. Belotskii, E.D. and Tomchuk, P.M., *Int. J. Electr.*, 73, 955, 1992.
104. Halté, V., Guille, J., Merle, J.-C., Perakis, I., and Bigot, J.-Y., *Phys. Rev. B*, 60, 11738, 1999.
105. Hamanaka, Y., Nakamura, A., Omi, S., Del Fatti, N., Vallée, F., and Flytzanis, C., *Appl. Phys. Lett.*, 75, 1712, 1999.
106. Liu, D., He, P., and Alexander, D.R., *Appl. Phys. Lett.*, 62, 249, 1993.
107. Zhang, J.Z., *Acc. Chem. Res.*, 30, 423, 1997.
108. Takagahara, T., *J. Lumin.*, 70, 129, 1996.
109. Guillon, C., Langot, P., Del Fatti, N., and Vallée, F., *New J. Phys.*, 5, 13, 2003.
110. Beaurepaire, E., Merle, J.C., Daunois, A., and Bigot, J.-Y., *Phys. Rev. Lett.*, 76, 4250, 1996.
111. Hohlfeld, J., Matthias, E., Knorren R., and Bennemann, K.H., *Phys. Rev. Lett.*, 78, 4861, 1997.
112. Ju, G., Vertikov, A., Nurmikko, A.V., Canady, C., Xiao, G., Farrow, R.F.C., and Cebollada, A., *Phys. Rev. B*, 57, R700, 1998.
113. Rhie, H.-S., Dürr, H.A., and Eberhardt, W., *Phys. Rev. Lett.*, 90, 247201, 2003.
114. Klar, T., Perner, M., Grosse, S., Von Plessen, G., Spirkl, W., and Feldmann, J., *Phys. Rev. Lett.*, 80, 4249, 1998.
115. Arbouet, A., Christofilos, D., Del Fatti, N., Vallée, F., Huntzinger, J.-R., Arnaud, L., Billaud, P., and Broyer, M., *Phys. Rev. Lett.*, 93, 127401, 2004.

116. Shalaev, V.M., *Phys. Rep.,* 272, 61, 1996.
117. Taleb, A., Petit, C., and Pileni, M.P., *Chem. Mater.,* 9, 950, 1997.
118. Krenn, J.R., Dereux, A., Weeber, J.C., Bourillot, E., Lacroute, Y., Goudonnet, J.P., Schider, G., Gotschy, W., Leitner, A., Aussenegg, F.R., and Girard, C., *Phys. Rev. Lett.,* 82, 2590, 1999.
119. Lamprecht, B., Schider, G., Lechner, R.T., Ditlbacher, H., Krenn, J.R., Leitner, A., and Aussenegg, F.R., *Phys. Rev. Lett.,* 84, 4721, 2000.
120. Silly, F., Gusev, A.O., Taleb, A., Charra, F., and Pileni, M.P., *Phys. Rev. Lett.,* 84, 5840, 2000.

6 Generation and Propagation of Monochromatic Acoustic Phonons in Gallium Arsenide

Anthony J. Kent and Nicola M. Stanton

CONTENTS

6.1 Introduction and Background 143
6.2 Basic Properties of Semiconductor Superlattices 146
6.3 Experimental Details 151
6.4 Experimental Results: Introduction 156
    6.4.1 Time of Flight Techniques: Detecting Propagating Phonons Following Ultrafast Excitation 156
    6.4.2 Determining the Monochromatic Nature of the Propagating Modes 159
    6.4.3 Resolving Phonon Modes: Transverse Mode Enhancement 163
    6.4.4 Studying the Transverse Contribution: (311) and (211) Substrates ... 166
6.5 Demonstrating Phonon Optics: Measuring the Mean Free Path of Terahertz (THz) Phonons 167
6.6 Other Materials: Gallium Nitride and Its Alloys 170
6.7 Summary and Outlook 173
References 174

6.1 INTRODUCTION AND BACKGROUND

Spectroscopic techniques employing terahertz (THz) acoustic phonons have proved extremely valuable in studies of condensed matter systems and, in particular, semiconductor nanostructures; for a recent review, see Challis.[1] A major advantage of phonons compared to terahertz electromagnetic radiation is their ability to directly probe transitions that are indirect in momentum space. Acoustic phonon spectroscopy can also probe regions of phase-space inaccessible to neutron scattering techniques. Phonon studies of carrier-phonon scattering and hot carrier energy relaxation complement the more widely used electronic transport and optical measurements.

Phonon techniques give very direct information about the relative strengths of the different carrier-phonon coupling processes (e.g., deformation potential and piezo-electric).

The ideal source of phonons for spectroscopy applications would generate short (nanosecond) pulses of monochromatic longitudinal (LA) and transverse (TA) polarized acoustic phonons in the frequency range from about 100 GHz to a few terahertz. Short pulses are useful because they allow separation of effects due to LA and TA polarizations, owing to the difference in their speeds of propagation. However, in most of the experiments described in Challis,[1] the heat pulse source was used. This consists of a thin metal film deposited on the sample and heated either electrically or optically. Such a source can generate short pulses, but the frequency spectrum is far from monochromatic, being approximately Planckian (black-body). Over the years, considerable effort has been put into developing alternative sources for phonon spectroscopy. Superconducting tunnel junctions have been used to generate beams of quasi-monochromatic phonons,[2] but these were usually cw (continuous wave) and superimposed on a large broad-spectrum background. Another possible source of quasi-monochromatic phonons is emission of cyclotron energy phonons by a two-dimensional electron system in a strong perpendicular magnetic field.[3] However, in GaAs (gallium arsenide), the form factor for electron-phonon coupling restricts the frequency to less than 150 GHz and, due to Landau level broadening, the generated spectrum is not particularly narrow.

Recently, progress in ultrafast pulsed laser technology has led to the development of more promising methods of phonon generation. The so-called pico(second)-ultrasonics techniques use femtosecond pulsed laser excitation of thin metal films or layered semiconductor structures. Absorption of an intense ultrashort laser pulse in a metal film deposited on a crystalline substrate leads to fast heating of the film. The resulting rapid thermal expansion of the film launches a strain pulse into the crystal. The dynamics of the strain pulse were studied using Brillouin scattering[4,5] or a second, weaker laser (probe) pulse to detect the strain-modulated reflectivity of a metal film.[6,7] This may be a second film on the opposite end of the crystal, or, after reflection of the pulse from the end of the crystal, the same film as used for its generation. The maximum phonon frequency associated with the strain pulse is related to its duration which, in turn, depends on the optical penetration depth in the film. Typically, frequency components up to about 100 GHz can be present in the pulse. Excitation with much higher optical power densities gives rise to a very intense strain pulse. As a result of the nonlinear acoustic properties of the crystal, the strain pulse sharpens up as it propagates into the crystal and develops into a shock front containing phonon frequency components of up to a terahertz. Due to phonon dispersion, this shock front ultimately develops into trains of acoustic solitons (considered elsewhere in this volume).[8,9] Pico-ultrasonic pulses have also been generated by femtosecond laser excitation of semiconductor quantum wells and detected through surface deflection[10] or the strain modulation of the surface reflectivity.[11] The frequency of the phonon oscillations depended on the width of the quantum well, and was approximately 100 GHz. Excitation of the modes has been attributed to deformation potential coupling.[10]

Pico-ultrasonic pulses are very short, approximately 10 nm long, and thus can be ideal for probing solid-state nanostructures[12] (e.g., semiconductor low-dimensional structures[13]). However, it is implicit in the shortness of the pulses that their spectral width is quite large. Furthermore, as mentioned above, unless the pulses are very intense (nonlinear regime), the maximum frequency is limited to about 100 GHz. Ultrafast optical excitation of higher-frequency modes is, however, possible using semiconductor superlattices (SLs). The artificial periodicity of the acoustic impedance leads to folding of the acoustic phonon dispersion into a mini-Brillouin zone (BZ) extending out to the boundary at $q = \pi/d_{SL}$, where d_{SL} is the period of the SL and q is the phonon wavevector. Therefore, high-frequency acoustic modes corresponding to where the folded dispersion meets the zone center ($q = 0$) can couple efficiently with photons. Such modes have, in the past, been extensively studied by Raman scattering; for a review, see Ruf.[14]

Excitation of terahertz coherent phonons in AlAs/GaAs SLs was first demonstrated by Yamamoto et al.[15] using ultrafast pump-probe measurements. The SLs studied consisted of 80 (60) periods, each of 15 (18) monolayers of GaAs and 15 (18) monolayers of AlAs. The first interband transition in the GaAs wells was excited using 1.68 eV, 130-fs pump pulses from a Ti:sapphire oscillator. Delayed probe pulses were used to measure the transient changes in surface reflectivity due to modulation of the refractive index by the strain oscillations. Observed reflectivity changes were approximately $10^{-4}R_0$, where R_0 is the normal reflectivity. The oscillatory component of the reflectivity was observed at frequencies of 0.613 (0.507) THz, corresponding to the first-order zone center modes at $v = c_s/d_{SL}$. Careful analysis of the data led to the conclusion that modes of B_2 symmetry were generated (i.e., modes having a node at the interface between layers of the SL). This was suggestive of an excitation mechanism in which optically excited carriers produce a stress in the GaAs quantum well layers.

Since the first observations, resonant photoexcitation of coherent zone-folded modes in AlAs/GaAs SLs have been studied in greater detail.[16,17] Bartels et al.[16] observed two triplets of modes, the center frequencies corresponding to the first- and second-order zone center ($q = 0$) modes with A_1 symmetry (i.e., with a node at the center of the layers). Each of these had a pair of sidebands with frequencies corresponding to $q = 2k_{laser}$. The dominant mode was the first-order zone center mode of frequency $v = c_s/d_{SL}$. These observations were suggestive of a Raman generation mechanism involving forward ($q = 0$) and backscattering ($q = 2k_{laser}$) of photons, and the process was attributed to impulsive stimulated Raman scattering (ISRS).

Coherent terahertz phonon generation has also been observed in other SL systems, notably in InGaN/GaN by Sun et al.[18] The 14 period SLs, each period consisting of 4.3 nm GaN barriers and InGaN wells in the range 1.2 to 6.2 nm, were excited with 180-fs, 3.2 eV pump pulses from a frequency-doubled Ti:sapphire oscillator. The coherent acoustic phonon oscillations were detected via the change in transmission through the SL of a delayed probe pulse. Very strong strain-induced modulation of the transmission, approximately 1%, was observed in the InGaN/GaN system. The process was attributed to screening by the photoexcited carriers of the strain-induced internal piezoelectric fields at the InGaN/GaN interfaces.

Coherent phonon oscillations in AlAs/GaAs SLs were observed to decay on a time scale of approximately 100 ps, and it was suggested that this was due to dephasing. However, the decay could also be due to the modes "leaking out" of the SL into propagating monochromatic phonons. The latter would be very interesting from the point of view of applications in phonon optics and phonon spectroscopy. Mizoguchi et al.[19] observed propagation of the coherent SL modes through a ~0.5 µm AlGaAs cladding layer above the AlAs/GaAs SL. They used two-color pump-probe measurements, with an infrared pump pulse to excite the SL and a UV probe pulse to measure the reflectivity change in a region within 20 nm of the surface of the AlGaAs cladding layer. However, after propagation over 0.5 µm, the phonon oscillations were already much weaker than measured at the SL. For this reason, and because the optical delay line would have to be abnormally long, it is unlikely that pump-probe techniques will be useful for detecting the phonons after propagation over ~1 mm, which is of order the mean free path of terahertz phonons in high-quality GaAs at liquid helium temperatures.

A much more sensitive method of phonon detection is to use superconducting bolometers, as in heat pulse experiments.[20] However, this is an incoherent detection technique and must be combined with other methods (e.g., frequency-selective phonon mirrors, etc.) to determine the spectral distribution of the phonons. In this chapter we review recent work in which we have combined coherent phonon generation in SLs and incoherent detection to study the propagation of terahertz monochromatic phonons over macroscopic distances in GaAs.

In Section 6.2 we briefly discuss the relevant electronic, photonic, and phononic properties of semiconductor SLs. We then proceed in Section 6.3 to describe the experimental arrangements for the generation and detection of monochromatic phonons. In Section 6.4, our results for longitudinal polarized and transverse polarized phonons are reviewed and discussed in detail. The measurement of the mean free path of terahertz phonons in GaAs is discussed in Section 6.5 as an example of an application of the technique. In Section 6.6, we consider some preliminary measurements on AlN/GaN systems, which offer the possibility of a much more intense source of monochromatic phonons. Finally, in Section 6.7, we conclude with a discussion of outstanding problems to be addressed and future applications of the techniques.

6.2 BASIC PROPERTIES OF SEMICONDUCTOR SUPER LATTICES

Superlattices consist of alternating layers *ABABAB*... of two different materials *A* and *B* (see Figure 6.1). A single period of the SL is made from *N* monolayers (ML) of material *A* and *M* ML of material *B*. Superlattices are fabricated using epitaxial crystal growth techniques (e.g., molecular beam epitaxy [MBE] and metal-organic vapor phase epitaxy). Using MBE, it is possible to produce SLs with abrupt, atomically flat interfaces between the layers when grown on an appropriate substrate.[21] In most of the work described in this chapter, we consider SLs based on the III–IV semiconductors AlAs and GaAs. Over the past two decades, this materials system has found widespread technological application. Improvements in growth have resulted in the ability to produce near-perfect SL structures. Both AlAs and GaAs have a cubic crystal structure and the lattice constant is very similar, which means

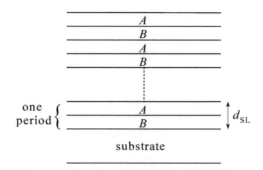

FIGURE 6.1 Superlattice structure made from materials A and B. The period of the super-lattice is d_{SL}.

TABLE 6.1
Relevant Parameters of AlAs and GaAs Semiconductor Materials at 4K[22]

Material	a (nm)	E_g (eV)	c_{LA} (ms^{-1})	c_{TA} (ms^{-1})	ρ (kgm^{-3})	Z_{LA} (kgm^{-2}s^{-1})	Z_{TA} (kgm^{-2}s^{-1})
AlAs	0.5661	2.250	5650	3960	3760	2.12×10^7	1.49×10^7
GaAs	0.5653	1.518	4730	3350	5317	2.52×10^7	1.78×10^7

Note: Phonon speeds are given for propagation along [001]; speeds are different for different propagation directions due to acoustic anisotropy.[23]

that there is little interfacial strain. The SL period $d_{SL} = Na_A/2 + Ma_B/2$, where a_A and a_B are, respectively, the lattice constants of AlAs and GaAs. Interesting electronic, photonic, and phononic properties of AlAs/GaAs SLs arise because of the materials' different electronic band gaps (E_g), speeds of sound for LA and TA modes (c_{LA} and c_{TA}), density (ρ), and acoustic impedance ($Z = \rho c_s$) (see Table 6.1).

The electronic band structure of an unbiased AlAs/GaAs SL is shown in Figure 6.2. Owing to the difference in E_g, the GaAs layers behave as quantum wells for electrons and holes between AlAs barriers. The electronic properties of the structure strongly depend on the thickness of the well and barrier regions. Provided the barriers and/or the wells are not too thick, carriers are able to tunnel between neighboring wells. For d_{SL} less than the phase coherence length of the carriers, the artificial periodicity of the lattice potential leads to a splitting of the conduction and valence bands into a series of minibands and minigaps for carrier motion in the z or growth direction (perpendicular to the plane of the layers). The electronic properties of such an SL were considered in detail by Esaki.[24] Essentially, the electronic minibands have an energy width Δ that is approximately proportional to $(d_W)^{-2}$, where d_W is the width of the well. If $\Delta > h/\tau$, where τ^{-1} is the mean scattering rate, the minibands are well defined. However, in the limit $\Delta < h/\tau$, most of the states will be localized in the quantum wells. The energy dispersion $E(k_z)$ of the electrons can be calculated

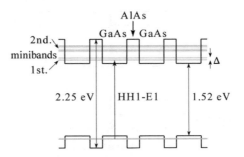

FIGURE 6.2 Electronic structure of an AlAs/GaAs SL. The optically driven HH1–E1 transition is shown.

using the Kronig-Penney model or the **k.p** method. For weak coupling (wide barriers), the energy dispersion of the lowest miniband is given by

$$E(k_z) = \frac{\Delta_1}{2}\{1 + \cos(k_z d_{SL})\}.$$

The optical response of semiconductor SLs has been extensively studied owing to their application in opto-electronics (e.g., lasers). Interband radiative transitions between the two-dimensional electron and hole states dominate the measured spectra. However, intersubband transitions can also be important (e.g., in FIR laser devices). Photocurrent measurements[25] have revealed peaks for photon energies corresponding to interband transitions involving the E1 band and higher, E2, etc., bands. Such measurements enable the experimental determination of the energy and width of the electronic minibands. The photoluminescence (PL) emission of SLs is dominated by excitonic transitions between the electron and heavy hole (HH) states, with the main peak corresponding to the E1-HH1 transition. The optical absorption edge occurs where the photon energy is resonant with the HH1–E1 transition.

As well as giving rise to the formation of electronic minibands, the artificial periodicity of the SL structure leads to folding of the phonon dispersion. Owing to the difference in acoustic impedance of the two materials making up one period of the SL, partial reflection of phonons occurs at the interfaces. For an SL containing a sufficient number (typically more than 20) of periods, narrow gaps (stop bands) open up in the phonon dispersion at wavevectors corresponding to the conditions for Bragg reflection of phonons. Here we consider the situation; relevant to the experiments described later, where the phonons are propagating in the z direction (i.e., normal to the interfaces between the SL layers) and that this is a direction of high symmetry (e.g., [100] in a cubic crystal or along the c-axis in a hexagonal crystal). The more complicated case of phonon propagation in other directions has been considered by Tamura et al.[26]

The phononic properties of SLs can be readily calculated in the framework of continuum elasticity theory using the transfer matrix method. Boundary conditions applicable to sound propagation across the interface between two solids are that the lattice displacement U and the stress S are continuous across the boundary. Consider

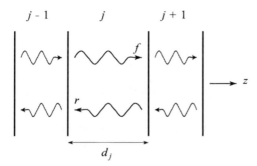

FIGURE 6.3 Model of propagation of phonons in the z, or growth, direction through layer j of a SL, as used in the transfer matrix calculation. The forward wave is denoted f and the reverse wave, r.

phonons traveling in layer j, of thickness d_j, in a direction toward the interface between layers j and $j + 1$ as shown in Figure 6.3,

$$U_j(z) = a_j^f e^{iq_j z} + a_j^r e^{-iq_j z} ,$$

and

$$S_j(z) = i\omega Z_j \left(a_j^f e^{iq_j z} - a_j^r e^{-iq_j z} \right) .$$

Here, a_j^f and a_j^r are, respectively, the amplitudes of the forward and reverse waves, and Z_j is the acoustic impedance of layer j. These equations can be written in a more convenient matrix form:

$$\mathbf{W}_j(z) = \underline{\mathbf{h}}_j(z)\mathbf{A}_j ,$$

where

$$\mathbf{W}_j(z) = \begin{pmatrix} U_j(z) \\ S_j(z) \end{pmatrix}; \quad \mathbf{A}_j = \begin{pmatrix} a_j^f \\ a_j^r \end{pmatrix}$$

and

$$\underline{\mathbf{h}}_j(z) = \begin{pmatrix} e^{iq_j z} & e^{-iq_j z} \\ i\omega Z_j e^{iq_j z} & -i\omega Z_j e^{-iq_j z} \end{pmatrix} .$$

The boundary conditions applied at the interface between layer $j - 1$ and layer j require

$$\mathbf{W}_{j-1}(z) = \mathbf{W}_j(z) ,$$

and, at the interface between layers j and $j + 1$,

$$\mathbf{W}_j(z + d_j) = \mathbf{W}_{j+1}(z + d_j) .$$

Therefore, the change in $\mathbf{W}$ after passing through layer j is given by

$$\mathbf{W}_{j+1}(z + d_j) = \underline{\mathbf{t}}_j(z + d_j)\mathbf{W}_{j-1}(z) ,$$

where

$$\underline{\mathbf{t}}_j(z + d_j) = \underline{\mathbf{h}}_j(z + d_j)[\underline{\mathbf{h}}_j(z)]^{-1} .$$

Setting at the interface between layers $j - 1$ and j, we obtain

$$\underline{\mathbf{t}}_j(d_j) = \begin{pmatrix} \cos(q_j d_j) & \dfrac{1}{\omega Z_j} \sin(q_j d_j) \\ -\omega Z_j \sin(q_j d_j) & \cos(q_j d_j) \end{pmatrix} .$$

A similar 2×2 matrix, $\underline{\mathbf{t}}_{j+1}(d_{j+1})$, can be defined for propagation through the $(j + 1)^{\text{th.}}$ layer. These two matrices can be used to give the change in $\mathbf{W}$ after propagation through a single period of the SL

$$\mathbf{W}_{j+2}(d_{\text{SL}}) = \underline{\mathbf{T}}\mathbf{W}_j ,$$

where $\underline{\mathbf{T}} = \underline{\mathbf{t}}_{j+1}\underline{\mathbf{t}}_j$ and $d_{\text{SL}} = d_j + d_{j+1}$. Finally, after passing through p equal periods of the SL,

$$\mathbf{W}_p(pd_{\text{SL}}) = (\underline{\mathbf{T}})^p \mathbf{W}(0) .$$

The calculated LA phonon transmission of an SL consisting of 40 periods of 7 ML AlAs and 7 ML GaAs is shown in Figure 6.4. Strong dips (stop bands) in the transmission at frequencies corresponding to the conditions for Bragg reflection, are clearly observed.

Following Mizuno and Tamura,[27] it is also possible to obtain analytic expressions for the center frequencies and widths of the stop bands at the BZ center ($q = 0$) and the BZ boundary ($q = \pi/d_{\text{SL}}$)

$$\omega_n = n\pi \frac{1}{\left(\dfrac{Na_A}{2c_A} + \dfrac{Ma_B}{2c_B} \right)}$$

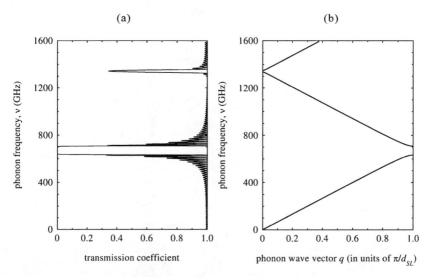

FIGURE 6.4 (a) The transmission coefficient of a 40-period 7-ML AlAs/7-ML GaAs SL as a function of the phonon frequency, calculated using the transfer matrix method; and (b) folded phonon dispersion of the same SL as in (a).

and

$$\Delta\omega_n = \frac{2\omega_n}{n\pi}\sqrt{2\gamma}\left|\sin\left(n\pi\frac{Na_A/c_A}{Na_A/c_A + Ma_B/c_B}\right)\right|,$$

where $\gamma = (Z_B - Z_A)^2/2Z_A Z_B$. The folded LA phonon dispersion relation for the 7 ML:7 ML AlAs/GaAs SL is also shown in Figure 6.4.

Experimental evidence for the zone-folded modes in AlAs/GaAs SLs has been observed in Raman scattering measurements.[28] In the backscattering geometry, pairs of modes are observed corresponding to $q = 2k_{laser}$ (see Figure 6.5). Resonant enhancement of the Raman cross-section was observed when the photon energy matched the HH1–E1 excitonic transition in the SL.[29]

The phononic properties of SLs have been exploited in phonon optics elements (e.g., band-stop phonon filters,[30] phonon mirrors, and phonon cavities[31]). In the experiments described by Narayanamurti et al.,[30] superconducting tunnel junctions were used to generate quasi-monochromatic phonons and also to detect them after propagation through the SL. Dips in the phonon transmission at frequencies corresponding to the SL stop bands were observed. However, due to the inherent limitations of the tunnel junction technique, the dips were weak and superimposed on a large background signal. Phonon propagation through superlattice structures has also been extensively studied using the phonon imaging (see, for example Ref. 32).

6.3   EXPERIMENTAL DETAILS

Using the theory discussed in Section 6.2, it is straightforward to design AlAs/GaAs SL structures with the desired acoustical and optical properties for the generation

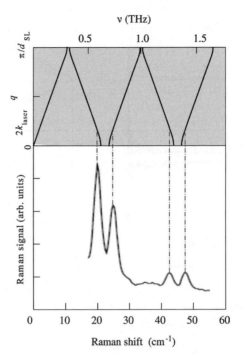

FIGURE 6.5 Typical backscattering Raman spectrum of a 40-period, 22-ML GaAs/4-ML AlAs SL. The peaks correspond to the excitation of near-zone center modes having $q = 2k_{laser}$.

of terahertz acoustic modes by femtosecond laser excitation. For example, in the 40-period 4-ML AlAs:22-ML GaAs SL, the first mini-zone center LA mode has a frequency of about 660 GHz and the HH1–E1 transition occurs at approximately 1.6 eV (780 nm), which is conveniently within the tuning range of commercial Ti:sapphire femtosecond oscillators.

The structures used for the experiments were grown by MBE on semi-insulating GaAs substrates of standard (0.4 mm) or 2-mm thickness. The back surface of the substrate was then polished and superconducting granular aluminum bolometers fabricated on it. Photolithographic techniques were used to define the 40×40 μm^2 active area bolometer pattern in a photoresist layer. This was followed by thermal evaporation of a 20-nm-thick layer of aluminum from an alumina crucible at a deposition rate of about 1 nm/s. Subsequent lift-off left behind the bolometers that had a room-temperature resistance of a few hundred ohms and a superconducting transition temperature 2K, (see Figure 6.6). In operation, a constant bias current of a few microamps was applied to the bolometer, which was then cooled through its superconducting transition until the resistance reached approximately 50 Ω. The temperature was stabilized at this point. Phonons incident on the bolometer caused a transient increase in its temperature and hence resistance, which produced a voltage pulse. The voltage pulse was amplified using gigahertz-bandwidth amplifiers and the signal captured using a high-speed (2 GHz) digitizer and signal averager (Perkin-Elmer Eclipse). The response of superconducting bolometers to phonon pulses is discussed in detail in Ref. 33–35.

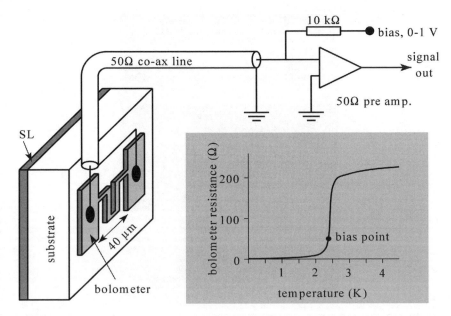

FIGURE 6.6 Detail of sample structure, showing 40×40 µm² active area superconducting bolometer and associated electrical arrangement. The inset shows a typical bolometer super-conducting transition and the optimum operating point is indicated.

The advantage of superconducting bolmeters, which makes them ideal for detecting the phonons after propagation across the GaAs substrate, is their very high sensitivity. With low-noise electronics and signal averaging sensitivities in the nanowatt range are achievable. The bolometers also have a fast response, typically on the order of nanoseconds. However, they are integrated energy detectors and give no information about the spectrum of the phonons detected. To obtain spectral information, they must be used in conjunction with a frequency-selective element. This took the form of another SL grown between the phonon generator and the bolometer. The second SL was designed to act as a notch filter, attenuating propagating phonons with frequency matching of the first mini-zone boundary stop band. A typical sample structure, including generator and analyzer SLs, is shown in Figure 6.7. The parameters of the various samples used in the experiments are detailed in Table 6.2.

All measurements were carried out with the sample mounted in an optical-access, liquid helium bath cryostat. The optical setup is shown in Figure 6.8. A tunable (690 to 1050 nm) mode-locked Ti:sapphire oscillator producing 100 fs-duration pulses of typical energy 10 nJ/pulse and repetition rate 82 MHz was used to excite the SL. The time between each laser pulse, 12 ns, is much less than the time required for the phonons to propagate across the GaAs substrate, typically 100 ns. Therefore, to allow the sample time to recover between excitation pulses, a pulse-picker was used to select laser pulses with a repetition rate less than or equal to 100 kHz. Under these conditions, the average power delivered was also sufficiently low that the sample did not heat up appreciably. The beam was focused onto the sample using an $f = 100$ mm lens, giving a spot diameter of approximately 100 µm. Taking into

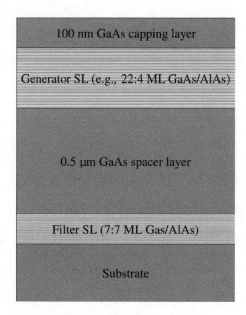

FIGURE 6.7 Layer structure typical of samples used in the experiments. This structure includes an analyzer (filter) SL below the generator SL.

TABLE 6.2
Parameters of the Samples Used in This Work

Sample	Well/Barrier Material	Substrate	Generator SL: Ratio of ML (well) to ML (barrier)	Filter SL: Ratio of ML (well) to ML (barrier)
A	GaAs/AlAs	0.4 mm, (001) SI GaAs	22:4	—
B	GaAs/AlAs	0.4 mm, (001) SI GaAs	22:4	7:7
C	GaAs/AlAs	0.4 mm, (001) SI GaAs	24:8	7:7
D	GaAs/AlAs	2 mm, (001) SI GaAs	22:4	—
E	GaAs/AlAs	0.4 mm, (311) SI GaAs	22:4	—
F	GaN/AlN	0.4 mm sapphire	16:12	—

account losses in the pulse-picker and optical components, this resulted in a power density on the sample of about 2 μJ cm^{-2}. Galvanometer-driven x-y scanning mirrors were used to position the laser spot at the desired position on the sample, normally directly opposite the bolometer. A pair of double-chirped mirrors was used to pre-compensate the laser pulse for the effects of dispersion of components within the optical path.

Figure 6.9 shows typical bolometer signals obtained after averaging about 1000 separate traces. The earliest feature is optical in origin (the bolometer is sensitive to photons as well as phonons) and probably due to stray light or PL reaching the

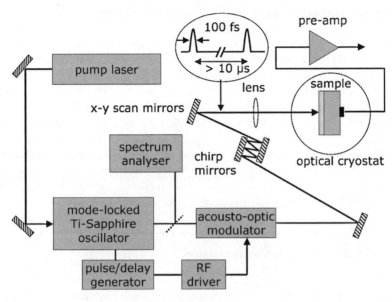

FIGURE 6.8 Optical setup for the generation of propagating monochromatic phonons.

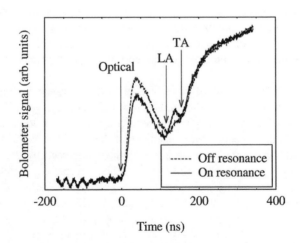

FIGURE 6.9 Typical bolometer traces, obtained for excitation wavelengths corresponding to on- and off-resonance conditions. The lines corresponding to the ballistic arrival times (c_{LA} = 4700 ms⁻¹ and c_{TA} = 3500 ms⁻¹ in [001]-orientated GaAs) are marked in the figure.

bolometer. After about 90 ns, the phonons arrive at the bolometer — first the LA and then the TA modes after a further 30 ns. These arrival times for phonons correspond to the ballistic propagation across the 0.4 mm-thick GaAs substrate. The rise time of the optical signal gives a measure of the speed of response of the bolometer and detection electronics, about 2 ns in this example.

6.4 EXPERIMENTAL RESULTS: INTRODUCTION

In this section, we concentrate on the results of measurements involving direct detection of phonons that have "leaked" from the SL and have propagated across the substrate material. Initially, in Section 6.4.1–2, the results of the first bolometric detection measurements using GaAs/AlAs SLs are reviewed.[36] We then describe measurements using novel structures, which showed that these "leaked" modes are monochromatic in nature.[37] In Section 6.5.4.3–4 we discuss the unexpected observation of emission of monochromatic TA phonons[38] and show results of experiments using SLs grown on different orientations of substrate. We then describe an experimental demonstration of "phonon optics" with a measurement of the terahertz phonon mean free path in GaAs.[39] Finally, in Section 6.5.4.5–6, we discuss measurements performed using other materials systems.[40]

6.4.1 TIME OF FLIGHT TECHNIQUES: DETECTING PROPAGATING PHONONS FOLLOWING ULTRAFAST EXCITATION

The first measurements attempting to directly detect the propagating phonons generated as a result of ultrafast optical excitation of a semiconductor SL using bolometric detection techniques were performed by Hawker et al.[36] In that work, a 40-period (001) GaAs/AlAs SL structure (22 ML GaAs, 4 ML AlAs) was excited with pulses of 100-fs duration from a Ti:sapphire laser. Pump-probe measurements using the same structure demonstrated coherent LA phonon generation under the same conditions.[16] By varying the excitation wavelength, it was possible to observe the onset of coherent phonon generation in the SL (using direct detection of phonons techniques) by measuring the "leaked" phonon intensity as a function of excitation wavelength.

Now consider the measured time of flight traces, shown in Figure 6.9. The different signals were obtained for excitation wavelengths corresponding to the off- and on-resonance regimes (that is, for excitation energy below and above the HH1–E1 transition, respectively). Consider first the off-resonance signal, obtained for =790nm. At this excitation wavelength, light is absorbed by the substrate and capping layer only, and the observed signal can be interpreted as follows: at $t \approx 0$ ns, the laser is incident on the sample and excites carriers in the GaAs capping layer and substrate. Relaxation of the photogenerated carrier density occurs with a time constant of several hundred picoseconds by emission of phonons. Phonon emission by optically excited GaAs has been discussed by a number of authors.[41–43] Initially, a cascade of high-energy optical (LO) phonons are emitted until the carrier excess energy is within one LO phonon energy (36.6 meV) of the GaAs band edge. Relaxation then continues via emission of deformation potential and piezoelectrically coupled LA and TA modes. Upon reaching the band edge, the electrons and holes can recombine radiatively, emitting a bandgap photon. Low-frequency (<1 THz) acoustic phonons emitted during the relaxation process will traverse the substrate ballistically and arrive at the detector at a time determined by the speed of sound in GaAs. The emitted LO phonons do not propagate through the crystal, but relax rapidly through a chain of down-conversions that ultimately results in propagating

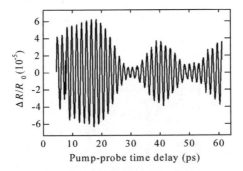

FIGURE 6.10 Oscillations in surface reflectivity as a function of pump-probe delay time for sample A. The period of the oscillations corresponds to a frequency of about 650 GHz.

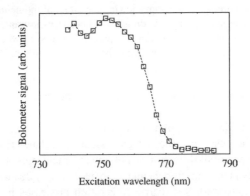

FIGURE 6.11 The LA mode intensity as a function of excitation wavelength for sample A. As the laser is tuned through the HH1–E1 resonance, we observe a strong increase in the LA signal.

high-frequency TA modes ($v \approx 1.5$ THz). These high-frequency modes are subject to scattering in the substrate, resulting in diffusive and dispersive (quasi-diffusive) propagation, giving rise to the characteristic slow decaying bolometer signal.

Excitation of coherent LA phonons was observed when the excitation wavelength was tuned to the first superlattice interband transition (HH1-E1), which in this sample corresponds to $\lambda = 767$ nm; see Figure 6.10. The on-resonance curve of Figure 6.9 corresponds to $\lambda = 760$ nm; and at this wavelength, carriers are excited in the SL as well as the substrate. The optical component of the signal intensity decreases, and a pronounced increase in LA phonon signal is visible. To better illustrate the effect of tuning the laser energy through the HH1–E1 resonance, the LA intensity as a function of excitation wavelength is plotted, shown in Figure 6.11. We see that negligible LA emission occurs for $\lambda > 770$ nm. As the wavelength is reduced, LA emission increases rapidly as carriers are excited in the SL. The enhancement in the LA signal we observe occurs over a narrow range of wavelength (rather than being a step at the HH1–E1 wavelength), due to the spectral width of the transform-limited

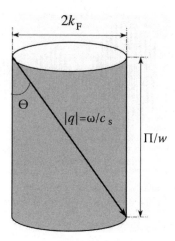

FIGURE 6.12 Momentum-space diagram showing the effect of momentum conservation on the emission of acoustic phonons by two-dimensional electrons in a quantum well of width w. Electronic transitions involving emission of a phonon must be between two points on the surface of the cylinder. The transition involving the emission of the highest possible energy phonon is indicated.

excitation pulse. Reducing the excitation energy further has no significant effect on the LA intensity.

Although the sudden onset of LA emission when the resonance condition is satisfied strongly suggests that the detected signal is related to coherent phonon generation, any other possibilities for signal enhancement must first be discounted. One such possibility is that the signal enhancement is simply due to carrier relaxation from the SL. Previous studies of carrier relaxation processes by two-dimensional carriers in heterostructures and quantum wells showed that the emitted acoustic phonon energy flux is concentrated in a cone of half angle θ around the direction perpendicular to the two-dimensional layer.[44] Due to the requirements of conservation of crystal momentum, for two-dimensional carriers in a quantum well of width w, the phonon emission is cut off for wavevector components $q_{perp}(\text{max}) \approx \pi/w$, and for $q_{parallel}(\text{max}) = 2k_F$ (where $k_F = \sqrt{(\pi n_s)}$); see Figure 6.12. The cone half angle, θ, is related to the well width (6.2 nm) and the induced carrier density per well, n_s. To estimate the latter, we can assume every photon absorbed creates an electron-hole pair, which gives $n_s \approx 10^{12}$ cm^{-2}. Therefore, the peak in acoustic phonon emission is expected at an angle of 30°. Hawker et al. measured the LA signal intensity for a range of different angles by moving the laser spot across the sample surface, as shown in Figure 6.13. The results showed that the LA emission was strongly peaked in a direction parallel to the SL growth direction. No strong emission was observed at an angle of 30°, as would be expected if hot carrier energy relaxation was responsible for the enhancement in the LA signal, providing additional evidence for the increase in LA emission being related to coherent phonon generation.

These initial measurements provided strong circumstantial evidence that the enhancement of LA phonon emission we observed when the wavelength corresponded to the HH1–E1 separation was due to propagating modes, produced as

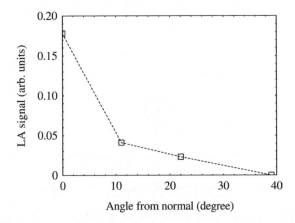

FIGURE 6.13 The angular dependence of LA intensity, as measured in sample A.

coherent phonons leak from the SL, reaching the detector. However, these measurements were unable to determine whether the propagating modes were of the same frequency as the coherent modes generated, because the superconducting aluminum bolometers used had no spectral resolution. To determine whether the phonons that leaked from the SL were monochromatic, we exploited the phonon stop bands of novel SL structures. The results of the experiment are discussed in the next subsection.

6.4.2 DETERMINING THE MONOCHROMATIC NATURE OF THE PROPAGATING MODES

As previously discussed in Section 6.2, the periodic structure of the SL gives rise to a folding of the Brillouin zone, and long wavelength phonons satisfying the Bragg condition are reflected at the interfaces of the SL. This gives rise to gaps in the dispersion curves, or phonon stop bands, where the transmission rate through the SL falls to zero. By designing novel SL structures, we have been able to show that the propagating modes are indeed monochromatic, and in addition, we have placed an upper limit on the bandwidth of the monochromatic phonon beam. The novel structures incorporated two SLs, with the layers designed so that the SL nearest the substrate would act as a reflective stop band filter for phonons that leaked from the generator SL grown on top of the filter SL (as shown schematically in Figure 6.7).

The experimental results described in this subsection are based on three samples, all grown by MBE on 0.4-mm GaAs substrates (see Table 6.2). The samples were designed such that the generator SLs had a narrow linewidth of monochromatic phonons, while the filter SL had a wide stop band. The dispersion relations for the generator and filter layers are shown in Figure 6.14.

Under resonant photoexcitation conditions, coherent LA phonons are generated with a frequency corresponding to the first Brillouin zone center mode (i.e., $q = 0$, and $n = 2$). For the generator layer used in samples A and B, this corresponds to $\nu = \omega_2/2\pi = 0.66$ THz, and in sample C, $\nu = 0.54$ THz. The sidebands at $q = 2k_{laser}$

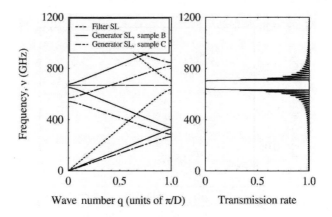

FIGURE 6.14 The dispersion relations for samples B and C, and the transmission rate of the filter SL, as calculated following the method described in Section 6.2.

and the higher-order modes ($n = 4, 6, \ldots$) are very much weaker than the fundamental and can be neglected. The frequency dependence of the transmission coefficient of the filter SL used in samples B and C is also shown in Figure 6.14. This curve was calculated using the transfer matrix approach as discussed in Section 6.2. The filter SL has a stop band corresponding to the first Brillouin zone boundary ($n = 1$) centered at $v = 0.67$ THz. The width of the stop band is 80 GHz. So, for our samples, the stop band frequency overlaps the generator frequency in sample B but not sample C.

With the theoretical background revisited briefly, we now turn our attention to the experimental results. As expected from consideration of previous work, in sample A we observed a strong enhancement in the LA signal intensity when the short laser energy was equal to or larger than the ground-state exciton (HH1–E1) energy, as shown in Figure 6.15. The increase occurred over a narrow range of excitation energy (from 1.59 to 1.63 eV), which was due to the spectral linewidth of the laser pulse. We also observed a strong TA signal, largely due to phonons emitted as a result of carriers relaxing excess energy in the GaAs (we return to consider this in a later subsection). The weak LA signal observed off resonance can also be attributed to the carrier relaxation.

As an extension of the previous work (where signals were recorded for a few different positions on the sample), images of the LA signal for the on- and off-resonance conditions were obtained by raster scanning the excitation spot over the sample. These images are shown in Figure 6.16 and represent a scan area of 0.75 mm × 0.75 mm, with the detector at the center of the image. The image obtained when off-resonance shows features that are characteristic of phonon propagation in GaAs. There is an enhancement of phonon flux at an angle of 45° to [001], typical of the weak LA focussing in cubic crystals.[45] This image showed that the phonons generated had an isotropic distribution of wavevectors, as would be expected if the generation was due to hot carrier relaxation in bulk GaAs (substrate and capping layer). When this image is compared to the on-resonance case, we observed a strong increase in the measured intensity at the center of the image, due to phonons generated with wavevectors directions within about 20° of (001), consistent with

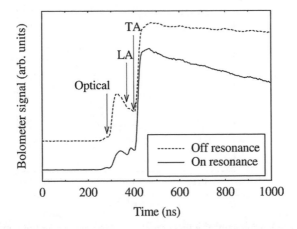

FIGURE 6.15 On- and off-resonance bolometer signals obtained using sample A. The enhancement of the LA mode on resonance is clearly visible.

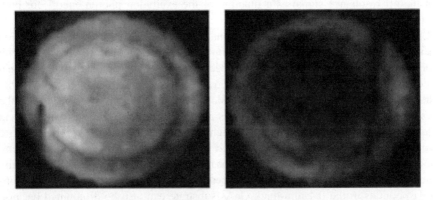

FIGURE 6.16 Images obtained using sample A, for excitation energies corresponding to the on- and off-resonance cases. The images cover an angular range of $\pm 45°$ in both (horizontal and vertical) directions, with the bolometer at the center, and the light-colored areas are regions of high phonon intensity. The enhancement of the LA mode at small emission angles on resonance is clearly visible.

coherent phonon generation occurring for this excitation energy. The stark difference in the appearance of the two images suggests that the phonons we detect on resonance are those coherent modes that have leaked from the SL. However, as before, the detection allows no spectral resolution; and although the evidence for a narrow angle monochromatic beam of phonons is strong, we could not exclude the possibility that the phonon emission anisotropy was not due to energy relaxation. In the case of hot carrier relaxation, LA phonons emitted at angles near (001) have a broad spectrum up to a cut-off. The cut-off is due to the effects of momentum and energy conservation and occurs at frequency $\nu_c = q_{perp}(\max)c_{LA}/2\pi \approx c_{LA}/2w$.[44]

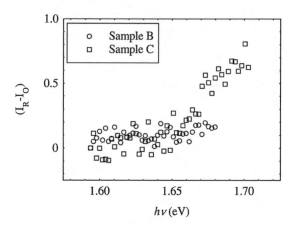

FIGURE 6.17 The LA intensity as a function of excitation energy for the samples containing the filter SL structures.

We now consider the results of ultrafast excitation of the samples containing the filter layers. In Figure 6.17, we show the LA signal intensity following subtraction of the off-resonance signal as a function of the excitation photon energy for both samples containing the "filter" SL structure. The results are quite striking. In sample B, there is no significant increase in signal strength when the resonance condition is satisfied. In this sample, the phonon stop band was designed to be coincident with the generator frequency; we therefore expect to see no enhancement in phonon signal because the propagating modes are not transmitted through the filter SL. We can discount the possibility that the filter SL blocks all phonons by considering the results obtained from sample C. There is a marked increase in the LA intensity in sample C when the laser is tuned through the resonance. However, in contrast to sample B, in sample C the generator and filter SL frequencies are not coincident, and the leaked propagating phonons can be transmitted through the filter SL, reaching the detector on the back surface of the crystal. Signal loss due to scattering of 0.66-THz phonons in the 0.4-mm-thick GaAs substrate can also be discounted because we observed a strong LA signal enhancement in sample A. The last process that must be considered is that the LA enhancement was due to photogenerated carriers relaxing their excess energy in the quantum wells. Previous work studying hot electron energy relaxation in QW structures showed that due to momentum conservation considerations, the phonons emitted perpendicular to the well are mostly of low frequency, up to the cut-off at v_c. For samples A and B v_c = 385 GHz and, for sample C, v_c = 353 GHz. Both these frequencies are much lower than the filter stop band frequencies, and thus carrier relaxation can be discounted as an explanation of the results.

The results described above provided additional and conclusive evidence that the longitudinal acoustic phonons leaking from the SL were indeed monochromatic. Fabrication of the filter SL caused 100% attenuation of the phonon beam when the stop band frequency matched the generator frequency. Also, because of the complete attenuation of the signal, we were able to conclude that the spectral width of the propagating monochromatic beam was less than 80 GHz (the width of the filter stop band).

6.4.3 RESOLVING PHONON MODES: TRANSVERSE MODE ENHANCEMENT

The discussion of the previous experimental measurements concentrated on the LA mode behavior. One of the advantages of using phonon spectroscopy measurements over other techniques is that individual phonon modes can be resolved by bolometers, owing to the differing times of flight of LA and TA phonons across the substrate. In addition, close inspection of bolometer signals often reveals features that are characteristic of down-conversion products and quasi-diffusive propagation. In all of the bolometer signals we have shown, a large transverse acoustic phonon signal was apparent, although thus far we have given little attention to this in our discussions. Originally, we attributed the large TA component to be the result of energy relaxation; as previously discussed, LO phonons down-convert into high wavevector TA modes, which arrive at the detector at later times than the ballistic peak. However, a more detailed analysis of the TA phonon signal revealed that the intensity of the ballistic component was also enhanced when the resonance condition was satisfied.

There are three possible explanations for the enhancement in the TA signal: (1) the TA signal enhancement is as a result of carrier energy relaxation; (2) the TA modes are due to anharmonic decay of the terahertz LA SL modes; and (3) coherent TA phonons were being generated and leaking into propagating modes, in addition to LA phonons. The similarity in behavior with the LA mode was unexpected, and coherent TA phonon generation has not been reported in pump-probe measurements. The proposed generation mechanism for coherent phonons is believed to be ISRS, and TA phonons propagating along (001) are not Raman active. That the TA signal was caused by (3) seemed unlikely. To determine conclusively the origin of the additional TA signal, we studied the temporal, angular, and excitation wavelength dependence of the transverse mode using the filter SL samples.

Figure 6.17 shows the effect of excitation wavelength on the LA signal; however, closer inspection of the data also shows that the TA component of the signal is enhanced on its rising edge and reaches a higher peak. At later times (not shown in the figure), the tail intensity of the on-resonance signal is reduced when compared to the off-resonance case. The enhancement of the signal at early times following the excitation pulse is more clearly seen in Figure 6.18, where the off-resonance signal has been subtracted from the signals taken for different excitation wavelengths. This effectively removes the large background that is due to carrier relaxation in the GaAs. The behavior of the TA component is very similar to that observed with the LA signal. There is a strong increase in TA intensity when the resonance condition is satisfied, and this increase occurs over a narrow range of wavelengths consistent with the spectral width of the laser pulse.

Let us first consider the off-resonance signal. This signal is well understood, as described in a previous section. At these wavelengths, there is little optical absorption in the SL. The phonon signal is largely due to energy relaxation by photoexcited carriers in the GaAs substrate. The excess energy, $h\nu - E_g$, is greater than the LO phonon energy ($\hbar\nu_{LO} = 36\,\text{meV}$) and thus most of the energy is lost through emission of LO phonons. These modes decay in a few picoseconds into large wavevector LA modes that subsequently undergo anharmonic decay to TA modes, which can propagate through the crystal. These large wavevector modes are subject to scattering

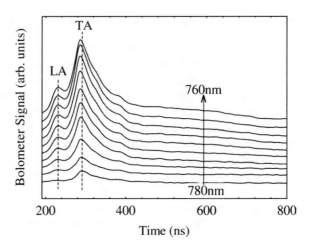

FIGURE 6.18 To show the increase in TA signal more clearly, the off-resonance curve has been subtracted from the traces. The lines are guides for the eye.

by isotope mass defects, impurities, and other crystalline defects. This results in diffusive propagation, and thus these scattered phonons arrive at the bolometer at times later than the ballistic time of flight across the crystal. The signal we observed shows a broad, delayed peak and a long tail, which is generally accepted as being characteristic of the arrival of the decay products of LO phonon emission.

The additional TA signal observed on resonance arrives at a time that corresponds to the ballistic time of flight (~120 ns). This suggests that the origin of the TA enhancement is due to sub-terahertz phonons that are not scattered in the substrate, rather than being the result of diffusive propagation. In addition, the peak in the TA signal shifts to earlier times as a larger proportion of the TA signal is due to the arrival of ballistic phonons. These observations effectively discount diffusive propagation as a possible cause of the observed signal.

Now consider the possible generation mechanisms for the enhanced TA signal. Initially, we limit the discussion to sample A. As with the LA mode, carrier relaxation from the QWs is subject to momentum conservation considerations, and thus there is a cut-off in the frequency of phonons that can be emitted. For TA phonons in sample A, $\nu_c = 200$ GHz. Next consider anharmonic decay of the coherent LA modes: there are two possible decay channels (LA $\rightarrow$ LA + TA and LA $\rightarrow$ TA + TA); and because we know that the LA frequency is 650 GHz, both decay channels would result in sub-terahertz TA phonons. Finally, it should be noted that if the increase in TA mode was due to leaking monochromatic TA phonons, their frequency would be $\nu = c_{TA}/d_{SL} = 450$ GHz.

It is clear from the above discussion that we are unable to determine the origin of the TA signal using sample A alone. By using the filter SL samples used in the phonon optics experiments, we can determine whether the TA mode was monochromatic. As described previously, the filter SL structure was chosen to block the propagation of phonons in one sample but not the other. As with the LA phonons, we calculated the generator frequencies of the SLs and the center stop band fre-

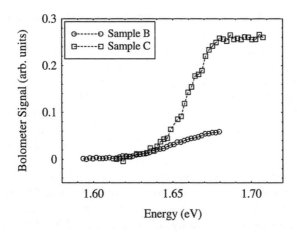

FIGURE 6.19 The TA intensity as a function of excitation energy for samples containing the filter SL structures. As previously observed with the LA mode, we see a pronounced increase when the resonance condition is satisfied only for the sample where the generator and filter frequencies did not coincide.

quency, as detailed in Figure 6.14. In sample B, TA phonons with $\nu = 450$ GHz were generated, whereas in sample C, the frequency was $\nu = 360$ GHz. The filter SL stop band was centered on 450 GHz. Figure 6.19 shows the dependence of phonon signal intensity on excitation wavelength for the two samples. In sample C (as in sample A), we see a strong increase in the TA intensity when the resonance condition is satisfied; whereas in sample B, we see no significant change in intensity across the wavelength range of interest. This result is to be expected if the TA phonons we detect when the pump photon energy is tuned onto resonance are monochromatic in nature. The notch filter results also ruled out the possibilities that the signal enhancement was due to anharmonic decay or carrier relaxation. Neither of those processes would result in monochromatic phonons, and samples B and C should have behaved in identical ways. However, that the decay of propagating LA phonons can account for the TA signal enhancement must be examined in more detail. Because the filter layer prevents propagation of the monochromatic LA phonons, the TA signal would be attenuated also. To achieve this required the fabrication of another sample, and the results are described in detail in Section 6.5.

Having discussed and subsequently discounted the various possible causes of the additional TA signal, the one that remains is that the signal is due to coherent TA modes leaking into propagating monochromatic modes. We now turn our attention to identifying the excitation mechanism responsible for this signal. It is not sufficient to reject this explanation simply because pump-probe measurements have not identified a TA mode. Transverse acoustic modes are not Raman active in (001) GaAs, and therefore, generation by ISRS is not expected. In addition, because the surface reflectance is proportional to the first-order Raman tensor, TA modes will not modulate the reflectance. We must then consider non-Raman processes as a possible explanation for the TA mode enhancement. A possible mechanism is displacive excitation of coherent phonons (DECP).[46] In this process, the basic idea is that the ions are set into motion via coupling to photoexcited carriers, or a sudden change in carrier temperature due to absorption of the optical pulse. If the optically induced changes occur on a time scale that is very short compared to the phonon period, then coherent modes are excited. However, this process excites only modes of A_1 symmetry (i.e., LA modes in [001] GaAs/AlAs). Another possibility is the effect of acoustic anisotropy giving rise to the excitation of TA modes. This process has been suggested previously as a possible explanation of the observation of Raman scattering by TA modes in (110) GaAs/AlAs.[47] Acoustic anisotropy has a significant effect on the mode and angular dependence of electron-phonon coupling in GaAs QWs and heterojunctions. Indeed, these effects are very strong for TA modes propagating in a direction close to, but not exactly along, (001) in GaAs and are responsible for the characteristic phonon focussing features observed.[45] In these directions, transverse modes cannot be described as "pure" transverse. The wavevector and polarization vector are not necessarily collinear, and as such, transverse modes may carry a small component of longitudinal polarization. This may permit the excitation of slightly off-axis TA modes by the ISRS process, or the piezoelectric coupling mechanism described by Winterling et al.[48] In pump-probe measurements, the weak TA mode would be easily obscured by the stronger LA signal, and so it is not surprising that TA modes were not observed using this technique.

6.4.4 STUDYING THE TRANSVERSE CONTRIBUTION: (311) AND (211) SUBSTRATES

Using bolometric detection techniques, we have been able to resolve the phonon modes and have observed the generation of monochromatic transverse acoustic phonons, in addition to monochromatic LA modes, when the resonance condition is satisfied. Further information on the generation of monochromatic TA modes was obtained by studying SLs grown on substrates of different crystallographic orientation.

The basic idea behind these measurements is to determine whether acoustic anisotropy was responsible for monochromatic TA phonon generation. We have already stated that TA modes propagating close to (001) in GaAs can carry a component of longitudinal polarization due to mode mixing effects. By changing the substrate orientation, we change the axis for phonon propagation. TA phonons propagating near (311) are not as strongly affected by acoustic anisotropy, and therefore, if the mechanism allowing the generation of monochromatic TA modes is mode mixing, then in (311)-orientated GaAs, there should be no enhancement in TA intensity when tuning through the resonance. The LA mode, however, is only weakly affected by acoustic anisotropy in any direction of propagation, and so its behavior should be unaltered by changes in crystal orientation.

Figure 6.20 shows a series of signals obtained for different excitation energies for a 40-period 22-ML:4-ML GaAs/AlAs SL grown on (311)A GaAs. As with the (001) orientation, we observe the decrease of the optical component, and corresponding increase in the LA mode intensity when the excitation energy is tuned through resonance. However, the transverse mode shows no appreciable increase when compared to the (001) case. Further tests involved measuring transverse

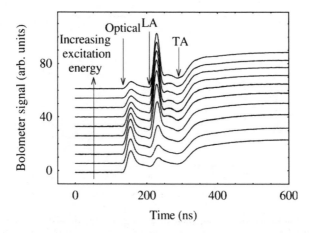

FIGURE 6.20 Bolometer signals obtained using sample D. For this crystal orientation, the LA mode enhancement is even more pronounced, while the TA contribution is not significantly altered across the wavelength range of interest.

phonon intensities for propagation along (211), and we found a similar behavior to the (311) case. These results suggest that acoustic anisotropy is allowing the generation of transverse monochromatic phonons.

6.5 DEMONSTRATING PHONON OPTICS: MEASURING THE MEAN FREE PATH OF TERAHERTZ (THZ) PHONONS

The SL filter experiment demonstrated that the phonons leaking from the SL were monochromatic, and for the SL structures used, the monochromatic beam had a relatively narrow linewidth. Having a source of monochromatic, and possibly coherent, phonons for phonon spectroscopy measurements is analogous to the advent of the laser for optical spectroscopy, and opens up a wealth of possibilities in phonon optics.

The aim of the work described in this section was to measure the mean free path of terahertz phonons, obtaining spectral resolution of the propagating LA modes using the GaAs substrate as a filter of phonons. At liquid helium temperatures, terahertz acoustic phonons can propagate ballistically over macroscopic distances in high-quality GaAs. There still remains, however, some question as to the relative strengths of the various phonon scattering processes. For point defect (Rayleigh type) phonon scattering, the scattering rate increases as the fourth power of the frequency, that is, $\tau^{-1} = A\nu^4$.[49] Long-wavelength LA phonons can also undergo anharmonic decay into lower-energy LA and TA modes, and for this process, $\tau^{-1} = B\nu^5$.[49-51] In addition to these processes, trace impurities (e.g., Cr ions)[52] and crystalline defects can also act as scattering centers. In either case, for phonons having a narrow frequency spectrum, the ballistic LA signal should be proportional to $e^{-l/\bar{\ell}}$, where l is the propagation distance and $\bar{\ell}$ is the phonon mean free path. However, for frequency-dependent scattering, the decay is nonexponential if the spectral width is significant. Having a source of monochromatic phonons therefore allows us to answer one of

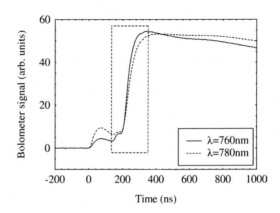

FIGURE 6.21 Bolometer traces obtained at a source-detector distance of 0.7 mm. The enhancement of the LA signal on-resonance is clearly visible, as is the corresponding increase in the TA mode intensity (as indicated by the dashed box).

the long-running questions in phonon physics regarding the dominant scattering mechanism in GaAs. Heat pulse measurements using conventional broadband metal film phonon sources have been used previously to study the scattering processes for terahertz phonons in GaAs, but the results were difficult to interpret due to the strong wavelength dependence of both scattering mechanisms. Using a beam of monochromatic phonons, we would expect to observe a strong exponential decay of the signal with increasing propagation distance, and therefore be able to determine the dominant scattering process in GaAs.

The experiment used the same SL structure as described in Section 6.4.1. To allow variation in the source-detector distance, while retaining the same direction of phonon propagation (i.e., along [001]), the SL was grown on a thick (2 mm) GaAs substrate and was subsequently polished at an angle such that the sample had a wedge shape. A line of aluminum bolometers was fabricated on the back surface of the substrate, and by scanning the laser spot along the sample, we could measure the phonon signal for different source-detector distances.

Figure 6.21 shows the signals obtained when the excitation spot was 0.7 mm from the bolometer. Both on- and off-resonance curves are shown, corresponding to excitation energies of 1.63 and 1.60 eV, respectively. As expected from earlier results, three peaks are again clearly visible in the signals. The peak at $t = 0$ ns is due to direct optical excitation of the bolometer by the laser and also PL from the substrate reaching the detector. The two later arriving peaks are consistent with the expected time of flight for LA and TA phonons. As previously observed, when the photon energy is resonant with the HH1–E1 transition, the optical signal intensity is reduced and there is a corresponding marked increase in the LA signal, as expected for this SL.

By looking at the propagation distance dependence of the LA mode intensity, it was possible to measure the mean free path of the propagating phonons, and to determine the dominant scattering mechanism in GaAs. In addition, the measurement provided yet more evidence that the leaked SL phonons were monochromatic.

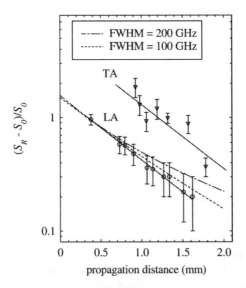

FIGURE 6.22 Normalized LA and TA signal intensity as a function of propagation distance. The solid lines correspond to a mean free path of 0.8 mm. Also shown (dashed lines) are simulations of the LA signal for point defect (isotope) scattering and two different spectral distributions centered on 650 GHz, one having a full width at half maximum (FWHM) of 200 GHz and the other a FWHM of 100 GHz.

To properly study the dependence, we needed to correct for differing detector sensitivity and also for geometrical (e.g., $1/r^2$) effects. To achieve this, the signals were normalized, and a plot of $(I_R - I_0)/I_R$ against propagation distance is shown in Figure 6.22, where I_R is the on-resonance LA peak intensity and I_0 is the off-resonance (background) LA intensity. We found that $(I_R - I_0)/I_R \propto e^{-l/\bar{\ell}}$, with $\bar{\ell} = 0.8$ mm; and extrapolating to $l = 0$, we obtain the ratio of LA signal on and off resonance to be $I_R/I_0 = 2.5$. That the decay is exponential indicates that the on-resonance signal we observed in previous measurements was largely due to phonons of a narrow frequency range. In addition, the value of the mean free path obtained is considerably lower than would be expected if the signal was due to carrier relaxation (where the phonons emitted would be of a relatively low frequency). This measurement provides conclusive evidence that the leaked propagating modes are monochromatic.

To estimate the spectral linewidth, the propagation distance dependence of $(I_R - I_0)/I_R$ for different width frequency distributions centered on 650 GHz was simulated for isotope scattering ($\tau^{-1} \sim \nu^4$), and the results are also shown in Figure 6.22. The predicted responses for spectral linewidths of 200 GHz and 100 GHz both show marked deviations from the experimental data, and therefore it was possible to conclude that the spectral width of the quasi-monochromatic phonons was ~50 GHz. (Roughly the same value of linewidth is obtained when assuming decay is dominant over isotope scattering.)

In addition to studying the LA signal dependence, the effect of phonon propagation distance on TA signal enhancement was also studied, as shown in Figure 6.22. The signals have been analyzed using the same method as for LA modes. One

TABLE 6.3

**Comparison of Measured and Calculated Values
of the LA Phonon Mean Free Path at 660 GHz**

	LA Mean Free Path (mm)
Measured, this work and Ref. 39	0.8
Calculated, Guseinov and Levinson[53]	4.4
Calculated, Tamura[54]	3.3
Calculated, Lax et al.[55]	12.0

Note: The calculations assume that isotope scattering dominates.

of the questions that was not fully answered by the SL filter experiment described in Section 6.4.3 was that the TA mode enhancement was due to anharmonic decay products of the leaked monochromatic LA modes reaching the detector.

In the previous study of LA phonons, we observed the LA phonon intensity decreased as the propagation distance increased. If the TA signal were due to the decay of LA modes, then we would expect to see the TA intensity increase with increasing propagation distance. However, we see a decrease of the TA intensity probably due to scattering in the substrate and consistent with a mean free path of ~0.8 mm. This not only discounts decaying monochromatic LA phonons as a source of the TA signal enhancement, but also allows us to comment on the dominant scattering mechanism in GaAs. Because we measure the same mean free path for both modes, we conclude that point defect type scattering dominates. This is because, despite their different frequencies, both LA and TA modes have the same wavelength equal to the period of the generator SL.

Assuming isotope scattering, the phonon mean free path is given by $\bar{\ell} = c_s \tau = c_s A^{-1} v^{-4}$. In Table 6.3, the measured value of ℓ is compared with that predicted using the various theoretical calculations of the scattering coefficient A.[53-55] We find that even the shortest of the calculated values is significantly longer than 0.8 mm, suggesting that defects other than natural isotopes may be dominating the scattering in our samples.

6.6 OTHER MATERIALS: GALLIUM NITRIDE AND ITS ALLOYS

So far, the discussion of experimental measurements has been limited to the GaAs/AlAs system. Recently, ultrafast excitation of InGaN/GaN multi-quantum-well (MQW) systems has been studied, and coherent phonon generation demonstrated using pump-probe techniques.[18,56] GaN epilayers are commonly grown on sapphire substrates; and due to lattice mismatch and c-axis growth, wurtzite GaN-based structures have large internal electric fields, of the order of several MV/cm. Owing to the different crystal structure and substantial piezoelectric nature of GaN, the observed reflection modulation is orders of magnitude higher than in GaAs-based SLs. Indeed, the acoustic coherent phonon oscillations were so strong that it

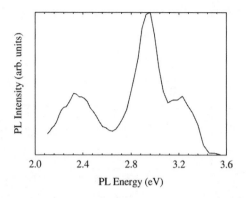

FIGURE 6.23 Photoluminescence (PL) spectrum of 40-period AlN/GaN SL (sample F). The main peak is at 2.95 eV (420 nm), which is well below the GaN bandgap due to the effect of the internal fields (quantum confined Stark effect).

was possible to measure in transmission, rather than reflection, geometry. Although there are several reports of studies of InGaN MQWs, there are (as yet) no reports of coherent phonon generation in AlGaN or AlN systems. The InGaN/GaN fundamental gap is readily accessible using a frequency-doubled Ti:sapphire, and thus is the obvious material choice for studying coherent phonon generation. However, growth of InGaN is complicated by the possibilities of phase separation, in segregation, and the possible formation of quantum dots, all of which could complicate the interpretation of results.[57–59] A more suitable system for use in these types of experiments would be AlGaN or AlN MQWs or SLs. Because the ability to study coherent phonon generation is ultimately limited by the tuning range of the laser, it might be expected that resonant excitation of the structures could be achieved only for very low Al fractions, and the lower the Al fraction, the lower the reflection modulation because the acoustic impedances of the well and barrier material are more closely matched. However, as previously mentioned, because of the large internal electric fields, the PL emission of AlGaN- and AlN-based structures is red shifted due to the quantum confined Stark effect. Although this is detrimental to device development (as it holds the emission wavelength much higher than would be expected), it allows us to study the generation and propagation of monochromatic phonons in a nitride-based alloy that is not subject to material variations to the same extent as InGaN.

To this end, a AlN/GaN SL, grown by MBE on a GaN template layer on sapphire, was excited by pulses from a frequency-doubled Ti:sapphire laser. The SL, consisting of 40 periods of 4.2-nm GaN and 3.2-nm AlN, was characterized by x-ray diffraction and PL measurements. The peak PL emission energy was measured at $\lambda = 435$ nm, well below the GaN band edge energy, as shown in Figure 6.23. A typical set of bolometer traces is shown in Figure 6.24. The most striking difference between the signals obtained from an AlAs/GaAs SL and those in Figure 6.9 is the intensity of the optical component when compared to the phonon signals. Consider the trace obtained for an excitation $\lambda = 450$ nm. At this wavelength, light passes through the sample and through the sapphire substrate. The observed signal is due to the laser

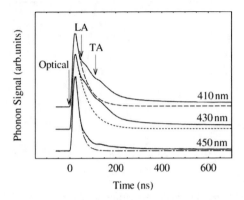

FIGURE 6.24 A series of bolometer signals from sample F. The LA intensity can be seen to increase strongly when the resonance condition is satisfied.

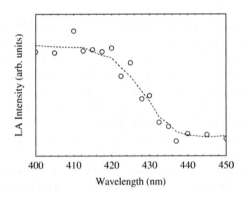

FIGURE 6.25 As with the GaAs-based SL samples, we observe a strong increase in the LA signal intensity when the excitation wavelength is tuned through the resonance.

pulse reaching the detector. When the excitation wavelength is coincident with the peak PL emission energy, light is absorbed by the SL and we observe the sudden appearance of a second and third peak in the bolometer traces. Because of the intense optical component, and the high sound velocities in sapphire ($c_{LA} = 11\ 100\ ms^{-1}$ and $c_{TA} = 6000\ ms^{-1}$), the phonon signals are not clearly resolved in these traces.

To better resolve the phonon modes, the optical decay was subtracted from the bolometer traces. As the resonance condition is satisfied, the phonon signal intensity shows a substantial increase. To obtain the curve of Figure 6.25, the signals were scaled to account for changes in the optical component intensity. This has the effect of normalizing the signals and is valid because across the wavelength range of interest, the laser power varies by an order of magnitude, due to changes in the frequency-doubling efficiency of the BBO crystal used. It is clear that the observed signals are not simply due to variations in laser power. Also, the nature of the signals measured following ultrafast excitation is characteristically different from those observed as a result of carrier relaxation. Studies of the radiative and nonradiative processes in the samples showed that TA phonon emission was the dominant relaxation mechanism.[60] In this measurement, the contribution to the bolometer signal from LA modes is much more significant, indicating a different origin for the observed signals. As the excitation wavelength is further reduced, the signal appearance does not vary until a wavelength of 355 nm. At this wavelength, light is absorbed in the GaN template, and the optical signal is greatly reduced, with a corresponding increase in the TA component. These results provide strong evidence that the enhancement in LA signal intensity we detect is due to monochromatic phonons that have leaked out of the SL. The excitation mechanism is probably similar to that proposed for InGaN/GaN SLs.[18] That is, absorption of photons in the GaN quantum wells photoexcites carriers that screen the internal electric fields, and this couples with the strain modes via the strong piezoelectric effect.

6.7 SUMMARY AND OUTLOOK

In this chapter we have described methods of generation of propagating pulsed beams of monochromatic acoustic phonons by ultrafast laser excitation of semiconductor SL structures. The method used to analyze the spectrum of the phonons generated, based on SL phonon filters, is itself an example of one application of the phonon pulses: phonon optics. Another application discussed was the measurement of the mean free path of terahertz (THz) frequency range phonons in GaAs at low temperatures. Although convincing evidence has been presented for the phonon pulses being monochromatic, it is still not clear whether they are coherent or, if they are, what the coherence length might be. An attempt to observe interference between two beams generated with a small time delay τ (pump-τ-pump experiment) failed to reveal any fringes. However, it is possible that this was simply due to the bolometer being too large and hence averaging the signal over a number of fringes. Further attempts to measure the mean coherence length of the phonons are presently underway.

The process by which the phonons leak out of the SL could have important consequences for the coherence of the propagating modes. The SL modes have $q = 0$; whereas, for propagating phonons $q = 2v/c_s$. Therefore, it is difficult to envisage a process by which an SL mode can convert to a propagating mode of the same energy without involving some elastic scattering process. Decay into two oppositely propagating phonons is unlikely for frequencies below 1 THz, and their frequency would be half that of the SL modes. If scattering is involved, this would lead to a loss of coherence. Clearly, more work, aimed at understanding the leakage mechanism, is required.

A surprising result of this study was the observation of monochromatic transverse polarized phonons. A possible explanation of this based on the effects of acoustic anisotropy was proposed. However, further theoretical work is required to confirm this hypothesis.

Despite the outstanding questions, it is clear that pulses of propagating terahertz monochromatic acoustic phonons can be generated by femtosecond laser excitation of SLs. Possible future applications of these nano-ultrasound pulses (ultrasonic pulses with nanometer wavelength) include spectroscopy and imaging of solid-state nanostructures. An important example is semiconductor quantum dots (QDs), which have applications in opto-electronics and, possibly, quantum information processing. It has already been shown that terahertz phonons can be used to probe, nondestructively, the structure of buried QD layers.[61] The wavelength of terahertz phonons is comparable to the QD size, giving rise to scattering and interference effects. Fringes were observed in the phonon reflection signal due to interference between phonons reflected from the top and bottom interfaces of the InAs dots with the GaAs matrix. However, the work described by Hawker et al.[61] used heat pulse sources, producing a black-body phonon spectrum. Significantly improved resolution should be obtained with a source of monochromatic phonons. Another application related to semiconductor nanostructures is the spectroscopy of carrier states and investigation of the carrier-phonon interaction. The latter is important in QDs for quantum information processing as it is a cause of dephasing of the carrier states. Related to this is the possibility that controlled beams of nano-ultrasound could also be used to "clock" the states in QD-based QUBITs.

Another area in which monochromatic terahertz phonons might be useful is in the generation and control of terahertz electromagnetic radiation. Currently, ultrafast laser-based methods of generating terahertz electromagnetic waves (e.g., the Auston switch[62]) produce very short pulses with an inherently broad frequency spectrum. For some applications (e.g., spectroscopy and imaging), nanosecond-duration pulses containing ~1000 cycles of electromagnetic radiation would be useful. One way of producing such narrow-spectral line terahertz radiation is to use intersubband lasers; another possibility would be to use piezoelectric films to couple the monochromatic phonons to EM radiation.

The work discussed in this chapter is mostly concerned with the AlAs/GaAs system. However, in Section 6.6, the first results on AlN/GaN SLs were described. Using bolometric techniques, we have recently observed evidence for the excitation of zone-folded modes that leak out of the SL and propagate across the sapphire substrate. However, as these results are preliminary in nature, to confirm that the signals we detect occur as a result of coherent SL modes leaking into propagating monochromatic phonons requires the use of pump-probe measurements and filter layers (as in the case of GaAs). These experiments are currently underway.

REFERENCES

1. L.J. Challis, Ed. *Electron-Phonon Interactions in Low-Dimensional Structures*, Oxford University Press, 2004.
2. H. Kinder. Spectroscopy with phonons on Al_2O_3 V^{3+} using phonon beam bremsstrahlung of a superconducting tunnel junction. *Phys. Rev. Lett.*, 28,1564–1567, 1972.
3. G.A. Toombs, F.W. Sheard, D. Neilson, and L.J. Challis. Phonon emission by a hot two-dimensional electron gas in a quantizing magnetic field. *Solid State Commun.*, 64, 577–581, 1987.
4. E.P.N. Damen, D.J. Dielman, A.F.M. Arts, and H.W. de Wijn. Generation and propagation of coherent phonon beams. *Phys. Rev. B*, 64, 174303-1-12, 2001.
5. D.J. Dielman, A.F. Koenderink, A.F.M. Arts, and H.W. de Wijn. Diffraction of coherent phonons emitted by a grating. *Phys. Rev. B*, 60, 14719–14723, 1999.

6. H.Y. Hao, H.J. Maris. Study of phonon dispersion in silicon and germanium at long wavelengths using picosecond ultrasonics. *Phys. Rev. Lett.,* 84, 5556–5559, 2000.

7. H.Y. Hao and H.J. Maris. Dispersion of the long-wavelength phonons in Ge, Si, GaAs, quartz, and sapphire. *Phys. Rev. B,* 63, 224301-1-10, 2001.

8. H.Y. Hao and H.J. Maris. Experiments with acoustic solitons in crystalline solids. *Phys. Rev. B,* 64, 064302–064308, 2001.

9. O.L. Muskens and J.I. Dijkhuis. High amplitude, ultrashort, longitudinal strain solitons in sapphire. *Phys. Rev. Lett.,* 89, 285504, 2002.

10. J.J. Baumberg, D.A. Williams, and K. Kohler. Ultrafast acoustic phonon ballistics in semiconductor heterostructures. *Phys. Rev. Lett.,* 78, 3358–3361, 1997.

11. O. Matsuda, I. Ishii, T. Fukui, J.J. Baumberg, and O. Wright. Wavelength selective photoexcitation of picosecond acoustic phonon pulses in a triple $GaAs/Al_{0.3}Ga_{0.7}As$ quantum well structure. *Physica B,* 316, 205–208, 2002.

12. C. Rossignol, B. Perrin, S. Laborde, L. Vandenbulcke, M.I. De Barros, and P. Djemia. Nondestructive evaluation of micrometric diamond films with an interferometric picosecond ultrasonics technique. *J. Appl. Phys.,* 95, 4157–4162, 2004.

13. B. Perrin. Private communication.

14. T. Ruf. Phonon Raman Scattering in Semiconductors, Quantum Wells and Superlattices: Basic Results and Applications, *Springer Tracts in Modern Physics,* Vol. 142, Springer, Berlin, 1998.

15. A. Yamamoto, T. Mishina, Y. Masumoto, and M. Nakayama. Coherent oscillation of zone folded phonon modes in GaAs-AlAs superlattices. *Phys. Rev. Lett.,* 73, 740–743, 1994.

16. A. Bartels, T. Dekorsy, H. Kurz, and K. Kohler. Coherent zone folded longitudinal acoustic phonons in semiconductor superlattices: excitation and detection. *Phys. Rev. Lett.,* 82, 1044–1047, 1999.

17. T. Dekorsy, A. Bartels, H. Kurz, and K. Kohler. Coherent acoustic phonons in semiconductor superlattices. *Phys. Stat. Sol. (b),* 215, 425–430, 1999.

18. C.K. Sun, J.C. Liang, and X.Y. Yu. Coherent acoustic phonon oscillations in semiconductor multiple quantum wells with piezoelectric fields. *Phys. Rev. Lett.,* 84, 179–182, 2000.

19. K. Mizoguchi, M. Hase, S. Nakashima, and M. Nakayama. Observation of coherent folded acoustic phonons propagating in a GaAs/AlAs superlattice by two-color pump-probe spectroscopy. *Phys. Rev. B,* 60, 8262–8266, 1999.

20. R.J. von Gutfeld and A.H. Nethercot. heat pulses in quartz and sapphire at low temperatures. *Phys. Rev. Lett.,* 12, 641–646, 1964.

21. J.J. Harris and C.T. Foxon. Growth and fabrication of semiconductor devices for hot-electron research, in *Hot Electrons in Semiconductors: Physics and Devices,* N. Balkan, Ed., Oxford University Press, 1998.

22. See, for example, http://www.ioffe.rssi.ru/SVA/NSM/Semicond/, and references therein.

23. B. Taylor, H.J. Maris, and C. Elbaum. Phonon focusing in solids. *Phys. Rev. Lett.,* 23, 416–419, 1969.

24. L. Esaki and R. Tsu. Superlattices and negative differential conductivity in semiconductors. *IBM J. Res. Dev.,* 14, 61, 1970.

25. R. Tsu, L.L. Chang, G.A. Saihalasz, and L. Esaki. Effect of quantum states on photocurrent in a superlattice. *Phys. Rev. Lett.,* 34, 1509–1512, 1975.

26. S. Tamura, D.C. Hurley, and J.P. Wolfe. Acoustic phonon propagation in superlattices. *Phys. Rev. B,* 38, 1427–1449, 1988.

27. S. Mizuno and S. Tamura. Transmission and reflection times of phonon packets propagating through superlattices. *Phys. Rev. B*, 50, 7708–7718, 1994.
28. C. Colvard, R. Merlin, M.V. Klein, and A.C. Gossard. Observation of folded acoustic phonons in a semiconductor superlattice. *Phys. Rev. Lett.*, 45, 298–301, 1980.
29. P. Manuel, G.A. Saihalasz, L.L. Chang, C.A. Chang, and L. Esaki. Resonant Raman scattering in a semiconductor superlattice. *Phys. Rev. Lett.*, 37, 1701–1704, 1976.
30. V. Narayanamurti, H.L. Stormer, M.A. Chin, A.C. Gossard, and W. Wiegmann. Selective transmission of high frequency phonons by a superlattice: the "dielectric" phonon filter. *Phys. Rev. Lett.*, 43, 2012–2016, 1979.
31. P. Lacharmoise, A. Fainstein, B. Jusserand, and V. Thierry-Mieg. Optical cavity enhancement of light-sound interaction in acoustic phonon cavities. *Appl. Phys. Lett.*, 84, 3274–3276, 2004.
32. D.C. Hurley, S. Tamura, J.P. Wolfe, and H. Morkoc. Imaging of acoustic phonon stop bands in superlattices. *Phys. Rev. Lett.*, 58, 2446–2449, 1987.
33. S.C. Edwards, H. Binrani, and J.K. Wigmore. The use of superconducting bolometers for detecting nanosecond heat pulses. *J Phys. E Sci. Instrum.*, 22, 582–586, 1989.
34. B.A. Danilchenko, C.Z. Jasiukiewicz, T. Paszkiewicz, and S. Wolski. Response of superconductor bolometer to phonon fluxes. *Acta Phys. Pol. A*, 103, 325–338, 2003.
35. W.A. Sherlock and A.F.G. Wyatt. The effect of self-heating on the dynamical response of a bolometric detector. *J. Phys. E: Sci. Instrum.*, 16, 669–672, 1983.
36. P. Hakwer, A.J. Kent, L.J. Challis, A. Bartels, T. Dekorsy, H. Kurz, and K. Kohler. Observation of coherent zone folded acoustic phonons generated by Raman scattering in a superlattice. *Appl. Phys. Lett.*, 77, 3209–3211, 2000.
37. N.M. Stanton, R.N. Kini, A.J. Kent, M. Henini, and D. Lehmann. Terahertz phonon optics in GaAs/AlAs superlattice structures. *Phys. Rev. B*, 68, 113302, 2003.
38. N.M. Stanton, R.N. Kini, A.J. Kent, and M. Henini. Monochromatic transverse-polarized phonons from femtosecond pulsed optical excitation of a GaAs/AlAs super-lattice. *Phys. Rev. B*, 69, 125341, 2004.
39. A.J. Kent, N.M. Stanton, L.J. Challis, and M. Henini. Generation and propagation of monochromatic acoustic phonons in gallium arsenide. *Appl. Phys. Lett.*, 81, 3497–3499, 2002.
40. N.M. Stanton, C.E. Martinez, A.J. Kent, S.V. Novikov, and C.T. Foxon. Phonon generation by femtosecond pulsed laser excitation of an AlN/GaN superlattice. *Phys. Stat. Solidi (c)*, 1, 11, 2678 (2004).
41. M.E. Msall and J.P. Wolfe. Ballistic phonon production in photoexcited Ge, GaAs, and Si. *Phys. Rev. B*, 65, 195205-1-11, 2002.
42. R.G. Ulbrich, V. Narayanamurti, and M.A. Chin. Propagation of large wave vector acoustic phonons in semiconductors. *Phys. Rev. Lett.*, 45, 1432–1435, 1980.
43. R.G. Ulbrich. Generation, propagation and detection of terahertz phonons in gallium arsenide, in *Non-equilibrium Phonon Dynamics*, W.E. Bron., Ed., Vol. 124, 101–128, 1984.
44. A.J. Kent. Energy and momentum relaxation of hot electrons by acoustic phonon emission, in *Hot Electrons in Semiconductors: Physics and Devices,* edited by N. Balkan, Oxford University Press, 1998.
45. J.P. Wolfe. Imaging Phonons: Acoustic Wave Propagation in Solids. Cambridge University Press, Cambridge, 1998.
46. H.J. Zeiger, J. Vidal, T.K. Cheng, E.P. Ippen, G. Dresselhaus, and M.S. Dresselhaus. Theory for displacive excitation of coherent phonons. *Phys. Rev. B*, 45, 768–778, 1992.

47. Z.V. Popovic, M. Cardona, E. Richter, D. Strauch, L. Tapfer, and K. Ploog. Phonon properties of GaAs/AlAs superlattices grown along the [110] direction. *Phys. Rev. B,* 40, 3040–3050, 1989.
48. G. Winterling, E.S. Koteles, and M. Cardona. Observation of forbidden Brillouin scattering near an exciton resonance. *Phys. Rev. Lett.,* 39, 1286–1289, 1977.
49. P.G. Klemens. In *Solid State Physics,* edited by F. Seitz and D. Turnbull. Academic, New York, Vol. 7, Chapter 1, 1958.
50. N.M. Guseinov and Y.B. Levinson. Diffusion of non-decaying TA phonons. *Solid State Commun.,* 45, 371–374, 1983.
51. S. Tamura. Spontaneous decay rates of LA phonons in quasi-isotropic solids. *Phys. Rev. B,* 31, 2574–2577, 1985.
52. L.J. Challis, M. Locatelli, A. Ramdane, and B. Salce. Phonon scattering by Cr ions in GaAs. *J. Phys. C: Solid State Phys.,* 15, 1419–1432, 1982.
53. N.M. Guseinov and Y.B. Levinson. Propagation of non-decaying TA phonons. *Zh. Eksp. Teor. Fiz.,* 85, 779–794, 1983.
54. S. Tamura. Isotope scattering of large-wave-vector phonons in GaAs and InSb: deformation-dipole and overlap-shell models. *Phys. Rev. B,* 30, 849–854, 1984.
55. M. Lax, V. Narayanamurti, R. Ulbrich, N. Holzarth. In *Phonon Scattering in Condensed Matter,* edited by W. Eisenmenger, K. Lassman, and S. Dottinger. Springer, Berlin, 103, 1984.
56. Ü. Özgur, C.W. Lee, and H.O. Everitt. Control of coherent acoustic phonons in semiconductor quantum wells. *Phys. Rev. Lett.,* 86, 5604–5607, 2001.
57. E.J. Thrush, M.J. Kappers, P. Dawson, M.E. Vickers, J. Barnard, D. Graham, G. Makaronidis, F.D.G. Rayment, L. Considine, and C.J. Humphreys. GaN/InGaN quantum wells grown in a closed coupled showerhead reactor. *J. Cryst. Growth,* 248, 518–522, 2003.
58. A. Dussaigne, B. Damilano, N. Grandjean, and J. Massies. In surface segregation in InGaN/GaN quantum wells. *J. Cryst. Growth,* 251, 471–475, 2003.
59. P. Waltereit, O. Brandt, K.H. Ploog, M.A. Tagliente, and L. Tapfer. In surface segregation during growth of (In,Ga)N/GaN multiple quantum wells by plasma assisted molecular beam epitaxy. *Phys. Rev. B,* 66, 165322-1-6, 2002.
60. C.E. Martinez, N.M. Stanton, A.J. Kent, C.R. Staddon, S.V. Novikov, and C.T. Foxon. Influence of internal fields on radiative and non-radiative processes in AlN/GaN superlattices. *J. Appl. Phys.,* 95, 7785–7789, 2004.
61. P. Hawker, A.J. Kent, and M. Henini. Measuring the size of buried quantum dots using phonons. *Physica B,* 263, 514–516, 1999.
62. D.H. Auston, K.P. Cheung, and P.R. Smith. Picosecond photoconducting Hertzian dipoles. *Appl. Phys. Lett.,* 45, 284–286, 1984.

7 Optical Studies of Carrier Dynamics and Non-Equilibrium Optical Phonons in Nitride-Based Semiconductors

K.T. Tsen

CONTENTS

7.1 Introduction 180
7.2 Experimental Approach, Samples, and Experimental Technique 181
    7.2.1 Experimental Approach...181
        7.2.1.1 Raman Spectroscopy in Semiconductors 181
    7.2.2 General Considerations 189
        7.2.2.1 Light Source..189
        7.2.2.2 Spectrometer 189
        7.2.2.3 Detector and Photon-Counting Electronics.....................190
    7.2.3 Samples and Experimental Technique...191
7.3 Experimental Results 192
    7.3.1 Experimental Results for Carrier Dynamics
        in $In_xGa_{1-x}As_{1-y}N_y$: Analysis and Discussion 192
    7.3.2 Experimental Results for Transient Carrier Transport
        in $In_xGa_{1-x}N$/GaN 195
    7.3.3 Experimental Results for Transient Carrier Transport
        in InN/GaN: Analysis and Discussion..196
    7.3.4 Experimental Results on Non-Equilibrium Longitudinal
        Optical Phonons in InN: Analysis and Discussion 202
    7.3.5 Experimental Results on the Observation of Large Electron
        Drift Velocities in InN Thick Film Grown on GaN............................206
7.4 Conclusion...210
Acknowledgments...210
References..210

7.1 INTRODUCTION

Gallium nitride (GaN), aluminum nitride (AlN), indium nitride (InN), and their alloys have long been considered very promising materials for device applications.[1,2] Semiconductor alloys such as $In_xGa_{1-x}N$ have been successfully used in the fabrication of blue-green light emitting diodes (LEDs) and laser diodes (LDs).[2–7] Recently, growth of high-quality InN as well as $In_xGa_{1-x}N$ have been demonstrated.[8–10] In particular, progress in the manufacture of very high-quality, single-crystal InN thin films has opened up a new challenging research avenue in the III-nitride semiconductors.[11] InN, together with its alloys of GaN and AlN, enable the operation of LEDs and LDs ranging in spectral wavelength from infrared all the way down to deep ultraviolet. It has also been predicted that InN has the lowest electron effective mass among all the III-nitride semiconductors.[12] As a result, very high electron mobility and very large saturation velocity are expected. It was found by ensemble Monte Carlo (EMC) simulations that InN possesses extremely high transient electron drift velocity.[13–15] InN had been reported to have a bandgap of $\cong 1.89$ eV.[16] In contrast to this general belief, recent experimental results on high-quality InN samples have suggested that the bandgap of InN should be around 0.8 eV.[17–24] Some researchers argued that the bandgap of InN was about 1.89 eV as reported in the literature, and the narrow bandgap reported recently might be due to radiative emission from a defect level, like the well-known yellow luminescence of GaN. Others believed that the previous studies were carried out in poor-quality InN films and the films might be contaminated by impurities such as oxygen, which can result in a deep level state well above the conduction band of InN, and as a result they showed a higher bandgap energy value than the real one.

The successful fabrication of operational InGaAsN/GaAs laser diodes using gas-source molecular beam epitaxy (GSMBE),[25–27] chemical beam epitaxy (CBE),[28] and metal-organic chemical vapor deposition (MOCVD)[29,30] techniques has attracted a lot of attention. The quaternary InGaAsN alloy system, because of its ability to remain lattice-matched to other semiconductors such as GaAs, Ge, and InP,[31] has been predicted to have superior temperature characteristics compared to the InGaAsP alloy system. InGaAsN alloys have also been predicted to have great potential for multi-bandgap solar cells.[32,33] It was found that the minority-carrier diffusion length increases substantially after postgrowth annealing in a nitrogen ambient, and as a result the quantum efficiency in such solar cells can be as high as 70%. More recently, Mair et al.[34] reported the results of time-resolved photoluminescence spectroscopy studies of an InGaAsN epilayer. These authors concluded that the localized states in InGaAsN, which arose from alloy fluctuations, played an important role in the decay of the photoluminescence intensity. The next natural step is to manufacture high-performance InN, $In_xGa_{1-x}N$, and InGaAsN electronic devices. To improve the design of these devices, knowledge of their electron transport properties is indispensable. In this chapter, carrier dynamics, carrier transport, and non-equilibrium phonons in various nitride-based semiconductors are studied using ultrafast Raman spectroscopy (see Figure 7.1).

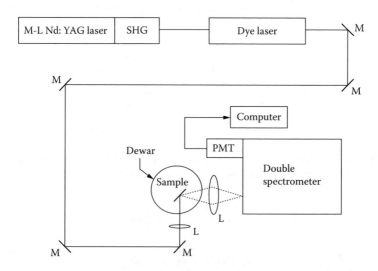

FIGURE 7.1 The experimental setup for transient (single-pulse) picosecond Raman scattering studies of electron transient transport in semiconductors. SHG: second harmonic generation system; M: mirror; L: lens; and PMT: photomultiplier tube. For transient experiments, the same laser pulse is used for both the excitation of electron-hole pairs and the detection of their transport properties.

7.2 EXPERIMENTAL APPROACH, SAMPLES, AND EXPERIMENTAL TECHNIQUE

7.2.1 EXPERIMENTAL APPROACH

7.2.1.1 Raman Spectroscopy in Semiconductors

We first present a brief theory of Raman scattering from carriers in semiconductors, which will be particularly useful for situations where electron distributions are non-equilibrium, such as in our current situation; and then we discuss the theory of Raman scattering by lattice vibrations in semiconductors.

7.2.1.1.1 Theory of Raman Scattering from Carriers in Semiconductors

(a) A Simple Model

To understand how Raman spectroscopy can be used to probe electron distribution function in semiconductors, we start with the simplest physical concept — Compton scattering. As shown in Figure 7.2, let us consider that an incident photon with wavevector $\vec{k}_i$ and angular frequency ω_i is interacting with an electron of mass m_e^* traveling at a velocity $\vec{V}$. After the scattering event, the scattered photon is characterized by wavevector $\vec{k}_s$ and angular frequency ω_s. The scattered electron is then moving at a velocity $\vec{V}'$. From the conservation of energy and momentum, we can write the following equations:

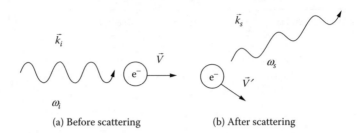

(a) Before scattering   (b) After scattering

FIGURE 7.2 A simple model: Compton scattering, demonstrating how the electron distribution in semiconductors can be directly probed by Raman spectroscopy.

$$\hbar\omega_i + \frac{1}{2}m_e^*\vec{V}^2 = \hbar\omega_s + \frac{1}{2}m_e^*\vec{V}'^2 \qquad (7.1)$$

$$\hbar\vec{k}_i + m_e^*\vec{V} = \hbar\vec{k}_s + m_e^*\vec{V}' . \qquad (7.2)$$

If we define the energy transfer and the wavevector transfer of the photon to be $\omega \equiv \omega_i - \omega_s$ and $\vec{q} \equiv \vec{k}_i - \vec{k}_s$ respectively, then from Equation 7.1 and Equation 7.2, we have

$$\omega = \vec{V}\cdot\vec{q} + \frac{\hbar\vec{q}^2}{2m_e^*} . \qquad (7.3)$$

This important equation states that the energy transfer of incident photon is (apart from a constant term: $\hbar\vec{q}^2/2m_e^*$) directly proportional to the electron velocity along the direction of wavevector transfer. That is, it implies that Raman scattering intensity, measured at an angular frequency ω, is proportional to the number of electrons that have a velocity component along the direction of wavevector transfer given by Equation 7.3, irrespective of their velocity components perpendicular to $\vec{q}$.

Therefore, if electron distribution function is Maxwell-Boltzmann-like, then the lineshape of the Raman scattering spectrum will be Gaussian-like, centered around $\omega \cong 0$, whereas a drifted Maxwell-Boltzmann distribution with an electron drift velocity $\vec{V}_d$ will result in a Raman scattering spectrum that is a shifted Gaussian centered around $\omega \cong \vec{q}\cdot\vec{V}_d$.

However, we note that, strictly speaking, this simple picture is only correct for a system of noninteracting electron gas in vacuum. For an electron gas in a semiconductor such as GaAs or GaN, many-body effects and the effects of band structure must be considered. The former is usually taken into account by the random-phase approximation (RPA)[40] and the latter by sophisticated band structure calculations such as $\vec{k}\cdot\vec{p}$ approximation[41] or pseudopotential calculations.[42]

(b) A Full Quantum-Mechanical Approach

We now use a quantum-mechanical method to calculate the Raman scattering cross-section or single-particle scattering (SPS) spectrum for a single-component plasma in a direct bandgap semiconductor such as GaAs, probed by an ultrafast laser having pulse width t_p. For simplicity, we assume that the probe pulse is a square pulse from $-t_p/2$ to $+t_p/2$ and that electron elastic scattering is the dominant scattering process in the solid-state system.

We start with a typical electron-photon interaction Hamiltonian, which has been shown in the equilibrium case to be[43]

$$H = H_0 + \sum_i \left[\frac{-e}{2m_e^* c} [\vec{p}_i \cdot \vec{A}(\vec{r}_i) + \vec{A}(\vec{r}_i) \cdot \vec{p}_i] + \frac{e^2}{2m_e^* c} \vec{A}^2(\vec{r}_i) \right]$$

$$\cong H_0 + H_1 + H_2 , \qquad (7.4)$$

where H_0 is the total Hamiltonian of the system in the absence of radiation field; e is the charge of an electron; c is the speed of light; $\vec{p}$ is the electron momentum; $\sum$ refers to summation over electrons; the second and third terms describe the interactions of electrons with radiation field; and $\vec{A}$ is the vector potential of radiation field:

$$\vec{A}(\vec{\gamma}_i) \cong \frac{1}{\sqrt{V}} \sum_j \left(\frac{2\pi hc^2}{w_j} \right)^{1/2} \left(e^{i\vec{k}_j \cdot \vec{\gamma}_i} b_{\vec{k}_j} + e^{-i\vec{k}_j \cdot \vec{\gamma}_i} b_{\vec{k}_j}^+ \right) \hat{e}_j \qquad (7.5)$$

where V is the volume of the semiconductor; and $b_{\vec{k}_j}$, $b_{\vec{k}_j}^+$ are photon annihilation and creation operators, respectively. Because the vector potential $\vec{A}$ is a linear combination of photon creation and annihilation operators, and the Raman scattering process involves the annihilation of an incident photon and the creation of a scattered photon, the $\vec{p} \cdot \vec{A}$ and $\vec{A} \cdot \vec{p}$ terms in Equation 7.4 will contribute to the scattering matrix element in the second order and the $\vec{A}^2$ terms contribute in the first order in the perturbation-theory calculations of Raman scattering cross-section.

In general, the SPS cross-section has three parts: namely, charge-density fluctuations (CDF), energy-density fluctuations (EDF), and spin-density fluctuations (SDF). The spin of the electron is coupled with the incident light through the second-order $\vec{p}_i \cdot \vec{A}$ perturbation terms. SDF is the dominant contribution for higher electron concentrations. In addition, the EDF contribution dominates when the CDF contribution is screened at high electron densities. These three contributions are related to different polarization selection rules.[44]

The CDF and EDF contributions are relatively complicated. They contain both the screening effects of Coulomb interaction and the electron-phonon interaction through the dielectric constant. The SDF contributions, on the other hand, are proportional to $(\hat{e}_i \times \hat{e}_s)^2$, where $\hat{e}_i$, $\hat{e}_s$ are the polarization vectors of the incident and scattered photons, respectively, and are independent of either the Coulomb interaction or electron-phonon interactions. It is therefore much simpler than either CDF or EDF for analysis of electron distributions. For this reason, SPS spectra obtained from SDF is most conveniently used to probe electron distribution functions in semiconductors.

The single-particle scattering (SPS) cross-section associated with spin-density fluctuations (SDF) for a single-component plasma in a direct bandgap semiconductor such as GaAs, probed by an ultrafast laser having pulse width t_p and when elastic scattering is dominant, can be shown to be given by [44-46]

$$\left(\frac{d^2\sigma}{d\omega d\Omega}\right)_{SDF} = C \cdot \sum_{\vec{p}} -n(\vec{p})[1 - n(\vec{p}+\vec{q})](\hat{e}_i \times \hat{e}_s)^2 \cdot$$

$$\int_{-\infty}^{\infty} d\omega' \int_{-t_p/2}^{t_p/2} dt \int_{-t_p/2-t}^{t_p/2-t} dt' e^{i(\omega-\omega')t'} S_p(t,\omega_i) S_p^*(t'+t,\omega_i) \cdot$$

$$\operatorname{Im}\left\{ \frac{1}{\hbar\omega' + \varepsilon_{\vec{p}} - \varepsilon_{\vec{p}+\vec{q}} + i\hbar/\tau} \left[1 - \frac{i\hbar}{\tau} \left\langle \frac{1}{\hbar\omega' + \varepsilon_{\vec{p}} - \varepsilon_{\vec{p}+\vec{q}} + i\hbar/\tau} \right\rangle_{\Omega_{\vec{p}}} \right]^{-1} \right\}, \quad (7.6)$$

where C is a constant; $n(\vec{p})$ is the electron distribution function; $\hat{e}_i$, $\hat{e}_s$ are polarization vectors of the incident and scattered light, respectively; ω_i, ω_s are angular frequencies of the incident and scattered light, respectively; $\omega \equiv \omega_i - \omega_s$; t_p is the pulse width of the probe laser; $\varepsilon_{\vec{p}}$ is the electron energy at $\vec{p}$; τ is the electron collision time; $\Omega_{\vec{p}}$ represents an average over the solid angle in the momentum space; and

$$S_p(t,\omega_i) \equiv -\left(\frac{P^2}{3m_e^*}\right) \cdot$$

$$\sum_{n=1}^{3} A_s(n) \cdot \frac{\hbar\omega_i - e^{\frac{i}{\hbar}E_{g_n}\left(t+t_p/2\right)}\left\{\hbar\omega_i \cos\left[\left(\omega_i\left(t+t_p\right)/2\right)\right] - iE_{g_n}\sin\left[\omega_i\left(t+t_p/2\right)\right]\right\}}{\left(E_{g_n} - i\Gamma_n\right)^2 - \left(\hbar\omega_i\right)^2},$$

where m_e^* is the effective mass of electron on the conduction band; $A_s(1) = A_s(2) = 1$, and $A_s(3) = -2$; and Γ_1, Γ_2, Γ_3 are the damping constants involved in the Raman scattering processes. The E_{g1}, E_{g2}, E_{g3} denote the energy difference between the conduction band and the heavy-hole, light-hole, and split-off-hole bands evaluated at wavevector $\vec{k}$, respectively. $P \equiv -i\langle S|p_z|Z\rangle$ is the momentum matrix element between the conduction and valence bands at the $\Gamma-$point in Kane's notations.[41]

We note that in the limit of very long probe pulse ($t_p \to \infty$) and equilibrium electron distributions, our results can be shown to reduce to expressions previously given for the Raman scattering cross-section in the equilibrium case.[47]

It is very instructive to note that if we assume that the pulse width of the probe laser is sufficiently wide, collision effects are negligible, the electron distribution

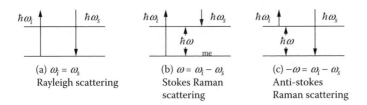

(a) $\omega_i = \omega_s$
Rayleigh scattering

(b) $\omega = \omega_i - \omega_s$
Stokes Raman
scattering

(c) $-\omega = \omega_i - \omega_s$
Anti-stokes
Raman scattering

FIGURE 7.3 A diagram showing (a) Rayleigh scattering process, (b) Stokes Raman scattering process, and (c) anti-Stokes Raman scattering process.

function is nondegenerate, and the term involving matrix elements $--S_p$ does not depend on the electron momentum, Equation 7.6 can be shown to become

$$\left(\frac{d^2\sigma}{d\omega d\Omega}\right)_{SDF} \propto \int d^3p \cdot n(\vec{p}) \cdot \delta\left[\omega - \vec{V}\cdot\vec{q} - \frac{\hbar q^2}{2m_e^*}\right];\qquad(7.7)$$

here, the δ-function in Equation 7.7 ensures that both the energy and momentum are conserved.

We note that Equation 7.7 shows that the measured Raman scattering cross-section at a given solid angle $d\Omega$ (which determines $\vec{q}$) provides *direct* information about the electron distribution function in the direction of wavevector transfer $\vec{q}$, in agreement with the simple classical picture.

One intriguing feature for probing carrier distributions with Raman spectroscopy is that because the Raman scattering cross-section is inversely proportional to the square of the effective mass of the carrier, it preferentially probes electron distribution even if holes are simultaneously present. This unique feature makes the interpretation of electron distribution in Raman scattering experiments much simpler than those of other techniques.

7.2.1.1.2 Theory of Raman Scattering by Lattice Vibrations in Semiconductors

Consider an incident laser beam of angular frequency ω_i that is scattered by a semiconductor and the scattered radiation is analyzed spectroscopically, as shown in Figure 7.3. In general, the scattered radiation consists of a laser beam of angular frequency ω_i accompanied by weaker lines of angular frequencies $\omega_i \pm \omega$. The line at an angular frequency $\omega_i - \omega$ is called a Stokes line; whereas, at an angular frequency, $\omega_i + \omega$ is usually referred to as an anti-Stokes line. The important aspect is that the angular frequency shifts ω are independent of ω_i. In this way, this phenomenon differs from that of luminescence, in which it is the angular frequency of the emitted light that is independent of ω_i. The effect just described is called the Raman effect. It was predicted by Smekal[48] and is implicit in the radiation theory of Kramers and Heisenburg.[49] It was discovered experimentally by Raman[50] and by Lansberg and Mandel'shtam[51] in 1928. It can be understood as an inelastic scattering of light in which an internal form of motion of the scattering system is either excited or absorbed during the process.

(a) A Simple Classical Theory

Imagine that we have a crystalline lattice having an internal mode of vibration characterized by a normal coordinate

$$Q = Q_0 \cos \omega t \; ; \tag{7.8}$$

the electronic polarizability α is generally a function of Q and, because in general, $\omega \ll \omega_i$, at each instant we can regard Q as fixed compared with the variation of the external field $\vec{E}$; that is, at angular frequency ω_i, the induced dipole moment $\vec{P}$ is

$$\vec{P} = \alpha \vec{E} = \alpha(Q) \vec{E} \; . \tag{7.9}$$

Let $\alpha_0 = \alpha(0)$ be the polarizability in the absence of any excitation. We can write

$$\alpha(Q) = \alpha_0 + \left(\frac{\partial \alpha}{\partial Q} \right)_0 Q + \frac{1}{2} \left(\frac{\partial^2 \alpha}{\partial Q^2} \right)_0 Q^2 + \cdots$$

$$= \alpha_0 + \alpha_1 Q + \frac{1}{2} \alpha_2 Q^2 + \cdots , \tag{7.10}$$

where $\left[\dfrac{\partial \alpha}{\partial Q} \right]_0 \equiv \alpha_1$; $\left[\dfrac{\partial^2 \alpha}{\partial Q^2} \right]_0 \equiv \alpha_2$ and the derivative is to be evaluated at zero excitation field.

If we assume that $\vec{E} = \vec{E}_0 \cos \omega_i t$, we find that

$$\vec{P}(t) = \left(\alpha_0 \vec{E}_0 + \frac{1}{4} \alpha_2 Q_0^2 \vec{E}_0 \right) \cos \omega_i t + \frac{\vec{E}_0}{2} \alpha_1 Q_0 \left[\cos(\omega_i + \omega)t + \cos(\omega_i - \omega)t \right]$$

$$+ \frac{1}{8} \alpha_2 Q_0^2 \vec{E}_0 \left[\cos(\omega_i + 2\omega) + \cos(\omega_i - 2\omega) \right] + \dots . \tag{7.11}$$

For an oscillating dipole moment, the magnetic and electric fields of emitted electromagnetic wave are given by[52]

$$\vec{B} = \frac{1}{c^2 r} \left[\frac{\partial^2 \vec{P} \left(t - \dfrac{r}{c} \right)}{\partial^2 t} \right] \times \hat{n} \tag{7.12a}$$

and

$$\vec{E} = \vec{B} \times \hat{n} \ , \tag{7.12b}$$

where $\vec{r}$ is the position vector connecting the center of the dipole moment to the point of observation and $\hat{n} = \vec{r}/|\vec{r}|$.

Therefore, the first term in Equation 7.11 gives rise to Rayleigh scattering; the second term gives the anti-Stokes and Stokes first-order Raman lines, respectively; the third term takes into account the anti-Stokes and Stokes second-order Raman lines, etc. We notice in Equation 7.11 that the intensities of the Stokes and anti-Stokes lines are equal. This is because all classical theories neglect the possibility of spontaneous emission.

(b) A Quantum-Mechanical Theory

In the quantum mechanical treatment of the scattering of light by lattice vibrations, we consider the total Hamiltonian of the system, including the radiation field:

$$H = H'_0 + H_{el-ph} + H' \ , \tag{7.13}$$

where H'_0 includes contributions from the electronic system, lattice vibrations (or phonons), and radiation field; $H_{el-ph} = -e\varphi(\vec{r}_i)$ describes the interaction of electrons with phonons; $\varphi(\vec{r}_i)$ is the potential due to, say, deformation potential and/or Fröhlich interactions; and

$$
\begin{aligned}
H' &= \sum_i \frac{-e}{2m_e c} \left[\vec{A}(\vec{r}_i) \cdot \vec{p}_i + \vec{p}_i \cdot \vec{A}(\vec{r}_i) \right] + \sum_i \frac{e^2}{2m_e c^2} \vec{A}^2(\vec{r}_i) \\
&= \sum_i \frac{-e}{m_e c} \left[\vec{p}_i \cdot \vec{A}(\vec{r}_i) \right] + \sum_i \frac{e^2}{2m_e c^2} \vec{A}^2(\vec{r}_i) \\
&\equiv H'_1 + H'_2 \tag{7.14}
\end{aligned}
$$

takes into account the electron-photon interactions, where $\vec{A}(\vec{r}_i)$ is the vector potential of radiation field given by Equation 7.5.

We notice that for a typical Raman scattering process in which $\omega_i \gg \omega$, photons do not interact directly with phonons, but through electron-phonon interactions, that is, the H_{el-ph} term in the total Hamiltonian. Because the Raman scattering process involves the annihilation of an incident photon and the creation of a scattered photon, the $\vec{p} \cdot \vec{A}$ and $\vec{A} \cdot \vec{p}$ terms in Equation 7.14 will contribute to the scattering matrix element in the third order, and $\vec{A}^2$ terms contribute in the second order in the perturbation-theory calculations of the Raman scattering cross-section. If we neglect nonlinear processes, then only the $\vec{p} \cdot \vec{A}$ and $\vec{A} \cdot \vec{p}$ terms in Equation 7.14 are important and need be considered.

From the time-dependent perturbation theory and Fermi golden rule, we obtain for the scattering probability (which is proportional to Raman scattering cross-section) for a one-phonon Stokes Raman process[53]:

$$P(\omega_s) = \frac{2\pi}{\hbar} \left| \sum_{n,n'} \frac{\langle i|H_1'|n\rangle\langle n|H_{el-ph}|n'\rangle\langle n'|H_1'|i\rangle}{\left[\hbar\omega_i - (E_n - E_i)\right]\left[\hbar\omega_i - \hbar\omega - (E_{n'} - E_i)\right]} \right.$$

$$+ \sum_{n,n'} \frac{\langle i|H_1'|n\rangle\langle n|H_1'|n'\rangle\langle n'|H_{el-ph}|i\rangle}{\left[\hbar\omega_i - (E_n - E_i)\right]\left[\hbar\omega_i - \hbar\omega_s - (E_{n'} - E_i)\right]}$$

$$+ \sum_{n,n'} \frac{\langle i|H_1'|n\rangle\langle n|H_{el-ph}|n'\rangle\langle n'|H_1'|i\rangle}{\left[-\hbar\omega_s - (E_n - E_i)\right]\left[-\hbar\omega_s - \hbar\omega - (E_{n'} - E_i)\right]}$$

$$+ \sum_{n,n'} \frac{\langle i|H_1'|n\rangle\langle n|H_1'|n'\rangle\langle n'|H_{el-ph}|i\rangle}{\left[-\hbar\omega_s - (E_n - E_i)\right]\left[-\hbar\omega_i + \hbar\omega - (E_{n'} - E_i)\right]}$$

$$+ \sum_{n,n'} \frac{\langle i|H_{el-ph}|n\rangle\langle n|H_1'|n'\rangle\langle n'|H_1'|i\rangle}{\left[-\hbar\omega - (E_n - E_i)\right]\left[-\hbar\omega + \hbar\omega_i - (E_{n'} - E_i)\right]}$$

$$+ \left. \sum_{n,n'} \frac{\langle i|H_{el-ph}|n\rangle\langle n|H_1'|n'\rangle\langle n'|H'|i\rangle}{\left[-\hbar\omega - (E_n - E_i)\right]\left[-\hbar\omega - \hbar\omega_s - (E_{n'} - E_i)\right]} \right|^2$$

$$\times\ \delta(\hbar\omega_i - \hbar\omega_s - \hbar\omega)\,, \tag{7.15}$$

where $|i\rangle$ is the initial state of the system and E_i is its energy; and $|n\rangle$, $|n'\rangle$ are intermediate states with energies E_n, $E_{n'}$, respectively.

We note that there are three processes involved in one-phonon Raman scattering: (1) the incident photon is annihilated; (2) the scattered photon is emitted; and (3) a phonon is annihilated (or created). Because these three processes can occur in any time order in the time-dependent perturbation-theory calculations of scattering probability, we expect that there will be six terms or contributions to $P(\omega_s)$, which is consistent with Equation 7.15. The δ-function here ensures that energy is conserved in the Raman scattering process.

One important advantage of probing non-equilibrium excitations with Raman spectroscopy in semiconductors is that because it detects the Raman signal only when excitation photons are present, its time resolution is essentially limited by the pulse width of the excitation laser and not by the response of the detection system. This explains why our detection system has a time resolution of the order of nanoseconds whereas the time resolution in our Raman experiments is typically on the scale of a picosecond or shorter.

7.2.2 GENERAL CONSIDERATIONS

The measurement of a Raman spectrum requires at least the following equipment:

1. A collimated and monochromatic light source
2. A spectrometer to analyze the spectral content of the scattered radiation
3. A sensitive optical system to collect and detect the generally weak scattered radiation

Because the Raman signal is typically very small (as is clear from the fact that third- or higher-order time-dependent perturbation theory is involved in the calculation of Raman scattering cross-section), every component mentioned above must be optimized. We consider these components individually:

7.2.2.1 Light Source

In the days before the invention of lasers, the light source was typically a high-power discharge lamp. Discrete emission lines of a gas or vapor were used. In those days, only transparent samples could be studied because of their much larger scattering volumes. Because many common semiconductors are opaque, Raman scattering studies of semiconductors became feasible only after the advent of lasers. The continuous wave (cw) He-Ne laser was the first to be used in Raman spectroscopy. It was soon replaced by Nd:YAIG, Ar$^+$, and Kr$^+$ ion lasers. The latter two produce several high-power (>1W in a single line) discrete emission lines covering the red, yellow, green, blue, and violet regions of the visible spectrum. With these high average power cw lasers, it became feasible to obtain not only one-phonon, but also two-phonon Raman spectra in semiconductors. With continuously tunable cw lasers based on dyes, color-centers in ionic crystals, and more recently Ti-doped sapphire, it became possible to perform Raman excitation spectroscopy, that is, resonant Raman scattering in which one monitors the Raman signal as a function of the excitation laser wavelength. Because of very low noise and excellent stability, cw mode-locked Ti:sapphire laser has become standard equipment in time-resolved or transient optical spectroscopy.

7.2.2.2 Spectrometer

In most Raman experiments on semiconductors, the Raman signal is typically four to six orders of magnitude weaker than the elastically scattered laser light. At the same time, the difference in energy between the Raman scattered photons and the excitation laser photons is only about 1% of the laser energy. This percentage is even smaller when Raman spectroscopy is used to measure the SPS spectrum. To observe this very weak sideband in the vicinity of a strong laser line, the spectrometer must satisfy several stringent conditions. First it must have a good spectral resolving power. Modern Raman spectrometers typically have a resolving power of $(\lambda/\Delta\lambda) \geq 10^4$, which can be obtained easily with diffraction gratings. It is, however, important that these gratings do not produce ghosts and/or satellites, which can be confused with Raman signal. A Raman spectrometer must have an excellent stray light rejection ratio. This is defined as the ratio of the background stray light (light at all wavelengths other than the nominal one specified by the spectrometer) to the signal. Stray light can be produced either by the imperfections in the optical system or by the scattering of light off walls and dust particles inside the spectrometer. Most simple spectrometers have a rejection ratio of 10^{-4} to 10^{-6}. As a result, the background stray light can be orders of magnitude larger than the Raman signal. This issue can be resolved by (1) making the sample surface as smooth as possible and therefore minimizing the elastically scattered laser light; (2) employing a notch filter to block out the elastically scattered laser light; (3) putting two or even three simple spectrometers in tandem. A properly designed double monochromator, in which two simple spectrometers are placed in series, can have a rejection ratio as small as 10^{-14}, equal to the product of the ratio for the two simple monochromators. This rejection ratio is adequate for Raman studies in most semiconductors. Currently, triple spectrometers have become popular for use with multichannel detectors (to be described next). In these spectrometers, two simple monochromators are put in tandem for use as a notch filter. The third spectrometer provides all the dispersion required for separating the Raman signal from the elastically scattered laser light.

7.2.2.3 Detector and Photon-Counting Electronics

Raman used photographic plates to record the weak inelastically scattered light in his pioneering experiment in 1928. These detectors actually possess many of the desirable characteristics of modern systems. They have the sensitivity to detect individual photons; they are multichannel detectors in that they can measure many different wavelengths at the same time; and they can integrate the signal over long periods of time, from hours to even days. They have one big advantage when compared to modern detection systems: they are relatively inexpensive. However, they also have some serious drawbacks, including that their outputs are not linear in the light intensity and it is also difficult to convert the recorded signal into digital form for analysis. The first major advance in photoelectric recording of Raman signals was the introduction of photon counting methods. Instead of integrating all the photocurrent pulses arriving at the anode of a photomultiplier tube as the signal, a discriminator selects and counts only those pulses with large enough amplitude to have originated from the photocathode. The background pulses (noise) remaining in such systems are those generated by thermionic emission of electrons at the photocathode. This can be minimized by cooling the entire photomultiplier tube to $-20°C$ (through thermoelectric coolers) or to $-78°C$ (using dry-ice coolers). One of the most popular photomultipliers for Raman scattering experiments has a GaAs photocathode cooled to $-20°C$. When coupled to properly designed pulse-counting electronics, such a detection system has a background noise or dark counts of a few counts per second and a dynamic range of 10^6.

The above-mentioned detector system has one major disadvantage compared with the photographic plate. It counts the total number of photons emerging from the spectrometer without spatially resolving the positions (and hence the wavelengths) of the photons. Consequently, the Raman spectrum is obtained only after scanning the spectrometer output over a wavelength range containing the spectral range of interest. Recently, several multichannel detection systems have become available commercially. These systems are based on either charge-coupled devices (CCDs) or position-sensitive imaging photomultiplier tubes.

7.2.3 SAMPLES AND EXPERIMENTAL TECHNIQUE

The InN sample reported in this work is a thin film grown on a HVPE GaN template by the conventional MBE technique.[35] RF-remote plasma nitrogen is supplied by an EPI Unibulb source operating at 260 W with 0.7-sccm nitrogen flow. Prior to InN growth, the substrate was directly heated to 525°C, measured by thermocouple, and then InN growth started. The InN growth rate is about 0.7 μm/h, with final film thickness around 7.5 μm. The HVPE GaN template has been compensated with Zn introduced during growth to suppress unintentional n-type conductivity. Its thickness is around 16 μm with 300K resistivity up to 10^9 Ω.cm and dislocation density around 5×10^8/cm^2. The InN film is n-type and has an electron density of ~5×10^8/cm^3 and electron mobility of 1300 cm^2/Vsec.

The Si-doped In$_x$Ga$_{1-x}$N ($x \cong 0.4$) epilayer of about 0.15-μm thick used in this work was grown on top of a 1.5-μm GaN epilayer by metal organic chemical vapor deposition (MOCVD). Prior to GaN growth, a 25-nm-thick GaN buffer layer was grown on c-plane sapphire at 550°C. Subsequent epilayer growth was carried out at 1050°C for GaN and 710°C for In$_x$Ga$_{1-x}$N. Trimethylgallium (TMGa) and trimethylindium (TMIn) were used as the precursors. Nitrogen and hydrogen were used as carrier gases for In$_x$Ga$_{1-x}$N and GaN, respectively. High-purity ammonia was used as the active nitrogen source. To vary the In content in In$_x$Ga$_{1-x}$N, the TMIn flow rate was varied while other growth parameters remained constant. The In$_x$Ga$_{1-x}$N epilayer was doped by Si at a flow rate of 0.25 sccm of 10-ppm silane to improve the materials quality as well as to enhance the emission efficiency. The GaN and In$_x$Ga$_{1-x}$N growth rates were 3.6 μm/hr and 0.3 μm/hr, respectively. The typical room-temperature electron concentration and mobility of In$_x$Ga$_{1-x}$N alloy is 2×10^{12}/cm^3 and 160 cm^2/Vsec, respectively, as determined by Hall effect measurements.

The In$_x$Ga$_{1-x}$As$_{1-y}$N$_y$ epilayer studied in this work was grown by MOCVD on a semi-insulating GaAs substrate and terminated with a 5-nm-thick GaAs cap. The nominal In and N molar fractions were 0.03 and 0.01, respectively. The In/N incorporation ratio of 3 has been shown to provide lattice match to GaAs. As grown, the undoped InGaAsN film was p-type. After growth, the sample was annealed at 600°C for 30 min in ambient nitrogen to improve the electrical as well as optical properties of the material.

The excitation source was a double-jet DCM dye laser[36] synchronously pumped by the second harmonic output of a continuous wave (cw), mode-locked yttrium-aluminum-garnet (YAIG) laser operating at a repetition rate of 76 MHz, which is shown in Figure 7.1. A variety of photon energies can be derived from such a setup. The pulse width of the dye laser can be tuned almost continuously from 0.6 to 5 ps

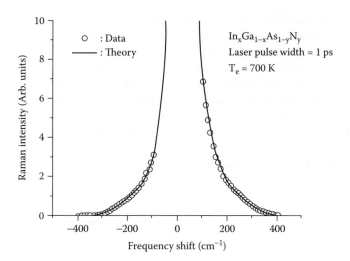

FIGURE 7.4 SPS spectrum (open circles) for an InGaAsN sample, taken with an ultrafast laser having 1 ps pulse width and photon energy $\hbar\omega_i = 1.931$ eV. The photoexcited electron-pair density is $n \cong 1 \times 10^{18}/cm^3$. The SPS spectrum is fit by Equation 7.6 (solid curve). The electron distribution is assumed to be a Fermi-Dirac function. The best fitting parameter set is: T_e (700 ± 35) K, = τ (10 ± 1) fs, and Γ = 15 meV.

by changing the concentration of the saturable absorber and the birefringence filter. Single-particle scattering (SPS) spectra were taken in the $Z(X,Y)\bar{Z}$ scattering geometry so that only the SPS spectra associated with spin-density fluctuations (SDF) were measured.[37,38] Here, X = (100), Y = (010), and Z = (001). The Raman scattering spectra for longitudinal optical phonons are taken in the $Z(X,X)\bar{Z}$ configuration. The photoexcited electron-hole pair density was estimated by fitting the luminescence spectrum of the E_0 bandgap of the respective semiconductor under study[39] or from the excitation laser power density. The backward-scattered Raman signal was collected and analyzed by a standard Raman system consisting of a double spectrometer and a low-background-count photomultiplier. All of the SPS spectra presented here were taken at T = 300K.

7.3 EXPERIMENTAL RESULTS

7.3.1 EXPERIMENTAL RESULTS FOR CARRIER DYNAMICS IN $In_xGa_{1-x}As_{1-y}N_y$: ANALYSIS AND DISCUSSION[54]

Figure 7.4 shows a typical SPS spectrum for an InGaAsN sample taken with photon energy $\hbar\omega_i = 1.931$ eV, an electron-hole pair density of $n \cong 1 \times 10^{18}/cm^3$, a laser pulse width of 1 ps, and at T = 300K.

Using Equation 7.6 and assuming that electron distribution is a Fermi-Dirac function with an effective electron temperature much higher than the lattice temperature, we will be able to fit the SPS of Figure 7.4 pretty well. This is shown in Figure 7.4.

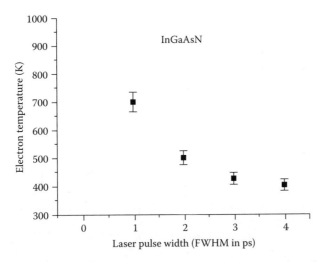

FIGURE 7.5 The deduced electron temperature is plotted as a function of the laser pulse width for an InGaAsN sample. From the data, the electron energy loss rate is estimated at (64 ± 6) meV/ps.

The parameter set that best fits experimental results is: $T_e = (700 \pm 35)$K, $\tau = (10 \pm 1)$fs, $\Gamma_1 = \Gamma_2 = \Gamma_3 = 15$ meV. The quality of the fit suggests that under our experimental conditions, electron distributions can be very well-described by Fermi-Dirac functions with an effective electron temperature much higher than the lattice temperature. We observe that the damping constants involved in the Raman scattering processes (Γ) are very close to the value (13 meV) that has been obtained from the analysis of resonance Raman profiles in the equilibrium case for GaAs.[55] We notice that, alternatively, a quantum-mechanical Monte Carlo simulation can be performed under our experimental conditions to obtain the non-equilibrium electron distributions and then compare them with our experimental results; however, the simulation requires input parameters such as electron-phonon scattering rates, effective masses of the electron and hole, and the band structure of the semiconductor. This information is not usually readily available for a new class of semiconductors such as that investigated here. Therefore, the simulation, in principle, can be performed on such a system but the results will not be very useful because of the requirements of several input parameters.

Figure 7.5 shows the effective electron temperature obtained in this way as a function of the pulse width of the excitation laser, ranging from 1 to 4 ps. This plot demonstrates that the effective electron temperature decreases from 700K to 400K when we increase the laser pulse width from 1 to 4 ps. Using the result of Kim and Yu,[56,57] which states that the electron effective temperature determined with a single pulse of full-width-at-half-maximum (FWHM) of δt is equal to the temperature of the electron after cooling for an equivalent duration $0.4\ \delta t$ when excited by an infinitely short pulse, we estimate that the electron cooling rate (energy loss rate) is equal to $[3/2(700 - 500)k_B]/(0.4)$ ps, where k_B is Boltzmann constant, or about (64 ± 6) meV/ps.

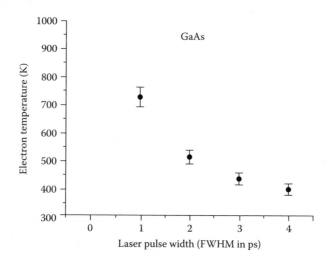

FIGURE 7.6 The deduced electron temperature is plotted as a function of the laser pulse width for a high-quality, MBE-grown GaAs sample. The energy loss rate under similar experimental conditions is found to be (63 ± 6) meV/ps, indicating that electrons lose energy primarily through electron-LO phonon interactions in InGaAsN.

To get better insight on our experimental results, we have also carried out similar experimental measurements on a molecular beam epitaxy (MBE)-grown GaAs sample. The results are shown in Figure 7.6. These results indicate that within our experimental uncertainty, the energy loss rate for an InGaAsN sample is the same as that for a GaAs sample, suggesting that an electron loses its energy primarily to the GaAs-like LO phonons in an InGaAsN sample. On the other hand, we have found that for a given laser pulse width, the electron collision time is about two times larger for a GaAs sample than for an InGaAsN sample. We attribute this finding to the much larger defect density in the InGaAsN sample, as a result of, say, alloy fluctuations. This interpretation is consistent with recent experimental results on this material.[34,58]

We note that for GaAs, our measured energy loss rate is substantially lower than that measured at much lower carrier densities ($\cong$36 meV/250 fs). We attribute this difference to the reabsorption of hot phonons and/or screening of the polar-optical phonons[59] for electron density $n \cong 10^{18}/cm^3$ excited in our experiments.

To obtain more information from our data, let us consider the experimental results for 1-ps laser pulse excitation. We have found that electron collision times are 22 and 10 fs for GaAs and InGaAsN, respectively. Because the photoexcited electron-hole pair density is very high ($\cong 1 \times 10^{18}/cm^3$) and the defect density for the MBE-grown GaAs sample is very low ($\leq 10^{13}/cm^3$), the 22-fs electron collision time has to primarily come from the electron-electron scattering in GaAs. Using this information, the collision time attributed to the defects in InGaAsN can be calculated to be $\tau_{ele-defect} \cong 18$ fs, or the scattering rate due to defects is $\Gamma_{defect} = 5.45 \times 10^{13}/\text{sec}$. Because the average electron velocity in InGaAsN is given by $\bar{v} = \sqrt{3k_B T_e / m_e^*} = 6.74 \times 10^5 \, m/\text{sec}$, we obtain for InGaAsN $n_{defect} \sigma_{defect} = \Gamma_{defect}/\bar{v} = 5.45 \times 10^{13}/6.74 \times 10^5 = 8.07 \times 10^7 \, m^{-1}$, where n_{defect}

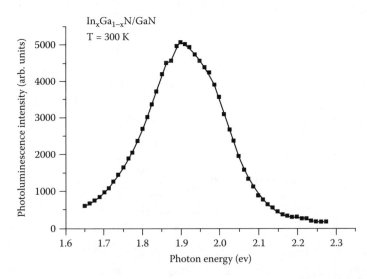

FIGURE 7.7 A typical electron-hole pair luminescence spectrum for an $In_xGa_{1-x}N/GaN$ sample. The photon energy of excitation laser is $\hbar\omega \cong 2.17$ eV. We deduce that In concentration in the sample is about x = 0.4 by comparing the data with those of Refs. 8, 9, and 10.

is the defect density and σ_{defect} is the microscopic scattering cross-section of the defect. Therefore, our experimental results can provide information about the defect density if the microscopic scattering cross-section of the defect is known.

7.3.2 EXPERIMENTAL RESULTS FOR TRANSIENT CARRIER TRANSPORT IN $IN_xGA_{1-x}N/GAN$[60]

Figure 7.7 shows a typical photoluminescence spectrum for an $In_xGa_{1-x}N$ sample. The E_0 bandgap luminescence of the $In_xGa_{1-x}N$ sample peaks at about 1.90 eV and has an FWHM of about 0.2 eV. Comparison of the spectrum with Refs. 8, 9, and 10 shows that our sample is of good quality and has an In concentration of about $x \cong 0.4$.

Figure 7.8a shows a typical SPS spectrum for an $In_xGa_{1-x}N$ ($x \cong 0.4$) sample taken at T = 300K and for an electron-hole pair density of $n \cong 1 \times 10^{18}/cm^3$. The SPS spectrum sits on a smooth background coming from the luminescence of E_0 bandgap of $In_xGa_{1-x}N$. Similar to previous studies on other III–V semiconductors such as GaAs, this background luminescence has been found to be fit very well by an exponential function.[61–64] The SPS spectrum is obtained by subtracting Figure 7.8a from this luminescence background. Following the procedure described in detail in Ref. 37, this subtracted spectrum (Figure 7.8b) can then be very easily transformed to electron distribution function. The electron distribution thus obtained is shown in Figure 7.8c. The intriguing feature worth pointing out is that the electron distribution function has been found to shift toward the wavevector transfer $\vec{q}$ direction, an indication of the presence of an electric field $\vec{E}$ parallel to $-\vec{q}$. The electron distribution has a cut-off velocity of around 1.5×10^8 cm/sec, indicative of the band

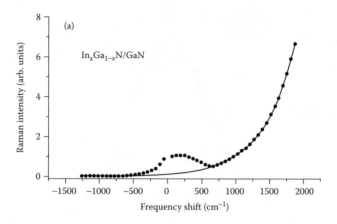

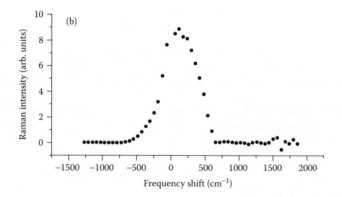

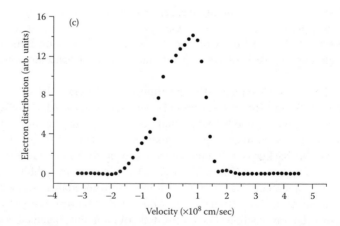

FIGURE 7.8 (a) A typical SPS spectrum for InGaN taken at T = 300K and a photoexcited electron-hole pair density of $n \cong 1 \times 10^{18}/$ cm^3. The SPS spectrum is found to lie on top of a luminescence background (solid curve) that can be fit very well by an exponential curve; (b) the SPS spectrum after the subtraction of the luminescence background; and (c) the electron distribution function obtained from (b).

structure effects and the onset of electron intervalley scattering processes in $In_xGa_{1-x}N$. The electron drift velocity deduced from the measured electron distribution (Figure 7.8c) is found to be $V_d \cong (3.8 \pm 0.4) \times 10^7 \, cm/sec$.

We have also carried out an ensemble Monte Carlo (EMC) simulation[65] for the transport of the photoexcited carriers in $In_xGa_{1-x}N$. Here, we treat polar optical phonons, acoustic phonons, inter-valley phonons, and dislocation scattering. For dislocation scattering,[66,67] we assumed that a defect density of $10^8/cm^2$ existed in the $In_xGa_{1-x}N$ layer; however, it is found that because of the presence of much more efficient inelastic scattering processes, this elastic defect scattering process can affect the low-field mobility but is not important for the high-field transient experiments carried out here. Disorder-induced scattering due to In concentration fluctuations in $In_xGa_{1-x}N$ is not included in the EMC simulation because it is an elastic scattering process that does not affect the high-field transport.[65] Within the Γ valley, which is the valley measured in the experiment, the high-field behavior is dominated by polar optical phonon scattering. In Figure 7.9a, we plot the velocity-field relationship for these carriers. The peak in the velocity occurs at around 150 kV/cm, beyond which carriers begin to rapidly transfer to the satellite valley. Nevertheless, even the carriers that remain in the Γ valley appear to show the onset of saturation and a weak negative differential resistance, both of which are due to this type of scattering in a non-parabolic band. In Figure 7.9b we plot the transient velocity of the Γ valley electrons at 200 kV/cm. Their initial velocity is determined by the excess energy of the photon excitation within the band. This gives rise to a non-zero velocity, as only a hemisphere (in momentum space) of carriers is excited at the surface. This directed velocity rapidly drops and the field-induced rise follows. We note that there is very little, if any, overshoot in the carrier velocity, which is a result of the fact that polar scattering is much better in relaxing energy than momentum. We notice that the EMC results are in reasonable agreement with the experimental values quoted above.

There has been experimental evidence[68-71] that an extremely large (of the order of MV/cm) electric field exists in the layer of $In_xGa_{1-x}N$ in the GaN/ $In_xGa_{1-x}N$ /GaN structures as a result of huge lattice mismatch between $In_xGa_{1-x}N$ and GaN. We believe that, similar to the previous studies, the source of the high electric field in the $In_xGa_{1-x}N$/GaN sample studied in this work arises from a combination of the spontaneous polarization field and the piezoelectric field. Although the direction of the net electric field is indicated in the shift of measured electron distribution (in this case, the net field points from GaN to $In_xGa_{1-x}N$ layers), the agreement between the experiment and the simulation is not sufficiently good to make a quantitative estimate of this field. Nevertheless, we can estimate the polarization field to be about 2.8 MV/cm, whereas the piezoelectric field is estimated to be some 50% smaller, although we do not have a good measure of the piezoelectric constants in this alloy.[72,73] We have extrapolated from the InN and GaN values, but this is not expected to be particularly accurate.

7.3.3 EXPERIMENTAL RESULTS FOR TRANSIENT CARRIER TRANSPORT IN InN/GaN: ANALYSIS AND DISCUSSION[74]

Figure 7.10a shows a typical SPS spectrum for an InN sample taken at T = 300K, with an excitation laser of photon energy $\cong$ 2.34 eV and pulse width $\cong$ 60 ps and

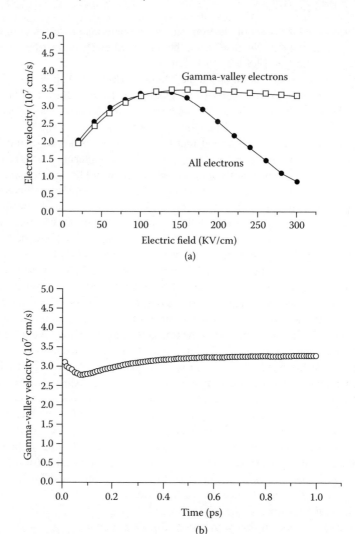

FIGURE 7.9 Ensemble Monte Carlo simulation for an $In_xGa_{1-x}N$ (x = 0.4) sample at T = 300K: (a) electron velocity as a function of the applied electric field intensity for Γ-valley electrons (open squares) and all electrons (open circles), as indicated; and (b) Γ-valley electron velocity as a function of time.

for a photoinjected electron-hole pair density of $n \cong 5 \times 10^{18}/cm^3$. Here, the spectral range between −200 and 200 cm^{-1} is not indicated because of the possible contributions from elastic scattering signals from the excitation laser. We have found that the observed SPS signal increases with laser intensity I in a quadratic way, as expected.[75] By considering conservation of energy and momentum, this SPS spectrum can be very easily transformed to an electron distribution function, which is shown in Figure 7.10b. However, we notice that Raman scattering cross-section for SPS is inversely proportional to the square of the electron effective mass and, because

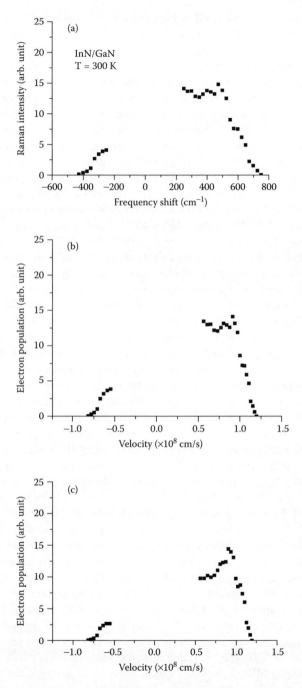

FIGURE 7.10 (a) A typical SPS spectrum for the InN sample taken at T = 300K, with an electron-hole pair density of $n \cong 5 \times 10^{18}/cm^3$; (b) electron distribution obtained from SPS spectrum of (a); and (c) electron distribution after the correction of effective mass effect on the band structure.

the electron velocity, as indicated in Figure 7.10b, extends up to 1.2×10^8 cm/sec, a proper correction for the electron distribution that takes into account the non-parabolicity on the conduction band of InN is needed. Unfortunately, information on the nonparabolicity of InN is currently not available in the literature. Therefore, to a good approximation, we have adopted the formula developed for GaAs with suitable InN parameters to correct for the effects of nonparabolicity on the conduction band.[76] The electron distribution thus obtained is shown in Figure 7.10c. The interesting feature worthwhile pointing out is that the electron distribution function has been found to shift toward the wavevector transfer $\vec{q}$ direction, an indication of the presence of an electric field $\vec{E}$ parallel to $-\vec{q}$ in our sample system. These experimental results suggest that the direction of net electric field points from GaN to InN layers. The electron distribution has a cut-off velocity of around 1×10^8 cm/sec, indicative of the band structure effects and the effect of excitation with a relatively long laser pulse width in InN. The electron drift velocity deduced from the measured electron distribution (Figure 7.10c) is found to be $V_d \cong (5.0 \pm 0.5) \times 10^7$ cm/sec.

We note that in the recent Raman work of Kasic et al.[77] and Inushima et al.[78] on highly n-type InN films, the dominant features are phonon spectra. On the other hand, the SPS spectrum dominates our Raman measurements. This is because of the nature by which the electrons are created. The former deals with equilibrium electron distributions; as a result, the Raman intensity for both the phonons and SPS spectra increases linearly with the laser intensity I; whereas, in the latter case, the electrons are photoinjected in a single laser-pulse pump/probe experiment, under such experimental conditions, the Raman intensity for phonons and SPS spectra is proportional to I and I^2, respectively. That is, for sufficiently large laser intensity, such as in our present work, contributions from SPS can overwhelm those from phonons.

The Monte Carlo simulation is a multianalytical-band formulation similar to that adopted for previous studies.[79] The basic program is known to give results in GaN that agree well with transport experiments[80] and in other III–V semiconductors with optical studies of the type discussed here.[81–83] We have assumed that the energy gap is approximately 0.7 eV, with satellite valleys lying some 2.7 eV above the main minimum in the conduction band. The Γ valley is assumed to be nonparabolic with a value of $\alpha = 1.33$. This is governed by a best fit to the 16×16 band $\mathbf{k} \cdot \mathbf{p}$ calculation of the bands.[76] The optical excitation creates carriers in the central Γ valley, and this excitation is assumed continuous throughout the simulation time, which is shorter than the actual laser pulse length. Thus, the properties measured are an average over the simulation time, in keeping with the experimental measurements themselves. A uniform electric field is considered here. In Figure 7.11, we plot the averaged velocity of the carriers remaining in the Γ valley as a function of the electric field. The peak velocity (within 10%) occurs near 75 kV/cm and is 4.7×10^7 cm/sec. Above this field, the averaged velocity begins to decrease, as only a fraction of the carriers is in the Γ valley; most have been scattered to the satellite valleys. Thus, while only a fraction is scattered back to the Γ valley, these enter isotropically and, because they are a comparable number to those already in this valley, lower the averaged velocity. From the comparison of drift velocities between experimental results (Figure 7.10c) and EMC simulation (Figure 7.11), we deduce that the built-in electric field intensity inside our InN thin film is about 75 kV/cm.

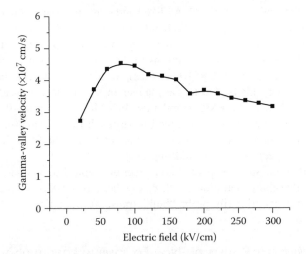

FIGURE 7.11 Electron drift velocity as a function of the electric field intensity for InN calculated under our experimental conditions by EMC simulation.

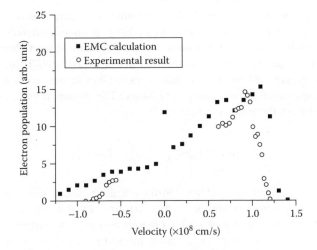

FIGURE 7.12 Comparison between the non-equilibrium electron distribution measured in the Raman experiments and that obtained by EMC simulation.

We also determine the V_z distribution function for the carriers in the Γ valley, where V_z is electron distribution along (001) or $-\vec{E}$ direction. Again, this is averaged over the simulation time, in keeping with the manner in which the experimental results are obtained. Figure 7.12 plots the V_z distribution function obtained from the simulation and compares it with that obtained in the experiment (after processing, as shown in Figure 7.10c). Qualitatively, the fit is reasonable except in the regions of -1×10^8 cm/sec and 1.2×10^8 cm/sec. The deviation close to 1.2×10^8 cm/sec is most probably due to the specific details of the band structure assumed in the

EMC calculations, as well as the GaAs-like band structure assumed in the correction of effective mass. Unfortunately, the band structure of InN is not well understood and, in particular, the rationale for the very low bandgap observed is not fully elucidated. The large number of electrons in the negative velocity region (between -0.5×10^8 cm/sec and -1.2×10^8 cm/sec), which are not seen in the Raman experiments, is very likely due to the manner in which electron scattering at the interface is handled in the EMC simulation. In EMC simulation, all the electrons that reach air–InN interface are assumed to suffer diffusive scattering. If some of these electrons were allowed to suffer specular scattering, the fit of the distribution function in this spectral range would improve.

The width of the surface-space charge layer in a typical semiconductor is ≤ 50Å. We estimate that the penetration depth of our laser is $\cong 500$Å. Consequently, our Raman experiments primarily probe "bulk" transport properties of InN film grown on GaN.

7.3.4 Experimental Results on Non-Equilibrium Longitudinal Optical Phonons in InN: Analysis and Discussion[84]

Figure 7.13 shows a typical Raman spectrum taken with an excitation laser having photon energy $\cong 2.3$ eV and an electron-hole pair density of $n \cong 2.5 \times 10^{18}$/cm^3. Similar results have been reported in the literature.[11] The sharp structure around 490 cm^{-1} comes from scattering of light from E_2(high) phonons, whereas the peak at 587 cm^{-1} is assigned to $A_1(LO)$ phonons in InN. The fact that both phonons exhibit very sharp peaks indicates that the InN used in our experiments is indeed of very high quality. Figure 7.14 shows anti-Stokes and Stokes Raman spectra for $A_1(LO)$ phonons for excitation photon energy $\cong 2.34$ eV. The phonon population $n(\omega)$ can be calculated from the following formula[85]:

$$n(\omega) = \left(I_{AS}\omega_S^4 \right) \Big/ \left(I_S\omega_{AS}^4 - I_{AS}\omega_S^4 \right), \qquad (7.16)$$

where I_S, I_{AS} are Stokes and anti-Stokes Raman intensities, respectively; $\omega_S \equiv \omega_L - \omega$, $\omega_{AS} \equiv \omega_L + \omega$ are the angular frequencies of Stokes and anti-Stokes Raman scattered light, respectively; ω is the phonon angular frequency; and ω_L is the angular frequency of the excitation laser. From Figure 7.14 we obtain $n(\omega) \cong 0.16 \pm 0.02$ for the population of $A_1(LO)$ phonon mode. We have also measured the population of E_2(high) phonon mode, which yields $n(\omega) \cong 0.10 \pm 0.01$, indicating that the lattice temperature of the excitation spot on the sample remains at T $\cong 300$K. Because the equilibrium phonon population for $A_1(LO)$ phonon mode at T $= 300$K is $n(\omega) \cong 0.063$, our experimental results suggest that non-equilibrium $A_1(LO)$ phonon population has been generated and detected in our Raman experiments. Interestingly, when we repeat exactly the same Raman experiments except that the excitation laser has photon energy $\hbar\omega_L \cong 1.96eV$, within our experimental accuracy, we do not observe any non-equilibrium $A_1(LO)$ phonon population ($n(\omega) \cong 0.07 \pm 0.02$), which is shown in Figure 7.15. These experimental results have the following important implications.

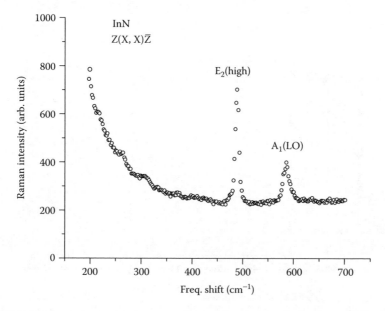

FIGURE 7.13 Raman scattering spectrum for an InN sample, taken at T = 300K with an excitation laser operating at photon energy of 2.34 eV.

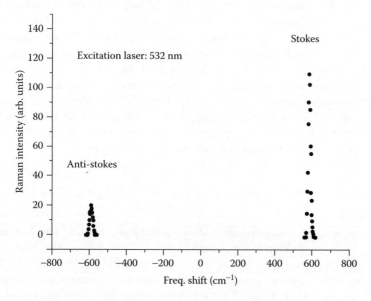

FIGURE 7.14 Raman scattering spectrum taken around the anti-Stokes and Stokes phonon lines of $A_1(LO)$ phonon mode and with excitation laser at 532 nm or 2.34 eV.

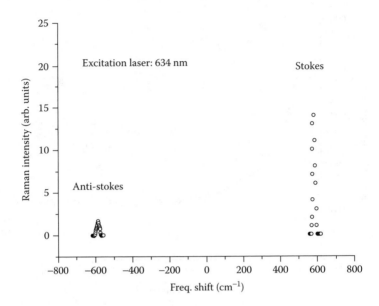

FIGURE 7.15 Raman scattering spectrum taken around the anti-Stokes and Stokes phonon lines of $A_1(LO)$ phonon mode and with excitation laser at 634 nm or 1.96 eV.

(1) Our experimental results indicate that the bandgap energy of InN cannot be 1.89 eV but they are consistent with a bandgap of 0.8 eV.

When a semiconductor is excited with a laser having photon energy larger than its bandgap, the electron and hole pairs are created in the conduction and valence bands, respectively. These energetic electrons/holes will very quickly relax toward the bottom/top of conduction/valence band by emitting phonons. For wurtzite structure InN used in our experiments, if we take $m_e^* = 0.14 m_e$[86]; $m_h^* = 1.63 m_e$[87]; $\hbar\omega_L = 2.34$ eV; Eq = 1.89 eV; the index of refraction n = 3.0[88]; $\hbar\omega_{LO} = 73.4$ eV (corresponding to $A_1(LO)$ phonon mode energy); then the phonon wavevector probed in our back-scattered Raman experiments is $q = 2nk_L = 7.08 \times 10^7 m^{-1}$, where k_L is wavevector of the excitation laser with photon energy 2.34 eV. Because of the much larger mass associated with holes, the emitted phonons by the holes can be ignored. If we assume a simple parabolic band, electron excess energy is then given by

$$\Delta E_e = \left(\hbar\omega_L - E_g\right)\left(\frac{m_h^*}{m_e^* + m_h^*}\right) \cong 0.41 eV .\tag{7.17}$$

This means that the energetic electrons are capable of emitting five LO phonons during their thermalization to the bottom of the conduction band. However, because of the conservation of both energy and momentum for the electron–LO phonon interaction process, there exists a range of LO phonon wavevectors that electrons can emit. As depicted in Figure 7.16, for an electron with wavevector $\vec{k}_e$ and excess energy ΔE_e, the minimum and maximum LO phonon wavevectors it can emit are given by[36]

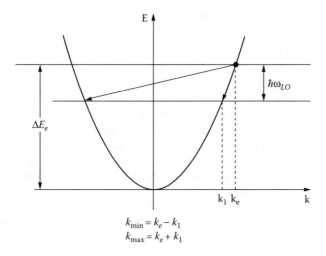

FIGURE 7.16 A simple two-band model showing the relaxation of energetic electrons through emission of optical phonons. The maximum and minimum phonon wavevectors that a given electron can emit are indicated.

$$k_{min} = \frac{\sqrt{2m_e^*}}{\hbar}\left(\sqrt{\Delta E_e} - \sqrt{\Delta E_e - \hbar\omega_{LO}}\right) \qquad (7.18)$$

and

$$k_{max} = \frac{\sqrt{2m_e^*}}{\hbar}\left(\sqrt{\Delta E_e} + \sqrt{\Delta E_e - \hbar\omega_{LO}}\right). \qquad (7.19)$$

Because of the nature of energy–wavevector relationship of the electron, the lower the electron's energy, the larger the k_{min} and the smaller the k_{max}. Therefore, at some electron energy, the k_{min} of the LO phonons will be larger than the wavevector probed by our Raman scattering experiments – $q = 7.08 \times 10^7$/m. When that happens during the relaxation process, the energetic electrons can no longer emit LO phonons with wavevector detectable in our Raman scattering experiments. By taking this into consideration, we have found that $k_{min} \cong 1.1 \times 10^8$/m and $k_{max} \cong 2.4 \times 10^9$/m for our experimental conditions. That is, although in principle the energetic electrons are capable of emitting five LO phonons during their thermalization to the bottom of the conduction band, none of them can be detected in our Raman experiments because of the conservation of both energy and momentum. The effect of non-parabolicity of the conduction band for InN is currently unknown. However, the general trend of the effect of nonparabolicity of the conduction band for a typical semiconductor has been to increase the effective mass of the electron on the conduction band. We expect that InN is no exception. By taking this into account we estimate that k_{min} will be larger than 1.1×10^8/m; and as a result no non-equilibrium $A_1(LO)$ phonon population should be detected in our Raman measurements when the effect of nonparabolicity of the conduction band is considered. Therefore, if bandgap energy of InN is 1.89 eV, then there will be no detectable non-equilibrium $A_1(LO)$ phonon population with laser source having photon energy 2.34 eV. This contradicts our experimental results of Figure 7.14 in which non-equilibrium $A_1(LO)$ phonons have been detected. Therefore, the detection of non-equilibrium $A_1(LO)$ phonons with excitation laser having photon energy of 2.34 eV in our experiments proves that the bandgap energy of InN cannot be 1.89eV. On the other hand, a similar argument, when applied to the scenario that the bandgap of InN is $\cong 0.8$ eV, indicates that six $A_1(LO)$ phonons can be detected by Raman scattering when InN is excited by a laser having photon energy of 2.34 eV and no $A_1(LO)$ phonons can be detected when it is excited by a laser with photon energy of 1.96 eV. All of these predictions are consistent with our experimental observations.

(2) Our experimental results suggest that the luminescence at 0.8 eV cannot be due to a deep level defect in InN.

It is well-known that in semiconductors the capture of electrons or holes by the deep levels can emit a photon (through a radiative relaxation process) whose energy is smaller than the bandgap. On the other hand, this capturing process can also emit phonons through a nonradiative relaxation process. Suppose that the bandgap of InN is 1.89 eV and the 0.8eV luminescence reported in the literature is due to such a capturing process by some unknown deep-level defects in InN, then our observation of non-equilibrium $A_1(LO)$ phonons in InN with excitation laser having photon energy 2.34 eV suggests that the nonradiative relaxation process also plays a role in the capturing. However, because the deep-level defects have very localized wave-functions, as a result their momenta are widely spread due to the uncertainty principle. This suggests that electron-phonon interaction during the capturing process does not need to conserve momentum. Therefore, phonon wavevectors of almost every magnitude can be emitted. That is, the defect model predicts that if you detect non-equilibrium $A_1(LO)$ phonons with the 532-nm (or 2.34-eV) excitation laser, then you should also detect non-equilibrium $A_1(LO)$ phonons with the 634-nm (or 1.96-eV) excitation source. This apparently is in contradiction with our experimental observation. We note that because of the much faster electron-LO phonon relaxation process occurring on the conduction band[36] than the electron-hole or electron-defect recombination process, the energetic electrons will primarily first relax to the bottom of the conduction band and then recombine with the defect. Therefore, no matter what excitation photon energy is used, the emitting electronic state in this defect scenario is always close to or at the bottom of the conduction band; that is, the above argument is applicable, irrespective of the excitation laser photon energy as long as it is larger than 1.89 eV. Consequently, we conclude that the 0.8-eV luminescence cannot be due to deep-level defects, just like yellow luminescence of GaN.

7.3.5 EXPERIMENTAL RESULTS ON THE OBSERVATION OF LARGE ELECTRON DRIFT VELOCITIES IN InN THICK FILM GROWN ON GaN

Figure 7.17 shows a typical non-equilibrium electron distribution for the InN sample excited by an ultrafast laser with a pulse width of 0.6 ps and taken with applied

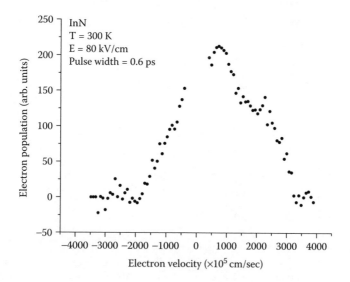

FIGURE 7.17 A typical non-equilibrium electron distribution for an InN film grown on GaN. The electric field intensity has been estimated in a way similar to that in Ref. 74. The photoexcited electron-hole pair density is $n \cong 5 \times 10^{18}/cm^3$. The excitation laser has a pulse width of $\cong 0.6$ ps. The excitation photon energy is $\hbar\omega_L = 1.92$ eV. The cut-off electron velocity is around 2×10^8 cm/sec.

electric field intensity of E = 80 kV/cm, and an injected electron-hole pair density of $n \cong 5 \times 10^{18}/cm^3$. The electric field intensity was deduced from a method similar to that described in Ref. 74. The electron distribution was observed to shift toward the $-\vec{E}$ direction, as expected. The appearance of a cut-off velocity around 2×10^2 cm/sec is indicative of the onset of electron scattering processes, the band structure effects, and the effect of excitation with an extremely short laser pulse width in InN. The most intriguing feature, however, is that this cut-off velocity is significantly larger than that found in other III–V compound semiconductors such as GaAs ($V_{cut-off} = 1.2 \times 10^8$ cm/sec) and InP ($V_{cut-off} = 1.4 \times 10^8$ cm/sec). The electron drift velocity deduced from Figure 7.17 has been determined as $V_d \cong (7.5 \pm 0.8) \times 10^7$ cm/sec.

We also carried out similar experimental measurements using an excitation laser with pulse width as long as 10 ps. A typical non-equilibrium electron distribution under these experimental conditions is shown in Figure 7.18. The electron drift velocity deduced in this case is $V_d \cong (5.1 \pm 0.5) \times 10^7$ cm/sec. Because experimental results for an excitation laser pulse width as long as 10 ps represent very closely the steady-state value, we conclude that transient electron velocity effects have been observed in InN samples in our shorter pulse experiments.

The different response of the electrons under these excitation laser pulse widths can be understood by the nature of the scattering processes undergone by the electrons. For 0.6-ps excitation, electrons are rapidly accelerated in the field to higher energies, subsequently scattered from the Γ valley to the satellite valleys of the conduction band. This process is very efficient and within a short time period (≤ 1.0 ps), those carriers remaining at higher-energy positions in the Γ valley can be

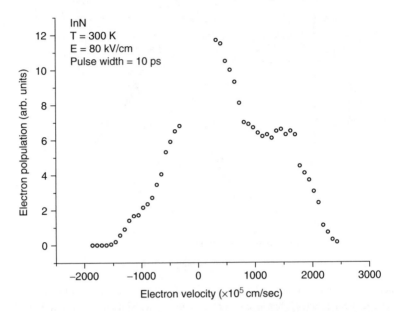

FIGURE 7.18 A typical non-equilibrium electron distribution for an InN film grown on GaN. The photoexcited electron-hole pair density is $n \cong 5 \times 10^{18}/cm^3$. The excitation laser has a pulse width of $\cong 10$ ps. The excitation photon energy is $\hbar\omega_L = 1.92$ eV.

monitored by the short Raman laser pulse. Experimentally, we observe a high velocity. On the other hand, for a longer period of time, the electrons in the Γ valley are mostly those that have returned from the satellite valleys and are cycling through the band and to/from these satellite valleys. Those electrons have been randomized efficiently due to the inter- and intra-valley scattering processes.[89] Hence, more low-energy electrons, compared with the energetic ones, weigh in the electron distribution, and we expect a lower quasi-steady-state drift velocity for the carriers in the Γ valley on this time scale.

In obtaining the electron distribution from the SPS spectrum, we have ignored the correction due to the nonparabolicity effects of the conduction band in InN. This is mainly because the precise band structure of InN is not known. Nevertheless, this omission will not affect our conclusion because normally this nonparabolicity effect enhances the high-energy electrons more than the low-energy ones; and because there are significantly more high-energy electrons in Figure 7.17 than in Figure 7.18, we expect that the consideration of nonparabolicity effects can only strengthen our conclusion.

We have tried to analyze the observed velocity and its distribution function through an ensemble Monte Carlo process. Here, we use a multivalley formalism for the conduction band, similar to that used in other III–V semiconductors.[90] Unfortunately, we do not know many of the important parameters of InN, which complicates the analysis considerably. Nevertheless, we can say some things about the likely structure of the conduction band. We use the latest estimate of the effective mass of $0.045m_0$ and the bandgap of 0.75 eV,[91] although the results are not sensitive to the exact value of this later parameter. The carriers are optically injected into the Γ conduction band valley from a photon source of 1.92 eV, and then evolve under the polarization-induced field of 80 kV/cm. Hence, they arrive in the Γ valley with about 1.0 eV excess energy. An important point for the long-time study of Raman scattering is that these carriers will move into the satellite valleys of the conduction band, then return to the Γ valley at a later time, at which point they can be reaccelerated in the field. We find that we are unable to reproduce the average velocity seen in the experiment, for the Γ valley carriers, *unless* the satellite valley is located *below* this injection point of 1.0 eV. We have looked at values of $\Delta E_{\Gamma-S} = 0.55$–$0.75$ eV, suggested by the similarities of InN to InGaAs lattice matched to InP, and the results are relatively insensitive to changes in this range. We have also varied the mass in the satellite valley m_s, although it is the product $m_s^{3/2}D^2$, where D is the coupling constant $(7 \times 10^8$ eV/cm) that is important; hence, this variation is shorthand for the overall coupling variation. Here, using the value of $m_s = 0.3m_0$, which is similar to that appropriate to GaAs and InGaAs, seems to give us the best fit to the experiment.

Another important observation is the spread to very high velocities that is seen in the velocity distribution function. This cannot be achieved with the nonparabolicity that would be expected for the small band gap. Here, we define the non-parabolicity parameter in the usual way[90]:

$$\frac{\hbar^2 k^2}{2m*} = E\left(1 + \alpha E\right) . \qquad (7.20)$$

Normally, one would expect a value of about $\alpha = 1.3$ from the Kane theory. Here, the observed spread in the velocity distribution function can only be achieved if α 0.6, or less than one half the expected value. When we adopt such values, and those discussed above, we find that the estimated average velocity from the ensemble Monte Carlo simulation is 5.7×10^7 cm/sec, close to the experimental results for 10-ps laser excitation. As discussed, of course, there is considerable uncertainty in this value, as we need much more information on the details of the band structure of InN and the corresponding parameters that must be used in such a simulation.

In conclusion, subpicosecond/picosecond Raman spectroscopy has been employed to study the transient transport properties of InN film grown on GaN. Non-equilibrium electron distributions, as well as electron drift velocities, have been directly measured. A much higher transient velocity is observed for the 0.6-ps, as opposed to the longer 10-ps excitation laser. This higher transient velocity, not to be confused with the phrase "overshoot velocity," which includes the role of velocities in the satellite valleys, probably arises from a number of effects, including the injection velocity inherent in the experimental conditions. The experimental results are compared with ensemble Monte Carlo simulations and reasonable agreement has been found for case of the 10-ps excitation. It is obvious from these experiments that the existence of carriers with velocities up to 2×10^8 cm/sec, found in the distributions plotted in Figures 7.17 and 7.18, is clear evidence that InN should provide considerably higher velocities, including overshoot velocity effects in devices, than other semiconductors. This result is in keeping with the earlier theoretical estimates, despite the newer (and significantly modified) band structure understanding.

7.4 CONCLUSION

Non-equilibrium electron distributions and energy loss rate in a MOCVD-grown $In_xGa_{1-x}As_{1-y}N_y$ (x = 0.03 and y = 0.01) epilayer on GaAs substrate have been studied by picosecond Raman spectroscopy. It is demonstrated that for electron density $n \cong 10^{12}/cm^3$, electron distributions can be described very well by Fermi-Dirac distributions with electron temperatures substantially higher than the lattice temperature. From the measurement of effective electron temperature as a function of the pulse width of excitation laser, the energy loss rate in $In_xGa_{1-x}As_{1-y}N_y$ is estimated to be 64 meV/ps. Within our experimental uncertainty, the electron energy loss rate has been found to be the same in InGaAsN as in GaAs, suggesting that the electron loses its energy primarily to the GaAs-like LO phonons in an InGaAsN sample. The electron collision time for GaAs has been found to be substantially larger than that for InGaAsN, indicating that the latter has much larger defect density than the former. We have studied field-induced electron transport in an $In_xGa_{1-x}N$ (x $\cong$ 0.4) sample grown on GaN by subpicosecond Raman spectroscopy. Non-equilibrium electron distribution and electron drift velocity due to the presence of piezoelectric and spontaneous fields in the $In_xGa_{1-x}N$ layer have been directly measured. The experimental results are compared with ensemble Monte Carlo calculations and reasonable agreements are obtained. Non-equilibrium electron transport in a high-quality, single-crystal InN sample has been studied by picosecond Raman spectroscopy. Both the non-equilibrium electron distribution and electron drift velocity were measured. These experimental results were compared with EMC calculations and good agreements were obtained. From the comparison, we also deduce the built-in electric field intensity in our sample system to be about 75 kV/cm. We have studied non-equilibrium optical phonons in a high-quality, single-crystal InN with picosecond Raman spectroscopy. Our experimental results not only strongly support a bandgap of around 0.8 eV, but also disprove the idea that the 0.8-eV luminescence observed recently is due to deep-level radiative emission in InN. From our transient Raman studies we have found the existence of carriers with velocities up to 2×10^8 cm/sec, significantly larger than other III–V semiconductors, which is clear evidence that InN should provide considerably higher velocities, including overshoot velocity effects in devices, than other semiconductors.

Acknowledgments

The author would like to thank D.K. Ferry for helpful discussions and EMC calculations, and William J. Schaff for providing InN samples. This work is supported by the National Science Foundation under Grant No: DMR-0305147 and Multi-Investigator Proposal Development Grant Program of College of Liberal Arts and Sciences at Arizona State University.

References

1. S. Strite and H. Morkoc, *J. Vac. Sci. Tech.*, B10, 1237, 1992.
2. S.N. Mohammad and H. Morkoc, *Prog. Quantum Electron.*, 20, 361, 1996.
3. S. Nakamura, M. Senoh, N. Iwasa, and S. Nagahama, *Jpn. J. Appl. Phys.*, 34, L797, 1995.

4. S. Nakamura, M. Senoh, N. Iwasa, S. Nagahama, T. Yamada, and T. Mukai, *Jpn. J. Appl. Phys.*, 34, L1332, 1995.

5. S. Nakamura, M. Senoh, S. Nagahama, N. Iwasa, T. Yamada, T. Matsushita, H. Kiyoku, and Y. Sugimoto, *Jpn. J. Appl. Phys.*, 35, L74, 1996.

6. L.J. Mawst, A. Bhattacharya, J. Lopez, D. Botez, D.Z. Garbuzov, L. De-Marco, J.C. Connolly, M. Jansen, F. Fang, and R. Nabiev, *Appl. Phys. Lett.*, 69, 1532, 1996.

7. E. Gregger, K.H. Gulden, P. Riel, H.P. Schweizer, M. Moser, G. Schmiedel, P. Kiesel, and G.H. Dohler, *Appl. Phys. Lett.*, 68, 2383, 1996.

8. J. Wu, W. Walukiewicz, K.M. Yu, J.W. Ager III, E.E. Haller, Hai Lu, William J. Schaff, Yoshiki Saito, and Yasushi Nanishi, *Appl. Phys. Lett.*, 80, 3967, 2002.

9. J. Wu, W. Walukiewicz, K.M. Yu, J.W. Ager III, E.E. Haller, Hai Lu, and William J. Schaff, *Appl. Phys. Lett.*, 80, 4741, 2002.

10. Takashi Matsuoka, Hiroshi Okamoto, Masshi Nakao. Hiroshi Harima, and Eiji Kurimoto, *Appl. Phys. Lett.*, 81, 1246, 2002.

11. For a review, see Ashraful Ghani Bhuiyan, Akihiro Hashimoto, and Akio Yamamoto, *J. Appl. Phys.*, 94, 2779, 2003.

12. S.N. Mohammad and H. Morkoc, *Prog. Quantum Electron.*, 20, 361, 1996.

13. S.K. O'Leary, B.E. Foutz, M.S. Shur, U.V. Bhapkar, and L.F. Eastman, *J. Appl. Phys.*, 83, 826 (1998).

14. E. Bellotti, B.K. Doshi, K.F. Brennan, J.D. Albrecht, and P.P. Ruden, *J. Appl. Phys.*, 85, 916, 1999.

15. B.E. Foutz, S.K. O'Leary, M.S. Shur, and L.F. Eastman, *J. Appl. Phys.*, 85, 7727, 1999.

16. T.L. Tansley and C.P. Foley, *J. Appl. Phys.*, 59, 3241, 1986.

17. V. Yu. Davydov, A.A. Klochikhin, R.P. Seisyan, V.V. Emtsev, S.V. Ivanov, F. Bechstedt, J. Furthmüller, H. Harima, A.V. Mudryi, J. Aderhold, O. Semchinova, and J. Graul, *Phys. Status Solidi*, B229, R1, 2002.

18. V. Yu. Davydov, A.A. Klochikhin, V.V. Emtsev, S.V. Ivanov, V.V. Vekshin, F. Bechstedt, J. Furthmüller, H. Harima, A.V. Mudryi, A. Hashimoto, A. Yamamoto, J. Aderhold, J. Graul, and E.E. Haller, *Phys. Status Solidi*, B230, R4, 2002.

19. J. Wu, W. Walukiewicz, K.M. Yu, J.W. Ager III, E.E. Haller, Hai Lu, William J. Schaff, Yoshiki Saito, and Yasushi Nanishi, *Appl. Phys. Lett.*, 80, 3967, 2002.

20. T. Matsuoka, H. Okamoto, M. Nakao, H. Harima, and E. Kurimoto, *Appl. Phys. Lett,.* 81, 1246, 2002.

21. M. Hori, K. Kano, T. Yamaguchi, Y. Saito, T Araki, Y. Nanishi, N. Teraguchi, and A. Suzuki, *Phys. Status Solidi*, B234, 750, 2002.

22. V. Yu. Davydov, A.A. Klochikhin, V.V. Emtsev, D.A. Kurdyukov, S.V. Ivanov, V.A. Vekshin, F. Bechstedt, J. Furthmüller, J. Aderhold, J. Graul, A.V. Mudryi, H. Harima, A. Hashimoto, A. Yamamoto, and E.E. Haller, *Phys. Status Solidi*, B234, 787, 2002.

23. Y. Saito, H. Harima, E. Kurimoto, T. Yamaguchi, N. Teraguchi, A. Suzuki, T. Araki, and Y. Nanishi, *Phys. Status Solidi*, B234, 796, 2002.

24. T. Miyajima, Y. Kudo, K.-L. Liu, T. Uruga, T. Honma, Y. Saito, M. Hori, Y. Nanishi, T. Kobayashi, and S. Hirata, *Phys. Status Solidi*, B234, 801, 2002.

25. M. Kondow, K. Uomi, A. Niwa, T. Kitatani, S. Watahiki, and Y. Yazawa, *Jpn. J. Appl. Phys., Part 1*, 35, 1273, 1996.

26. M. Kondow, T. Kitatani, M.C. Larson, K. Nakahara, K. Uomi, H. Inoue, *J. Cryst. Growth*, 188, 255, 1998.

27. H.P. Xin and C.W. Tu, *Appl. Phys. Lett.*, 72, 2442, 1998.

28. T. Miyamoto, K. Takeuchi, T. Kageyama, F. Koyama, and K. Iga, *Jpn. J. Appl. Phys., Part 1*, 37, 90 (1998).

29. S. Sato, Y. Osawa, and T. Saitoh, *Jpn. J. Appl. Phys., Part 1,* 36, 2671, 1997.
30. S. Sato and S. Satoh, *J. Cryst. Growth,* 192, 381, 1998.
31. L. Bellaiche, *Appl. Phys. Lett.,* 75, 2578, 1999.
32. J.F. Geisz, D.J. Friedman, J.M. Olson, S.R. Kurz, and B.M. Keyes, *J. Cryst. Growth,* 195, 401, 1998.
33. D.J. Friedman, J.F. Geisz, S.R. Kurz, and J.M. Olson, *J. Cryst. Growth,* 195, 409, 1988.
34. R.A. Mair, J.Y. Lin, H.X. Jiang, E.D. Jones, A.A. Allerman, and S.R. Kurz, *Appl. Phys. Lett.,* 76, 188, 2000.
35. Hai Lu, William J. Schaff, Lester F. Eastman, J. Wu, Wladek Walukiewicz, David C. Look, and Richard J. Molnar, *Prod. Mat. Res. Soc. Symp.,* Vol. 743, L4.10, 2003.
36. K.T. Tsen, in *Ultrafast Physical Processes in Semiconductors,* edited by K.T. Tsen, Vol. 67 of the series *Semiconductors and Semimetals,* Academic Press, New York, 2001, p. 109.
37. K.T. Tsen, in *Ultrafast Phenomena in Semiconductors,* edited by K.T. Tsen, Springer-Verlag, New York, 2001, p. 191.
38. M.V. Klein, in *Light Scattering in Solids I,* edited by M. Cardona and G. Guntherodt Springer-Verlag, New York, 1983, p. 151.39. D.S. Kim and P.Y. Yu, *Phys. Rev. B,* 43, 4158, 1991.
40. P.M. Platzman and P.A. Wolff, in *Waves and Interactions in Solid State Plasmas,* Vol. 13 of the series of *Supplement in Solid State Physics,* Academic Press, New York, 1973.
41. E.O. Kane, *J. Phys. Chem. Solids,* 1, 249, 1957.
42. M.L. Cohen and J. Chelikowsky, in *Electronic Structure and Optical Properties of Semiconductors,* 2nd edition, Vol. 75 of *Springer Series in Solid State Sciences,* Springer, Berlin, 1989.
43. S.S. Jha, *Nuovo Cimento,* 63B, 331, 1969.
44. C. Chia, O.F. Sankey, and K.T. Tsen, *Mod. Phys. Lett.,* B7, 331, 1993.
45. C. Chia, O.F. Sankey, and K.T. Tsen, *Phys. Rev. B,* 45, 6509, 1992.
46. C. Chia, O.F. Sankey, and K.T. Tsen, *J. Appl. Phys.,* 72, 4325, 1992.
47. D.C. Hamilton and A.L. McWhorter, in *Light Scattering Spectra of Solids,* edited by G. Wright, Springer, New York, 1969, p. 309.
48. A. Smekal, *Naturwissensch.,* 11, 873, 1923.
49. H.A. Kramers and W. Heisenburg, *Zeit. F. Physik,* 31, 681, 1925.
50. C.V. Raman, *Ind. J. Phys.,* 2, 387, 1928.
51. G. Lansberg and L. Mandel'shtam, *Naturwissensch.,* 16, 57; ibid, 16, 772, 1928.
52. J.D. Jackson, Classical Electrodynamics, 2nd edition, John Wiley & Sons, New York, 1975, pp. 391–397.
53. See, for example, P.Y. Yu and M. Cardona, Fundamentals of Semiconductors, Springer, New York, 1996.
54. Y. Chen and K.T. Tsen, *Appl. Phys. Lett.,* 78, 3094–3096, 2001.
55. A. Pinczuk, G.A. Abstreiter, R. Trommer, and M. Cardona, *Solid State Commun.* 30, 429, 1979.
56. D.S. Kim and P.Y. Yu, *Appl. Phys. Lett.,* 56, 1570, 1990.
57. D.S. Kim and P.Y. Yu, *Appl. Phys. Lett.,* 56, 2210, 1990.
58. D. Kwon, R.J. Kaplar, S.A. Ringle, A.A. Allerman, S.R. Kurtz, and E.D. Jones, *Appl. Phys. Lett.,* 74, 2830, 1999.
59. See, for example, *Hot Carrier in Semiconductors,* edited by K. Hess, J.-P. Leburton, and U. Ravaioli, Plenum Press, New York, 1995.
60. W. Liang, K.T. Tsen, D.K. Ferry, K.H. Kim, J.Y. Lin, and H.X. Jiang, *Appl. Phys. Lett.,* 82, 1413, 2003.

61. D.S. Kim and P.Y. Yu, *Phys. Rev. B,* 43, 4158, 1991.
62. E.D. Grann, K.T. Tsen, D.K. Ferry, A. Salvador, A. Botcharev, and H. Morkoc, *Phys. Rev. B,* 53, 9838, 1996.
63. E.D. Grann, S.J. Sheih, K.T. Tsen, O.F. Sankey, S.E. Guncer, D.K. Ferry, A. Salvador, A. Botcharev, and H. Morkoc, *Phys. Rev. B,* 51, 1631, 1995.
64. W. Liang, K.T. Tsen, C. Poweleit, J.M. Barker, D.K. Ferry, and H. Morkoc, *J. Physics: Condensed Matter,* 17, 1679, 2005.
65. For a review of ensemble Monte Carlo simulation, see for example, D.K. Ferry, *Semiconductor Transport,* Taylor & Francis, London, 2000, chap. 4.
66. N.G. Weimann, L.F. Eastman, D. Doppulapudi, H.M. Ng, and T.D. Moustakas, *J. Appl. Phys.,* 83, 3656, 1998.
67. D. Jena, A.C. Gossard, and U.K. Mishra, *Appl. Phys. Lett.,* 1707, 2000.
68. M.R. McCartney, F.A. Ponce, Juan Cai, and D.P. Bour, *Appl. Phys. Lett.,* 76, 3055, 2000.
69. D. Cherns, J. Barnard, and F.A. Ponce, *Solid State Commun.,* 111, 281, 1999.
70. T. Takeuchi, S. Sota, M. Katsuragawa, M. Komori, H. Takeuchi, H. Amano, and I. Akasaki, *Jpn. J. Appl. Phys.,* 36, L382, 1997.
71. T. Takeuchi, C. Wetzel, S. Yamaguchi, H. Sakai, H. Amano, and I. Akasaki, *Appl. Phys. Lett.,* 73, 1691, 1998.
72. F. Bernardini, V. Fiorentini, and D. Vanderbilt, *Phys. Rev. B,* 56, R10024, 1997.
73. T.-H. Yu and K. F. Brennan, *J. Appl. Phys.,* 89, 3827, 2001.
74. W. Liang, K.T. Tsen, D.K. Ferry, Hai Lu, and W. Schaff, *Appl. Phys. Lett.,* 84, 3681, 2004.
75. D.S. Kim and P.Y Yu, in *Light Scattering in Semiconductor Structures and Superlattices,* edited by D.J. Lockwood and J.F. Young, NATO ASI Series, Vol. 273, p. 383, 1992.
76. T. Ruf and M. Cardona, *Phys. Rev. B,* 41, 10747, 1990, and references therein.
77. A. Kasic, M. Schubert, Y. Saito, Y. Nanishi, and G. Wagner, *Phys. Rev. B,* 65, 115206, 2002.
78. T. Inushima, M. Higashiwaki, and T. Matsui, *Phys. Rev. B,* 68, 235204, 2003.
79. D.K. Ferry, *Semiconductor Transport,* Taylor & Francis, London, 2000.
80. J.M. Barker and D. K. Ferry, unpublished.
81. W. Liang, K.T. Tsen, D.K. Ferry, M.-C. Wu, C.-L. Ho, and W.-J. Ho, *Appl. Phys. Lett.,* 83, 1439 (2003).
82. E.D. Grann, K.T. Tsen, O.F. Sankey, D.K. Ferry, A. Salvador, A. Botcherev, and H. Morkoç, *Appl. Phys. Lett.,* 67, 1760, 1995.
83. K.T. Tsen, D.K. Ferry, J.S. Wang, C.-H. Huang, and H.S. Lin, *Appl. Phys. Lett.,* 69, 3575, 1996.
84. W. Liang, K.T. Tsen, D.K. Ferry, Hai Lu, and W. Schaff, *Appl. Phys. Lett.,* 84, 3849, 2004.
85. W. Hayes and R. Loudon, *Scattering of Light by Crystals,* John Wiley & Sons, New York, 1978.
86. A. Kasic, M. Schubert, Y. Saito, Y. Nanishi, and G. Wagner, *Phys. Rev. B,* 65, 115206, 2002.
87. Yong-Nian Xu and W.Y. Ching, *Phys. Rev. B,* 48, 4335, 1993.
88. The index refraction for our sample is determined by independent experiment.
89. R.P. Joshi, R.O. Grondin, and D.K. Ferry, *Phys. Rev. B,* 42, 5685, 1990.
90. D.K. Ferry, *Transport in Semiconductors*, Taylor and Francis, London, 2000, pp. 216–225.
91. J. Wu, W. Walukiewicz, K.M. Yu, J.W. Ager III, E.E. Haller, Hai Lu, William J. Schaff, Yoshiki Saito, Yasushi Nanishi, *Appl. Phys. Lett.,* 80, 3967, 2002.

8 Electromagnetically Induced Transparency in Semiconductor Quantum Wells

Mark C. Phillips and Hailin Wang

CONTENTS

8.1 Introduction 215
8.2 Electromagnetically Induced Transparency: Theory 217
    8.2.1 Density Matrix for the Λ-System 218
    8.2.2 Steady-State EIT Solutions 220
    8.2.3 Transient EIT Solutions 222
8.3 Experimental Methods 225
8.4 EIT via Intervalence Band Coherence 227
8.5 EIT via Exciton Spin Coherence 230
    8.5.1 Spin Coherence via Bound Biexciton States 233
    8.5.2 Spin Coherence via Unbound Two-Exciton States 234
8.6 EIT via Biexcitonic Coherence 239
8.7 Summary 245
Acknowledgments 247
References 247

8.1 INTRODUCTION

Nonradiative quantum coherences between states that are not directly dipole-coupled can lead to quantum interference in optical transitions. Destructive interference induced by nonradiative quantum coherence in a three-level system can render an otherwise opaque medium nearly transparent, leading to the phenomenon of electromagnetically induced transparency (EIT).[1–3] EIT and EIT-related processes provide a powerful and effective mechanism for controlling and manipulating light with nonradiative quantum coherences. Recent successes include the reduction of the group velocity of light by many orders of magnitude, the reversible storage of light as a stationary spin coherence, and the generation of entangled photon pairs with controllable delays.[4-13] Other well-known phenomena closely related to EIT include coherent population trapping, lasing without inversion, and stimulated Raman adiabatic passage.[1-3] While most EIT studies have thus far been carried out in atomic or atomic-like systems, the basic concept of destructive quantum interference induced by nonradiative quantum coherence can also be applied to extended optical excitations such as excitons in semiconductors.[14-24] EIT studies in semiconductors can further deepen our understanding of quantum coherence and interference in solids and can potentially open up new avenues for applications of EIT and EIT-related phenomena.

There are two major obstacles to realizing EIT and, more generally, to coherent manipulation of quantum coherences in semiconductors. First, coherent nonlinear optical processes in an excitonic system are profoundly modified by many-body Coulomb interactions between excitons, often in a detrimental manner.[25-27] While the similarities between atomic and excitonic transitions provide guidance for the realization of EIT in semiconductors, the differences must also be fully considered. As we discuss in detail in this chapter, under properly designed conditions, we can exploit and take advantage of these many-body Coulomb interactions to generate and manipulate nonradiative quantum coherences. In this regard, EIT studies also provide a powerful tool to probe how quantum coherences behave in an interacting many-particle system.

Second, quantum coherences in semiconductors are typically short lived. With the exception of electron spin coherence,[28,29] these coherences are extremely fragile against dynamic processes such as exciton-exciton and exciton-phonon scattering.[30] The rapid decoherence, however, can be overcome by carrying out measurements in a transient regime with ultrafast optical pulses. The use of ultrafast optical pulses enables us to complete an EIT measurement on a time scale comparable to the relevant decoherence time. By employing an intense ultrafast optical pulse, we can also generate a large Rabi frequency and thus attain a characteristic time scale for coherent optical processes that is short compared with the relevant decoherence time. Note that traditionally, EIT studies have been carried out in steady-state regimes or in an adiabatic limit. EIT measurements in a transient regime, in which adiabatic conditions are not satisfied, raise the important question of whether and how nonradiative quantum coherences induced in these measurements can lead to destructive quantum interference. Furthermore, we must determine how to probe and identify quantum interference processes in these transient measurements.

This chapter is organized as follows. Section 8.2 discusses, theoretically, the manifestation and behavior of EIT processes in transient regimes. The discussions are based on numerical solutions to the density matrix equations for a three-level system. The emphasis is on the understanding of EIT and especially the underlying destructive quantum interference in transient regimes.[31] Sections 8.3 through –8.6 discuss experimental demonstrations of EIT using excitons in GaAs quantum-well (QW) structures.[31-34] Three types of nonradiative quantum coherences — intervalence band coherence, exciton spin coherence, and biexcitonic coherence — are explored in these studies. These three EIT schemes are analogous to V-type, Λ-type, and cascaded-type 3-level EIT systems, respectively. We show that many-body Coulomb interactions between excitons can be exploited for the generation and manipulation of exciton spin coherence and biexciton coherence. These interactions, however, also greatly complicate the temporal evolution of these coherences. Pronounced EIT effects are realized only after these complications are properly compensated under carefully designed experimental conditions. A further emphasis of these sections is to elucidate the similarities and especially the essential differences in EIT processes between atomic and excitonic systems. Section 8.7 presents a summary and a brief discussion of future directions.

8.2 ELECTROMAGNETICALLY INDUCED TRANSPARENCY: THEORY

Electromagnetically induced transparency (EIT) is a process whereby destructive quantum interference causes an otherwise absorbing optical transition to become nearly transparent. This section analyzes, theoretically, EIT processes in an atomic-like three-level system in a transient regime. There are three different configurations for three-level systems: (1) Λ-type, (2) cascaded-type, and (3) V-type, as shown in Figure 8.1. This section focuses on the Λ-system with states $|a\rangle$, $|b\rangle$, and $|e\rangle$, where the $|a\rangle \leftrightarrow |e\rangle$ and $|b\rangle \leftrightarrow |e\rangle$ are dipole transitions coupled by electric fields ε_a and ε_b, respectively, as shown in Figure 8.2. The $|a\rangle \leftrightarrow |b\rangle$ transition is dipole-forbidden.

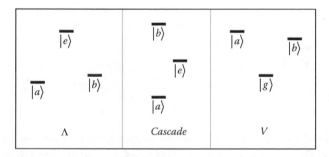

FIGURE 8.1 Classification of three-level systems.

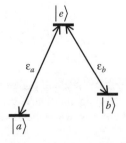

FIGURE 8.2 A Λ-type system with two external optical fields coupling to two respective dipole transitions.

8.2.1 DENSITY MATRIX FOR THE Λ-SYSTEM

Density matrix formulations for three-level systems have been discussed in quantum optics textbooks.[2] For a Λ-system coupling to two nearly resonant external optical fields, the Hamiltonian consists of two parts: (1) the atomic Hamiltonian H_0, describing the free evolution of the atomic states, given by

$$H_0 = \hbar \begin{pmatrix} \omega_{ee} & 0 & 0 \\ 0 & \omega_{aa} & 0 \\ 0 & 0 & \omega_{bb} \end{pmatrix} \tag{8.1}$$

and (2) an interaction Hamiltonian V, describing the interaction between the atoms and the applied electric fields, given by

$$V = - \begin{pmatrix} 0 & \mu_a \varepsilon_a & \mu_a \varepsilon_a \\ \mu_b \varepsilon_b & 0 & 0 \\ \mu_b \varepsilon_b & 0 & 0 \end{pmatrix} \tag{8.2}$$

where μ_a and μ_b are the relevant dipole matrix elements. For convenience, μ_a and μ_b are taken to be real. The applied electric fields are written as

$$\varepsilon_a = \tfrac{1}{2} E_a e^{-i\nu_a t} + c.c.$$
$$\varepsilon_b = \tfrac{1}{2} E_b e^{-i\nu_b t} + c.c. \tag{8.3}$$

where $E_a(t)$ and $E_b(t)$ are the slowly varying amplitudes of the probe and pump fields, respectively. The corresponding Rabi frequencies, Ω_a and Ω_b, are given by

$$\Omega_a(t) = \frac{\mu_a E_a(t)}{\hbar}$$
$$\Omega_b(t) = \frac{\mu_b E_b(t)}{\hbar} \tag{8.4}$$

The diagonal matrix elements ρ_{ee}, ρ_{aa}, and ρ_{bb} give the populations in states $|e\rangle$, $|a\rangle$, and $|b\rangle$, respectively. The off-diagonal matrix elements $\rho_{ea} = \rho_{ae}{}^*$ and $\rho_{eb} = \rho_{be}{}^*$ describe the relevant dipole coherences, while the matrix elements $\rho_{ba} = \rho_{ab}{}^*$ describe the nonradiative coherence.

The equations of motion for the density matrix elements are given by the Liouville equation. To make the rotating wave approximation (RWA) in the equations of motion, we make the substitutions $\rho_{ea} = \tilde{p}_a e^{-i\nu_a t}$, $\rho_{eb} = \tilde{p}_b e^{-i\nu_b t}$, $\rho_{ba} = \tilde{p}_{ba} e^{-i(\nu_b - \nu_a)t}$. For convenience, we also denote $\rho_{ee} = n_e$, $\rho_{aa} = n_a$, $\rho_{bb} = n_b$. With these assumptions, we arrive at the following equations of motion for the dipole and nonradiative coherences:

$$\dot{\tilde{p}}_a(t) = (i\delta_a - \gamma_a)\tilde{p}_a(t) - \frac{i\Omega_a(t)}{2}\big[n_e(t) - n_a(t)\big] + \frac{i\Omega_b(t)}{2}\tilde{p}_{ba}(t) \tag{8.5a}$$

$$\dot{\tilde{p}}_b(t) = (i\delta_b - \gamma_b)\tilde{p}_b(t) - \frac{i\Omega_b(t)}{2}\big[n_e(t) - n_b(t)\big] + \frac{i\Omega_a(t)}{2}\tilde{p}_{ba}(t)^* \tag{8.5b}$$

$$\dot{\tilde{p}}_{ba}(t) = \big[i(\delta_a - \delta_b) - \gamma_{ab}\big]\tilde{p}_{ba}(t) - \frac{i\Omega_a(t)}{2}\tilde{p}_b(t)^* + \frac{i\Omega_b(t)}{2}\tilde{p}_a(t) \tag{8.5c}$$

where $\delta_a = \nu_a - \omega_a$ and $\delta_b = \nu_b - \omega_b$ are the detunings of the applied fields from the dipole transitions with energies $\omega_a = \omega_{ee} - \omega_{aa}$ and $\omega_b = \omega_{ee} - \omega_{bb}$, respectively; γ_a and γ_b are the phenomenological decay rates for the relevant dipole coherences; and γ_{ab} is the phenomenological decay rate for the nonradiative coherence.

We are primarily interested in optical absorption for the $|a\rangle \leftrightarrow |e\rangle$ transition, and will therefore make a perturbative expansion in the probe field Ω_a, which we assume to be weak. We take state $|a\rangle$ to be the ground state, with $|b\rangle$ and $|e\rangle$ initially empty. To zeroeth order in $\Omega_a, n_a^{(0)} = 1$, and all other matrix elements are zero, which is a direct consequence of states $|b\rangle$ and $|e\rangle$ being unoccupied in the absence of Ω_a, so that Ω_b is pumping on an "empty" transition. Note that this would not be the case for a V-system. While we assume a weak probe beam, we place no limitation on the strength of the pump. To first order in Ω_a, we obtain the following coupled equations[2]:

$$\dot{\tilde{p}}_a^{(1)}(t) = (i\delta_a - \gamma)\tilde{p}_a^{(1)}(t) + \frac{i\Omega_a(t)}{2} + \frac{i\Omega_b(t)}{2}\tilde{p}_{ba}^{(1)}(t) \tag{8.6a}$$

$$\dot{\tilde{p}}_{ba}^{(1)}(t) = \big[i(\delta_a - \delta_b) - \gamma_{ab}\big]\tilde{p}_{ba}^{(1)}(t) + \frac{i\Omega_b(t)}{2}\tilde{p}_a^{(1)}(t) \tag{8.6b}$$

where we have replaced γ_a by γ because the dipole decoherence rate for the $|b\rangle \leftrightarrow |e\rangle$ transition does not enter the above equations. Note that in the absence of the pump field (i.e., $\Omega_b = 0$), Equation 8.6a reduces to the result for the linear response of a two-level system. Likewise, if we set $\tilde{p}_{ba} = 0$ so that there is no nonradiative coherence, we obtain the same result. Therefore, any changes in the absorption spectrum can be traced directly back to the effects of the nonradiative coherence.

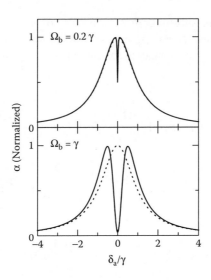

FIGURE 8.3 Dependence of the EIT spectrum on the pump intensity. Absorption spectra are plotted for the values of Ω_b given in the figure, and with $\gamma_{ab} = 0.01\,\gamma$ and $\delta_b = 0$.

8.2.2 STEADY-STATE EIT SOLUTIONS

Before considering EIT in a transient regime, it is helpful to review the steady-state analytic solutions for cw (continuous wave) excitations. The absorption spectrum, $\alpha(\delta_a)$, for the $|a\rangle \leftrightarrow |e\rangle$ transition is determined by $\mathrm{Im}\left[\tilde{p}_a^{(1)}\right]$. Solving Equation 8.6a and Equation 8.6b for $\tilde{p}_a^{(1)}$ in the steady-state limit yields:

$$\tilde{p}_a^{(1)} = \frac{-\frac{i\Omega_a}{2}\left[i\left(\delta_a - \delta_b\right) - \gamma_{ab}\right]}{\left(i\delta_a - \gamma\right)\left[i\left(\delta_a - \delta_b\right) - \gamma_{ab}\right] + \frac{|\Omega_{ab}|^2}{4}} \tag{8.7}$$

from which the absorption spectrum is readily calculated.

In Figure 8.3, we plot as the solid line the absorption spectra, with $\gamma_{ab} = 0.01\,\gamma$, $\delta_b = 0$, and with $\Omega_b = 0.2\,\gamma$ and also $\Omega_b = \gamma$ (the dashed line is the absorption spectrum in the absence of the pump). The dip in these absorption spectra at $\delta_a = 0$ arises from destructive quantum interference induced by the nonradiative coherence. For the purposes of this chapter, it will be helpful to understand the EIT process in the context of Equation 8.6a. Optical excitation for the $|a\rangle \leftrightarrow |e\rangle$ transition can take place via two pathways: (1) one is due to only the probe field (the term $i\Omega_a/2$ in Equation 8.6a and (2) the other is due to the coupling between the pump field and the nonradiative coherence (the term $i\Omega_b\tilde{p}_{ba}^{(1)}/2$ in Equation 8.6a). These two pathways can interfere with each other destructively if $\mathrm{Re}\left[\tilde{p}_{ba}^{(1)}\right] < 0$. It is this destructive interference that leads to the dip in the absorption spectra in Figure 8.3. The nonradiative coherence, which can be obtained from Equation 8.6b, features a resonance centered at $\delta_a = \delta_b$. In the limit of a weak pump field, γ_{ab} determines the spectral width of this two-photon resonance and therefore the spectral width of the EIT dip.

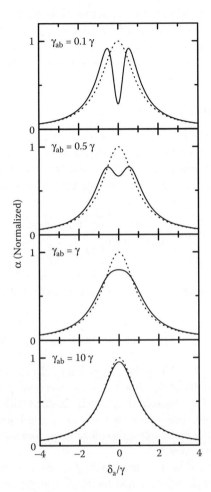

FIGURE 8.4 EIT spectra for different values of the decoherence rate γ_{ab}. For all curves, δ_b = 0, and $\Omega_b = \gamma$.

With increasing pump field amplitude, the EIT dip deepens and also broadens, as shown in Figure 8.3.

At a given pump field amplitude, the magnitude of the nonradiative coherence that can be generated is determined by the decoherence rate γ_{ab}. Figure 8.4 shows how the width and depth of the EIT dip in the absorption spectrum varies with γ_{ab}. In the limit of large γ_{ab}, no significant nonradiative coherence can be generated. As a result, the EIT dip vanishes and the absorption spectrum reverts to that of linear absorption for a two-level system. Note that a general condition for significant EIT effects can also be obtained from Equation 8.7. With $\delta_b = 0$, the relative reduction in absorption at the line center ($\delta_a = 0$) is given by the ratio $1 / [1 + |\Omega_b|^2 / (4\gamma \cdot \gamma_{ab})]$. To achieve greater than 50% reduction in absorption, we therefore need $|\Omega_b|^2 > 4\gamma \cdot \gamma_{ab}$.

For completeness, we should point out that a sharp EIT dip in the absorption spectrum is also accompanied by a steep variation in the refractive index with frequency. This steep variation in the refractive index has been used to demonstrate slowing down of the group velocity of light pulses.[5–8] Although the results discussed in this chapter focus on the changes in the absorption spectrum induced by EIT, these changes are also accompanied by corresponding changes in the refractive index.

8.2.3 TRANSIENT EIT SOLUTIONS

This section examines EIT processes in a transient regime. In this case, we solve Equation 8.6a and Equation 8.6b numerically, with Gaussian temporal pulse shapes for Ω_a and Ω_b. Fourier transforms of $\tilde{p}_a^{(1)}(t)$ and $\Omega_a(t)$ were taken to yield $\tilde{p}_a^{(1)}(\delta_a)$ and $\Omega_a(\delta_a)$. The absorption spectrum was then calculated using the relation

$$\alpha(\delta_a) \propto \mathrm{Im}\left[\frac{\tilde{p}_a^{(1)}(\delta_a)}{\Omega_a(\delta_a)}\right]$$

The resulting spectrum was normalized so that the peak value of the linear absorption equaled one.

For comparison, we first discuss the absorption spectrum of a two-level system measured by a short probe pulse with a duration of $0.1\,\gamma^{-1}$. We plot the numerical solutions for $\mathrm{Im}[\tilde{p}_a^{(1)}(t)]$ in Figure 8.5a and the corresponding absorption spectrum in Figure 8.5b, for $\Omega_b = 0$. The polarization $\mathrm{Im}[\tilde{p}_a^{(1)}(t)]$ excited by the short probe pulse decays exponentially at the dipole decoherence rate γ, yielding a Lorentzian absorption resonance. The absorption spectrum obtained with the short probe pulse agrees well with that obtained with a cw probe shown in Figure 8.5c. Note that the spectral bandwidth of the probe pulse should be large compared with the absorption linewidth we wish to measure.

We turn next to transient EIT processes. To induce a strong EIT response in systems with rapid decoherence, a first inclination might be to use a short pump pulse to achieve a large Rabi frequency. However, as we discuss below, in addition to a large pump Rabi frequency, a relatively long pump duration is also essential. In the context of Equation 8.6a, EIT arises from the term, $i\Omega_b(t)\tilde{p}_{ba}^{(1)}(t)/2$. In the limit of $\Omega_b(t) = 0$, this term will have no effect on the absorption for the $|a\rangle \leftrightarrow |e\rangle$ transition, regardless of the magnitude of the nonradiative coherence. To maximize the effect of the nonradiative coherence on the $|a\rangle \leftrightarrow |e\rangle$ transition, the pump pulse should exceed the decay time for the nonradiative coherence.

Figure 8.6 shows the absorption spectrum with $\gamma_{ab} = 0.01\,\gamma$, a pump duration of $10\,\gamma^{-1}$, and $\Omega_b = \gamma$ at its peak, and with the probe arriving at the peak of the pump pulse. The absorption spectrum features a distinct EIT dip similar to that observed for cw excitations. To better understand the transient EIT behavior, we also plot in Figure 8.7 the time evolution of both $\mathrm{Im}[\tilde{p}_a^{(1)}(t)]$ and $\mathrm{Re}[\tilde{p}_{ba}^{(1)}(t)]$. Note that the nonradiative coherence $\mathrm{Re}[\tilde{p}_{ba}^{(1)}(t)]$ is zero until both the pump and probe have arrived, and is negative for times thereafter, confirming that it leads to destructive interference.

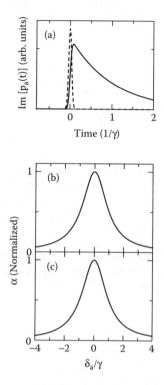

FIGURE 8.5 (a) Time dependence of $\mathrm{Im}\left[\tilde{p}_a^{(1)}(t)\right]$ (solid curve). The dashed curve gives the probe amplitude $\Omega_a(t)$. (b) Calculated linear absorption spectrum for a two-level system measured by a probe pulse with a duration of $0.1\,\gamma^{-1}$. (c) Calculated linear absorption spectrum for the two-level system measured by a cw probe.

As it was for the cw excitation, the depth of the EIT dip depends strongly on the decoherence rate of the nonradiative coherence. Figure 8.8 shows absorption spectra for various values of γ_{ab}. Note that decreasing γ_{ab} decreases the width of the EIT dip, but only up to a certain point. As discussed previously, the transient EIT process depends on the coupling between the nonradiative coherence and the pump field. Therefore, a short pump pulse also shortens the coherent interaction time. Decreasing γ_{ab} below the pump spectral bandwidth thus has little effect on the EIT dip.

One important parameter in the transient case, which is not present in the cw case, is the delay between the peaks of the pump and probe pulses. We denote this pump-probe delay by τ, with $\tau > 0$ indicating that the probe arrives after the peak of the pump. Figure 8.9 shows absorption spectra for various pump-probe delays. The depth of the EIT dip depends strongly on the pump-probe delay, and the results are not symmetric about $\tau = 0$. The EIT dip appears at relatively large negative delays, and is deepest when the probe slightly precedes the pump, but disappears rapidly for positive delays. This behavior can be understood by examining the temporal behavior of $\mathrm{Im}[\tilde{p}_a^{(1)}(t)]$ and $\mathrm{Re}[\tilde{p}_{ba}^{(1)}(t)]\cdot\Omega_b(t)$ shown in Figure 8.10. For negative delays large relative to γ^{-1}, $\mathrm{Im}[\tilde{p}_a^{(1)}(t)]$ decays significantly before the arrival of the pump. For negative delays $\sim\gamma^{-1}$, the coupling between the nonradiative coherence

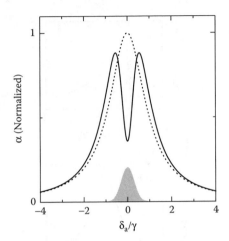

FIGURE 8.6 Calculated EIT absorption spectrum for transient excitation conditions. The probe and pump durations are $0.1\ \gamma^{-1}$ and $10\ \gamma^{-1}$, respectively, and the probe arrives at the peak of the pump. Other parameters are $\gamma_{ab} = 0.01\ \gamma$, $\Omega_b\,(0) = \gamma$, and $\gamma_b = 0$. The pump spectrum is shown as the shaded area.

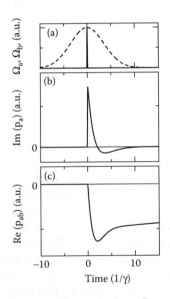

FIGURE 8.7 Numerical solutions for $\mathrm{Im}\left[\tilde{p}_a^{(1)}(t)\right]$ and $\mathrm{Re}\left[\tilde{p}_{ab}^{(1)}(t)\right]$ under transient excitation conditions of Figure 8.6. (a) Timing of pump and probe pulses. The probe Ω_a (t) is the solid curve and the pump Ω_b (t) is the dashed curve. (b) Time dependence of $\mathrm{Im}\left[\tilde{p}_a^{(1)}(t)\right]$. (c) Time dependence of $\mathrm{Re}\left[\tilde{p}_{ab}^{(1)}(t)\right]$.

and the pump field, characterized by $\tilde{p}_{ba}^{(1)}(t) \cdot \Omega_b(t)$, can maximally affect the temporal evolution of $\tilde{p}_a^{(1)}(t)$ over its entire decay, leading to a strong EIT process. When the probe arrives after the peak of the pump, the coupling between the nonradiative

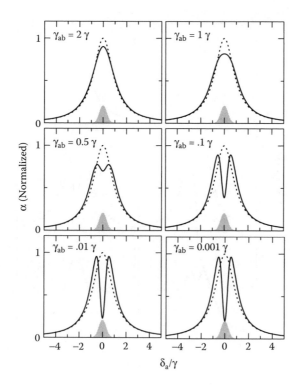

FIGURE 8.8 Effects of the decoherence rate γ_{ab} on the EIT absorption spectrum. The values of γ_{ab} are indicated in the figure, and other parameters are the same as for Figure 8.6, but with $\tau = -2\gamma^{-1}$.

coherence and the pump field becomes relatively weak because, in this case, the nonradiative coherence, which is induced by both the pump and probe, can only couple to the trailing edge of the pump pulse.

This section has discussed the basic properties of EIT in a Λ-system. These properties provide important guidance for understanding as well as designing transient experiments to realize EIT in semiconductors. While attaining the large Rabi frequency necessary to overcome rapid decoherence favors the use of very short pump pulses, the decay time for the relevant coherences also sets a lower limit on the duration of the pump pulses that are suitable for the transient EIT experiment. It should be added that transient behaviors of EIT in semiconductors are also strongly influenced by underlying many-body Coulomb interactions, as we discuss in more detail in subsequent sections.

8.3   EXPERIMENTAL METHODS

The experimental setup for transient EIT studies was based on a standard transient pump-probe configuration, but with different pump and probe pulse durations. Pump and probe pulses were split from the output of a mode-locked Ti:sapphire laser with a pulse repetition rate of 82 MHz, a spectral bandwidth of 7 to 8 nm, and a pulse

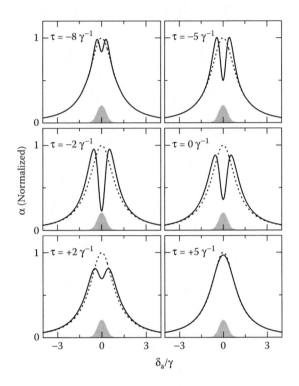

FIGURE 8.9 EIT absorption spectra for the pump-probe delays τ indicated in the figure. Other parameters are the same as for Figure 8.6.

duration of 150 fs. To obtain the spectrally narrow (and temporally long) pump pulses needed for the experiments, we used an external pulse shaper with which nearly Fourier-transform limited pulses can be obtained. The minimum filtered spectral bandwidth obtained with this setup was approximately 0.2 nm. For the experiments presented here, the filtered pump spectral widths were between 0.2 and 0.3 nm, with pulse durations of 5 to 6 ps, as measured by cross-correlation with the probe pulse. The probe pulse was unfiltered. The laser spot sizes at the sample were 3×10^{-5} cm^2 for the pump and 4×10^{-6} cm^2 for the probe. The energy flux of the probe pulse is less than 1% of the pump.

The EIT experiments were performed with three different GaAs/AlGaAs QW structures with thicknesses of 17.5 nm, 13 nm, and 10 nm. All were multiple QW structures, with sufficiently wide barriers so that the individual QWs behave independently, and all samples were grown by molecular beam epitaxy. The GaAs substrate of the samples was removed with chemical etching for transmission measurements. The samples were mounted on sapphire disks and were held at a temperature near 10K in a helium flow cryostat. The probe beam transmitted through the sample was collected and sent through a spectrometer with 0.05-nm resolution. The resulting spectrally resolved signal was detected using a photomultiplier tube. Absorption spectra in the presence or absence of the pump were calculated using the relationship:

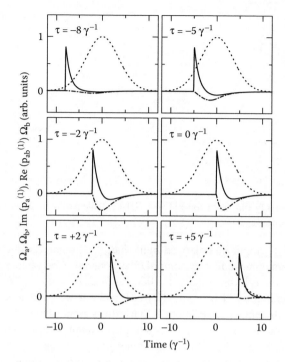

FIGURE 8.10 Time dependence of $\mathrm{Im}\left[\tilde{p}_a^{(1)}(t)\right]$ (solid), $\mathrm{Re}\left[\tilde{p}_{ab}^{(1)}(t)\right]\cdot\Omega_b(t)$ (dash-dot), and Ω_b (t) (dashed) for the same pump-probe delays as for Figure 8.9. Other parameters are the same as for Figure 8.6.

$$\alpha(\lambda)L = -\ln\left[\frac{I(\lambda)}{I_0(\lambda)}\right]$$

where $\alpha(\gamma)$ is the absorption coefficient as a function of the probe wavelength, L is the thickness of the absorbing region of the sample, $I(\lambda)$ is the measured intensity of the probe after passing through the sample, and $I_0(\lambda)$ is the intensity of the probe before passing through the sample, approximated by measuring the probe spectrum when passed through the sapphire disk only. We plot the absorbance αL when presenting experimental data.

8.4 EIT VIA INTERVALENCE BAND COHERENCE

A model system for inducing nonradiative coherence in semiconductors such as a GaAs QW is the heavy-hole (HH) and light-hole (LH) excitonic transitions. The band structure of a GaAs QW near the band edge consists of doubly degenerate conduction bands with $s_z = \pm 1/2$, and two doubly degenerate valence bands (the HH band with $J_z = \pm 3/2$ and the LH band with $J_z = \pm 1/2$). As shown in the band diagram in Figure 8.11a, an HH transition with σ^+ circularly polarized light and an LH

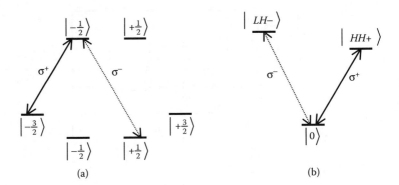

FIGURE 8.11 Energy diagrams for EIT via HH-LH valence band coherence: (a) electron states and (b) excitonic states.

transition with σ^- circularly polarized light share a common conduction band state. One can, in principle, induce a nonradiative coherence between the two valence bands via this three-level system. Earlier theoretical studies have suggested the possibility of realizing EIT using the HH-LH intervalence band coherence.[16] Observations of the coherently coupled optical Stark shifts associated with these transitions have been interpreted as a strong evidence for coherent coupling between the HH and LH valence bands.[19] Note that the HH and LH transitions do not form a Λ-type three-level system because both valence bands are initially occupied. These transitions should be viewed as a V-type three-level system, as shown schematically by the exciton energy diagram in Figure 8.11b.

To realize EIT from the intervalence band coherence, we applied a pump pulse with σ^+ polarization and an energy flux per pulse of 160 nJ.cm^{-2} to the HH exciton transition. The EIT signature should then appear as a dip in the LH exciton absorption resonance measured by a weak probe pulse with σ^- polarization. Experimental results presented in this section were all obtained in a 13-nm GaAs QW. Figure 8.12 shows the absorption spectrum measured by the probe in the presence (solid line) and absence (dashed line) of the pump. As shown in the figure, the pump induces a dip in the LH exciton absorption resonance.[33] Note that a pronounced absorption dip, which is due to EIT arising from the exciton spin coherence, also appeared in the HH exciton resonance. The additional absorption resonance below the HH exciton resonance is due to biexcitons. Both of these features are discussed in detail in the next section.

We have carried out additional experimental studies to show that the absorption dip in the LH exciton resonance satisfies the two-photon resonance condition expected for the intervalence band coherence. Figure 8.13 shows that the spectral position of the absorption dip varies with the central wavelength for the pump pulse. The experimental data is in general agreement with the two-photon resonance condition for the nonradiative coherence in a V-type three-level system.

One possible complication in the above experiment is spectral hole burning in the inhomogeneously broadened exciton absorption spectrum. In our experiment, the Rabi frequency of the pump (near the peak amplitude of the pump pulse)

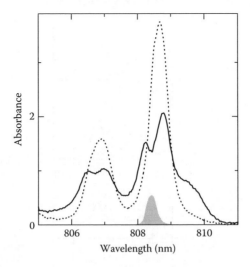

FIGURE 8.12 Absorption spectrum measured by a σ^- probe in the presence (solid line) and absence of (dashed line) a σ^+ pump. The pump-probe delay is −1 ps and the pump pulse energy flux is 160 nJ.cm⁻².

exceeded the inhomogeneous linewidth. Under these conditions, spectral hole burning is no longer possible.[35] Figure 8.14 shows absorption spectra measured by the probe as a function of the delay between the probe and pump. The absorption dips in both the HH and LH exciton resonances appeared when the probe-pump delay was near zero or when the probe slightly preceded the pump, but completely vanished when the probe arrived 10 ps after the pump. The temporal dependence confirms that the observed absorption dip is due to a coherent process such as EIT and is not due to incoherent spectral hole burning.

In the absence of spectral hole burning, the observed dip in the probe absorption spectrum can only arise from the nonradiative coherence. To confirm this, we performed numerical calculations of the transient optical response using a homogeneously broadened V-type three-level system. Similar to that discussed in the earlier section, the density matrix equations were solved to the first order of the probe field and all orders of the pump field. We plot in Figure 8.15 the calculated absorption spectra as we varied the decay rate for the intervalence band coherence γ_{ab} while keeping other parameters unchanged. In this case, increasing γ_{ab} to infinity is equivalent to artificially turning off the nonradiative coherence without affecting all other effects such as incoherent saturation inherent in a V-type three-level system. While the incoherent saturation is present regardless of the nonradiative coherence, the absorption dip appears only when γ_{ab} is sufficiently small and vanishes when γ_{ab} far exceeds the pump Rabi frequency. These calculations, although they do not provide a quantitative description of the experiments, confirm that the absorption dip observed in the LH exciton resonance arises from destructive interference associated with the intervalence band coherence and can thus be viewed as a signature of EIT.

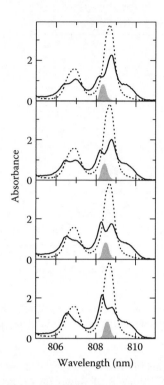

FIGURE 8.13 Dependence of the probe absorption spectrum on the pump wavelength. The pump spectrum is shown as the shaded area and the pump pulse energy flux is 160 nJ.cm^{-2}. Other conditions are the same as for Figure 8.12.

Although the above experimental results show that EIT due to the intervalence band coherence is possible, the EIT signature observed is rather weak. In such a V-type system, strong resonant excitation of excitons by the pump is necessary. In this case, the degree of transparency that can be achieved from the intervalence band coherence is limited by rapid decoherence induced by exciton-exciton scattering. To realize strong EIT processes in a semiconductor, systems analogous to Λ-type or cascade-type three-level systems, for which strong resonant excitation can in principle be avoided, are thus needed. In the next two sections, we discuss the use of nonradiative coherences induced via interactions or correlations between excitons to realize EIT in Λ-type and cascaded-type three-level systems.

8.5 EIT VIA EXCITON SPIN COHERENCE

While one can attempt to adapt and extend concepts and techniques developed in an atomic system to an excitonic system, a fundamental issue for manipulating excitonic quantum coherences is how many-body Coulomb interactions between excitons, which are inherent in an excitonic system but are absent in simple atomic systems, affect the creation, evolution, and manipulation of these coherences. We have taken advantage of the many-body Coulomb interactions between excitons and

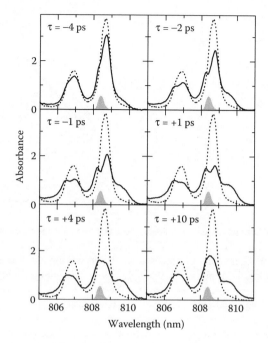

FIGURE 8.14 Dependence of the probe absorption spectrum on the pump-probe delay. The pump pulse energy flux is 160 nJ.cm^{-2}.

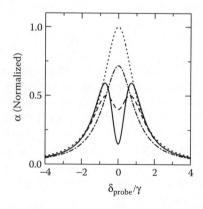

FIGURE 8.15 Calculated absorption spectra of a σ^- probe pulse in the presence of a σ^+ pump pulse, with δ being detuning relative to the LH exciton resonance, for $\gamma_{ab} = 0.1\ \gamma$ (solid), $\gamma_{ab} = 0.5\ \gamma$ (dashed), and $\gamma_{ab} = 10\ \gamma$ (dot-dashed), where γ is the decoherence rate for both dipole transitions. The dotted curve shows the probe absorption in the absence of the pump. The pump is resonant with the HH exciton and with a duration of $6/\gamma$. The peak Rabi frequency for the pump is set to equal γ. The probe duration is $0.1/\gamma$ and the pump-probe delay is $-1/\gamma$. The population relaxation rate is $0.01\ \gamma$.

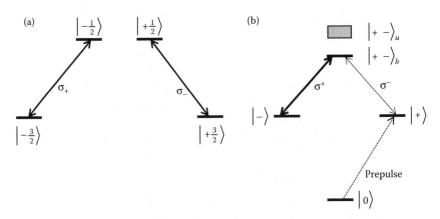

FIGURE 8.16 (a) Dipole optical transitions between the HH valence band states and the conduction band states. Transitions with opposite circular polarization share no common energy levels. (b) N-exciton energy eigenstates. The crystal ground state is denoted by $|0\rangle$. The one-exciton states $|-\rangle$ and $|+\rangle$ denote spin-down and spin-up excitons, excited by σ^- and σ^+ circularly polarized light, respectively. The two-exciton states are labeled by $|+-\rangle_b$ and $|+-\rangle_u$ for the bound and unbound biexciton states, respectively.

developed EIT schemes that are based on the use of exciton spin coherence and biexciton coherence in GaAs QW structures. Exciton spin coherence is a coherent superposition of the exciton spin states. Biexciton coherence is a coherent superposition of the crystal ground state and a bound state of two excitons. Both of these coherences arise from Coulomb interactions between excitons and have no counterparts in independent atomic systems. In this section, we discuss EIT arising from the exciton spin coherence. Experimental results presented in this section were all obtained in a 10-nm GaAs QW.

For the HH transitions, spin-up ($|+\rangle$) and spin-down ($|-\rangle$) excitons can be excited with σ^+ and σ^- circularly polarized light, respectively, as shown in Figure 8.16a. Although the $|+\rangle$ and $|-\rangle$ excitons share no common states, these excitons can interact with each other via Coulomb correlations, leading to the formation of bound and unbound two-exciton states (biexcitons), as shown by the N-exciton energy eigenstates in Figure 8.16b. Exciton spin coherence (i.e., coherent superposition of the $|+\rangle$ and $|-\rangle$ states) can then be induced via the one-exciton to two-exciton transition. We stress that the exciton spin coherence is a direct result of correlations between excitons with opposite spins. In the absence of these correlations, the two-exciton states can be factored into product states of single excitons and no exciton spin coherence can be induced.

Effects of Coulomb correlations on coherent nonlinear optical processes have been investigated extensively in earlier studies and can be described by microscopic theories based on dynamics-controlled truncation schemes and also on the use of N-exciton energy eigenstates.[36–44] These earlier studies, however, primarily focused on the effects of Coulomb correlations in a weakly nonlinear regime. In contrast, EIT involves nonlinear optical processes to all orders of the pump field.

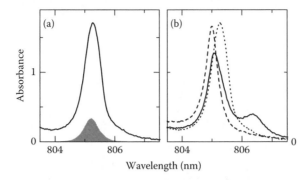

FIGURE 8.17 (a) Linear absorption spectrum of a σ^- probe in the absence of the prepulse. The shaded region also shows the prepulse spectrum. (b) Absorption of a σ^- probe (solid) and σ^+ probe (dashed) in the presence of a σ^+ prepulse. The dotted line shows the linear absorption for reference. The prepulse energy flux is 100 nJ.cm^{-2}, and the probe is delayed by 10 ps relative to the prepulse.

8.5.1 SPIN COHERENCE VIA BOUND BIEXCITON STATES

As shown in Figure 8.16b, we can form a Λ-type three-level system, consisting of the two exciton spin states $|+\rangle$ and $|-\rangle$, and the bound biexciton state $|+-\rangle_b$. To measure EIT in this system, we applied a strong pump pulse with σ^+ polarization to the $|-\rangle \leftrightarrow |+-\rangle_b$ transition and a weak probe pulse with σ^- polarization to the $|+\rangle \leftrightarrow |+-\rangle_b$ transition. In this configuration, the exciton spin coherence can lead to EIT in the optical transition between the one-exciton and bound two-exciton states (i.e., in the biexciton resonance).

Because both of the one-exciton states are initially unoccupied, we prepared an initial population in the $|+\rangle$ exciton state by applying a σ^+ polarized prepulse with 3-ps duration to the σ^+ HH excitonic transition. The prepulse was set to arrive 10 ps before the probe pulse, which is long compared to the exciton decoherence time so that dipole coherences induced by the prepulse do not interfere with the observation of EIT. The 10-ps delay, however, is short compared to the exciton spin relaxation time (50 to 100 ps)[45] so that only the $|+\rangle$ exciton state is occupied at the start of the EIT experiment. Figure 8.17a shows the linear absorption spectrum (obtained in the absence of the pump and prepulse). Figure 8.17b shows the absorption spectra obtained 10 ps after the prepulse (in the absence of the pump). The biexciton absorption resonance corresponding to the $|+\rangle \leftrightarrow |+-\rangle_b$ transition was observed when the probe and the prepulse had the opposite circular polarization. The biexciton resonance vanished when the probe and the prepulse had the same circular polarization.

Figure 8.18 shows absorption spectra of a σ^- probe when both the σ^+ prepulse and σ^+ pump were present. The pump was resonant with the $|-\rangle \leftrightarrow |+-\rangle_b$ transition and had an energy flux per pulse of 800 nJ.cm^{-2} (corresponding to a peak intensity of 150 MW.cm^{-2}). The presence of the pump induces a distinct dip in the biexciton absorption resonance. This dip is not due to incoherent spectral hole burning because

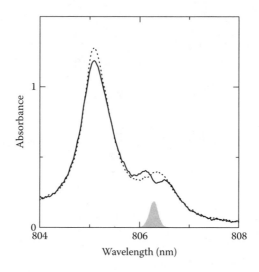

FIGURE 8.18 EIT via exciton spin coherence. The absorption of a σ^- probe in the presence of a σ^+ prepulse and σ^+ pump is shown by the solid line, while the dotted line shows the probe absorption in the absence of the pump. The pump and prepulse energy fluxes are 800 and 100 nJ.cm^{-2}, respectively. The probe arrives 10 ps after the prepulse and 3 ps before the peak of the pump.

the pump and probe had opposite circular polarization and also because only the $\left|+\right\rangle$ exciton state was populated by the prepulse. The absorption dip was most pronounced when the probe arrived 3 ps before the peak of the pump, as shown by the pump-probe delay dependence in Figure 8.19, which is consistent with the expected behavior of EIT in a transient regime discussed earlier. The dip disappeared when the probe arrived 6 ps after the pump, further confirming that the absorption dip is due to a coherent process. Figure 8.20 also shows that the spectral position of the dip follows the spectral position of the pump, thus satisfying the characteristic two-photon resonance condition expected for the exciton spin coherence in a Λ-type, three-level system.

From the polarization configuration and especially the characteristic temporal and spectral dependence of the absorption dip in the biexciton resonance, we can conclude that the origin of this absorption dip is EIT arising from an exciton spin coherence. Although EIT resulting from this spin coherence shares similar properties with EIT in an atomic system, the exciton spin coherence itself is the direct result of many-body exciton-exciton correlations. Without the exciton-exciton interactions, no exciton spin coherence could be induced.

8.5.2 Spin Coherence via Unbound Two-Exciton States

As discussed above, in addition to the bound two-exciton state, Coulomb interactions between excitons can also lead to a continuum of unbound (but still correlated) two-exciton states $\left|+-\right\rangle_u$. This section explores how these continuum states can be exploited to induce exciton spin coherence and to realize EIT at the exciton resonance.[32]

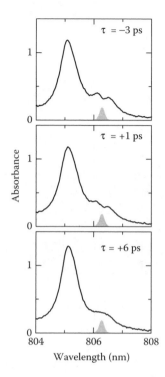

FIGURE 8.19 Dependence of EIT via exciton spin coherence on the pump-probe delay. The pump-probe delay τ is indicated in the figure, and other conditions are the same as for Figure 8.18.

For these experiments, the pump was σ⁺ polarized and was resonant with the excitonic transition (no prepulse was used). The absorption spectrum was measured with a σ⁻ polarized probe.

Figure 8.21 shows the absorption spectra obtained at various pump-probe delays. An absorption dip now occurs at the exciton resonance. The dip becomes visible when the probe precedes the peak of the pump by 6 ps and is the most pronounced when the probe precedes the peak of the pump by 3 ps. The absorption dip becomes strongly asymmetric near the zero pump-probe delay and disappears when the probe arrives after the peak of the pump by a few picoseconds. The absorption dip is not due to spectral hole burning because the σ⁺ and σ⁻ HH exciton transitions share no common states and also because the delay dependence in Figure 8.21 clearly shows that the dip is due to a coherent process. In fact, aside from the strongly asymmetric spectral lineshape near the zero pump-probe delay, the delay dependence is characteristic of the transient EIT process discussed earlier. The asymmetry in the spectral lineshape arises from exciton energy renormalization induced by the pump field,[46,47] as discussed in detail in Ref. 31. In Figure 8.22 we further show that the spectral position of the dip follows almost exactly the spectral position of the pump, indicating that the dip satisfies the two-photon resonance condition characteristic of exciton spin coherence and thus is associated with the exciton spin coherence.

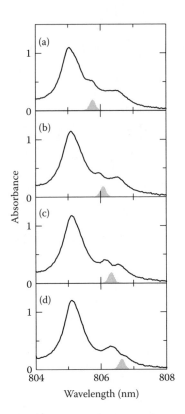

FIGURE 8.20 Dependence of EIT via exciton spin coherence on the pump wavelength. The pump spectrum is shown as the shaded area. Other conditions are the same as for Figure 8.18.

We have also performed experiments under conditions more closely corresponding to the adiabatic conditions typical of EIT experiments in atomic systems. To increase the duration of the probe pulse (and therefore reduce its spectral width), we added a pulse shaper to the probe beam. We obtained the absorption spectrum by scanning the center wavelength of the probe pulse instead of spectrally resolving the probe pulse after its transmission through the sample. For these measurements, the probe pulse was detected with a silicon photodiode, the probe duration was 8 ps, and the pump-probe delay was nominally zero. Figure 8.23 shows a representative absorption spectrum obtained in this manner, for a pump energy flux of 800 nJ.cm^{-2}. The absorption dip is clearly visible but less pronounced than that obtained in Figure 8.21. This is due, in part, to the fact that the spectral resolution of the above experiment is limited to 0.15 nm, set by the spectral width of the probe pulse. The relatively short duration for the probe is necessary to minimize the effects of rapid decoherence. In this regard, for systems with short decoherence times, the adiabatic experiments are less desirable as a spectroscopic tool for probing nonradiative coherences than the transient experiments discussed previously.

For the EIT experiments in this subsection, the pump pulse actually plays two distinct roles, as shown in Figure 8.24. Because the pump is resonant with the $|0\rangle \leftrightarrow |+\rangle$

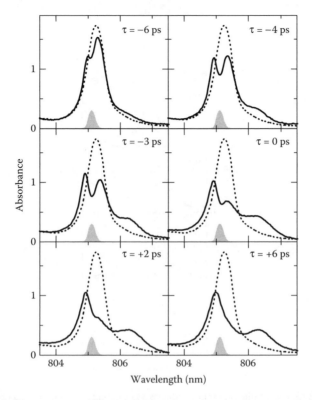

FIGURE 8.21 Exciton spin coherence induced via unbound biexcitons: dependence on the pump-probe delay τ. The dotted line shows the absorption of the probe in the absence of the pump. The pump spectrum is shown as the shaded area, and the pump pulse energy flux is 400 nJ.cm^{-2}.

transition, it plays the role of the prepulse in the EIT experiment discussed in Section 8.5.1, by exciting the $|+\rangle$ state. The pump, along with the probe, can also set up the exciton spin coherence via the unbound two-exciton states by coupling to the $|-\rangle \leftrightarrow |+-\rangle_u$ transition. This second role can be easily overlooked because both $|-\rangle$ and $|+-\rangle_u$ are initially unoccupied, but is made apparent by comparison with spin coherence induced via bound two-exciton states.

As discussed above, the exciton spin coherence can be induced via both bound and unbound two-exciton states. Regardless of the details of the different pathways, an order-by-order perturbation analysis using the N-exciton eigenstates in Figure 8.16b shows that the exciton spin coherence is at least to the fourth order of the applied optical fields. In this regard, the effects of the exciton spin coherence, while playing a prominent role in the strongly nonlinear regime, are not expected in third-order nonlinear optical measurements carried out at relatively low excitation levels. Further theoretical development on higher-order coherent nonlinear optical processes in interacting excitonic systems, however, is still needed for a satisfactory micro-scopic description of exciton spin coherence and its effects on coherent nonlinear optical responses.

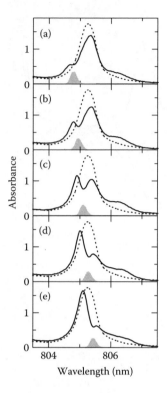

FIGURE 8.22 Exciton spin coherence induced via unbound biexcitons: dependence on the pump wavelength. The pump spectrum is shown as the shaded area. The probe arrives 3 ps before the pump. Other conditions are the same as for Figure 8.21.

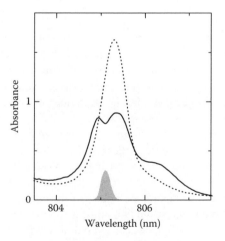

FIGURE 8.23 EIT at exciton resonance with a spectrally narrow probe. The probe absorption is shown in the presence (solid) and absence (dotted) of a pump with opposite circular polarization. The pump duration is 6 ps, the probe duration is 8 ps, and the peaks of the pump and probe pulses arrive at the same time. The pump energy flux is 800 nJ.cm^{-2}.

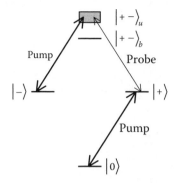

FIGURE 8.24 Dual role of pump for exciton spin coherence induced via unbound biexcitons.

Note that the EIT dip due to the exciton spin coherence is considerably stronger than that observed for the HH-LH intervalence band coherence. This suggests that the exciton spin coherence is more robust than the intervalence band coherence. In fact, the width of the observed EIT dip induced by the exciton spin coherence shown in Figure 8.21 may be limited by the pump spectral width, and not by the decoherence rate for the spin coherence. This relative robustness of the exciton spin coherence also makes it possible for the high-order Coulomb correlation underlying the exciton spin coherence to have such large effects on the overall optical response. It should be added that decoherence induced by exciton-exciton scattering still remains a major limiting factor for EIT from the exciton spin coherence. As shown in Figure 8.25, where we have increased the pump intensity in an effort to increase the degree of transparency, the depth of the EIT dip does not continue to increase for higher pump intensities. Instead, the overall absorption feature broadens due to increased decoherence induced by exciton-exciton scattering.

8.6   EIT VIA BIEXCITONIC COHERENCE

In the EIT schemes based on the use of exciton spin coherence, resonant optical excitations of excitons are still necessary. For spin coherence induced via the bound two-exciton state, one of the exciton states must be populated. For spin coherence induced via unbound two-exciton states, the pump pulse leads to strong resonant optical excitation. The degree of quantum interference that can be achieved in these EIT schemes is therefore severely limited by excessive decoherence due to exciton-exciton scattering induced by strong resonant optical excitations. In this section we discuss the use of biexciton coherence to realize EIT in a cascaded three-level system. The destructive quantum interference in this biexciton EIT scheme is set up by a pump pulse that couples to the biexciton absorption resonance. The biexciton EIT scheme thus avoids the resonant excitonic excitation that has hindered our other EIT schemes in semiconductors.

For the exciton and biexciton energy eigenstates shown in Figure 8.26, we can form a cascaded three-level system using the $|0\rangle$, $|-\rangle$, and $|+-\rangle_b$ states. By applying a σ^+ polarized pump to the $|-\rangle \leftrightarrow |+-\rangle_b$ transition and a σ^- polarized pump to the

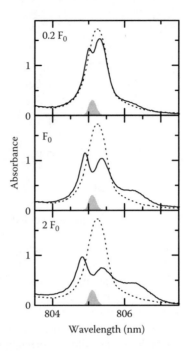

FIGURE 8.25 Exciton spin coherence induced via unbound biexcitons: dependence on the pump intensity. The pump energy flux is indicated in the figure, with $F_0 = 400$ nJ.cm^{-2}.

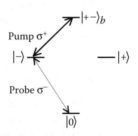

FIGURE 8.26 Cascaded three-level system for EIT via biexciton coherence. The ground state $|0\rangle$, the HH exciton states $|-\rangle$ and $|+\rangle$, and the bound biexciton state $|+-\rangle_b$ are shown. The pump and probe fields are also shown.

$|0\rangle \leftrightarrow |+\rangle$, we can then induce a nonradiative coherence between the ground and biexciton states (i.e., the biexciton coherence[48,49]) and can use destructive interference induced by this biexciton coherence to realize EIT. Experimental results presented in this section were all obtained in a 17.5-nm GaAs QW. To determine the spectral position of the biexciton resonance, we excited the sample with a pump pulse that was resonant with the HH exciton and had a circular polarization opposite of the probe. Figure 8.27 shows the absorption spectrum obtained with the probe pulse, from which we determine a biexciton binding energy of 1.6 meV.

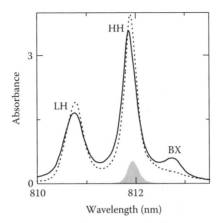

FIGURE 8.27 Biexciton absorption resonance. The curves show the probe absorption in the presence (solid) and absence (dotted) of the pump that is resonant with the exciton transition. The pump spectrum is shown as the shaded area. The pump pulse energy flux is 40 nJ.cm^{-2}, and the probe arrives 10 ps after the peak of the pump.

To demonstrate EIT using the biexciton coherence, we applied a 6-ps pump pulse with σ^+ polarization tuned to the vicinity of the exciton to biexciton transition and used a 150-fs probe pulse with σ^- polarization to measure the absorption spectrum. Based on the energy level diagram, we would expect to see an EIT dip appearing at the HH exciton resonance.

Figure 8.28 shows the absorption spectra measured by the probe with the pump pulse at two different spectral positions. When the pump pulse was at the biexciton resonance, a small absorption dip appears on the high-energy side of the exciton resonance (see Figure 8.28a). However, from the two-photon resonance condition for the biexciton coherence, we would expect the EIT dip to occur exactly at the center of the exciton resonance. When the pump was tuned to an energy slightly above the biexciton resonance, a much greater reduction in the exciton absorption was observed. The absorption dip now occurs at the center of the exciton resonance (see Figure 8.28b). In this case, the absorption is reduced by factor of $\exp(3.1)$ = 22 at the center of the HH exciton resonance.

The dependence of the dip in the absorption spectrum on the pump-probe delay is shown in Figure 8.29. The absorption spectrum obtained when the probe arrives 10 ps after the pump is nearly identical to the linear absorption spectrum obtained in the absence of the pump, indicating minimal real absorption of the pump pulse. This nearly complete recovery of the absorption spectrum shows that the reduction in the absorption of the exciton resonance in the above experiments is primarily a coherent process with only minimal contributions from incoherent bleaching or incoherent two-photon absorption processes.

For the above experiments, the absorption dip shifted to lower energy when the pump pulse was tuned to higher energy, consistent with the two-photon resonance condition expected for EIT. The spectral positions of the dips, however, do not exactly correspond to EIT dips expected for an atomic-like cascaded three-level

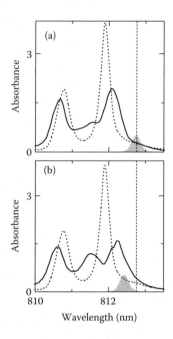

FIGURE 8.28 EIT via biexciton coherence: (a) pump resonant with the biexciton transition, and (b) pump slightly detuned from the biexciton resonance. The spectral position of the biexciton resonance is indicated by the dashed line. The pump spectrum is shown as the shaded area and the pump pulse energy flux is 2 μJ.cm^{-2}. The probe arrives 1 ps before the peak of the pump.

system. In Figure 8.30 we plot the absorption spectra obtained with increasing pump intensity, which reveals another unexpected behavior: the position of the EIT dip changes with the pump intensity. For the lowest intensity, the EIT dip appears on the low-energy side of the HH exciton. This is consistent with the two-photon resonance condition because the pump is on the high-energy side of the biexciton resonance. At higher pump intensities, however, the EIT dip shifts gradually to higher energy.

We have applied a microscopic many-particle theory to understand the physical processes underlying the biexciton EIT process.[34] That analysis, however, is rather involved. To make direct connections with the well-known concepts from atomic physics, here we discuss instead a phenomenological model that illustrates the essential physical mechanism of the observed EIT process.

For EIT due to the biexciton coherence, the spectral position of the EIT dip is set by the resonance frequency of the biexciton coherence and the spectral position of the pump. Formally, the two-photon resonance condition is $E_{EIT} = E_{bx} - E_{pump}$, where E_{EIT} is the energy position of the EIT dip, E_{bx} is the energy of the biexciton state, and E_{pump} is the energy position of the pump. The shift of the EIT dip with increasing pump intensity indicates that the two-photon resonance condition, and specifically the biexciton energy, varies with the pump intensity.

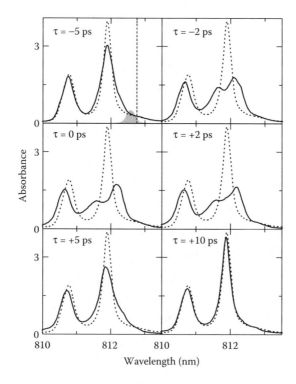

FIGURE 8.29 Dependence of EIT via biexciton coherence on pump-probe delay τ. The pump spectrum is indicated by the shaded area, and the pump pulse energy flux is 1.6 μJ.cm⁻².

Because a biexciton is a bound state of two excitons, the biexciton energy is determined by the sum of the σ+ and σ− exciton energies, minus the biexciton binding energy, all of which are affected by exciton-exciton interactions, leading to renormalization in both exciton and biexciton energies. At relatively low intensity for the σ+ polarized pump, we can simulate the effect of the biexciton energy renormalization by introducing phenomenologically an intensity-dependent biexciton energy $E_{bx} = E_{bx}^0 + \hbar\beta\left|\Omega_{pump}\right|^2$, where E_{bx}^0 is the energy of the biexciton in the absence of the pump and β is a phenomenological constant.

Figure 8.31 plots the EIT spectrum calculated using a cascaded three-level system with an intensity-dependent biexciton energy under the conditions of decoherence for the biexciton coherence $\gamma_{bx} = \gamma$ and the peak Rabi frequency for the pump $\Omega_0 = \sqrt{2}\gamma$. Figure 8.31a shows the spectrum for β = 0 so that there is no renormalization for the biexciton energy. As expected, the EIT dip is positioned at the center of the exciton resonance. Figure 8.31b shows the absorption spectrum when the biexciton energy renormalization is included by setting β = 0.2γ⁻¹. In this case, a pump resonant with the un-renormalized biexciton resonance becomes detuned with respect to the renormalized biexciton resonance. The EIT dip then appears on the high-energy side of the exciton peak. The EIT dip can be centered on the exciton peak by anticipating the biexciton energy shift and positioning the

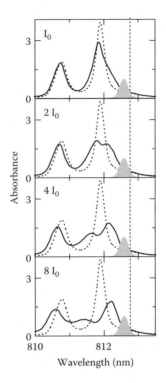

FIGURE 8.30 Dependence of EIT via biexciton coherence on pump intensity. The pump pulse energy flux is indicated in the figure, with $I_0 = 400$ nJ.cm^{-2}. The spectral position of the biexciton resonance is indicated by the dashed line. The pump spectrum is shown as the shaded area.

pump near the renormalized biexciton resonance. Figure 8.32 shows that the spectral position of the EIT dip varies with pump intensity, where the pump is positioned slightly above the un-renormalized biexciton resonance. At low pump intensity, the biexciton energy renormalization is small and the EIT dip occurs on the low-energy side of the exciton resonance. As the intensity increases, a blue shift of the biexciton energy leads to a corresponding shift of the EIT dip to the high-energy side of the exciton resonance, in general agreement with the experiment.

Coherent nonlinear optical processes in an exciton-biexciton system have been investigated extensively. Reports of exciton red optical Stark shifts and Coulomb memory effects due to biexcitons, which can be viewed as a precursor for the EIT process discussed here, can be found in Refs. 38 and 50. Autler-Townes splitting of biexcitons has also been observed in two-photon absorption spectra of biexciton.[51] Despite these extensive efforts, EIT occurring at the exciton resonance via biexciton coherence had not been reported previously. As we discussed above, to observe the strong EIT process at the exciton resonance, it is essential that the biexciton energy renormalization is properly compensated. In addition, it is also important that the pump pulse duration is long, or at least comparable to the biexciton decoherence time.

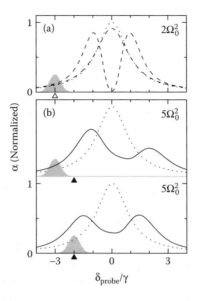

FIGURE 8.31 Effect of biexciton energy renormalization. Curves give the calculated probe absorption spectrum in the presence (solid) and absence (dotted) of the pump. The pump and probe durations are $10\gamma^{-1}$ and $0.1\gamma^{-1}$, respectively, with the probe arriving $1/\gamma$ before the pump. The peak pump intensity is indicated in the figure, with $\Omega_0^2 = 2\gamma^2$, and the shaded area shows the pump spectrum. The open and solid triangles mark the position of the un-renormalized biexciton resonance and the position of the normalized biexciton resonance at the peak of the pump pulse, respectively. (a) No biexciton energy renormalization ($\beta = 0$), for $\gamma_{bx} = 0.1\gamma$ (deep EIT dip) and 10γ (no dip); and (b) with biexciton energy renormalization ($\beta = 0.2\gamma^{-1}$), and $\gamma_{bx} = \gamma$.

8.7 SUMMARY

In this chapter we discussed the experimental realization of EIT in semiconductor QW structures using three different types of nonradiative coherences: (1) interval-ence band coherence, (2) exciton spin coherence, and (3) biexciton coherence. Coherent manipulation of quantum coherences in semiconductors in general has been stimulated by and, to a large extent, is based on the understanding of similar processes in simple atomic systems. Quantum coherences in semiconductors, how-ever, are typically very fragile against decoherence processes. In addition, coherent optical interactions in semiconductors are also strongly modified by the underlying many-body Coulomb interactions between optical excitations. Using transient optical techniques to overcome rapid decoherence processes and understanding the role of many-body Coulomb interactions, we have successfully demonstrated EIT in these semiconductor systems. These studies illustrate that, with an adequate understanding of the role of many-body interactions, we can harness and take advantage of these interactions for coherent manipulation of nonradiative quantum coherences.

Future progress in the manipulation of quantum coherences in semiconductors will certainly benefit from systems with smaller decoherence rates. Semiconductor

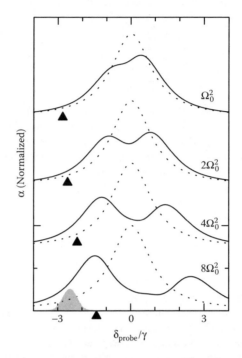

FIGURE 8.32 Effects of intensity-dependent biexciton energy renormalization. The peak pump intensity is indicated in the figure, with $\Omega_0^2 = 2\gamma^2$ and other parameters the same as for Figure 8.31.

quantum dot systems can have decoherence times for dipole transitions of hundreds of picosecond,[52,53] and the effects of exciton-exciton scattering are expected to be greatly reduced. Experiments showing coherent manipulation of dipole coherence in the form of Rabi oscillations[54–56] are a natural progression toward realizing quantum interference effects such as EIT. While the quantum dot system is conceptually more atomic-like than the quantum well system, theoretical treatments of EIT in quantum dots show that many-body interactions cannot be ignored.[57] Further study is also needed to explore the properties of nonradiative coherences in quantum dots.

Compared with the nonradiative coherences discussed in this chapter, electron spin coherence is exceptionally robust. Earlier studies on electron spin precession in semiconductors have shown that the electron spin coherence can last as long as microseconds and can also persist to room temperature.[28,29] Electron spin coherence is thus very promising for pursuing optical manipulation and control of quantum coherences in semiconductors. Recent experimental studies in a GaAs QW have already shown precursors of EIT from the electron spin coherenc.[58] A new scheme to induce electron spin coherence and realize EIT in a QW waveguide has also been proposed recently.[59] The successful realization of EIT based on the use of robust electron spin coherence can open up possibilities of applications of EIT in both classical and quantum information processing.

Acknowledgments

We wish to acknowledge Rolf Binder for fruitful collaborations on EIT studies in semiconductors, and J.E. Cunningham and D.G. Steel for supplying the QW structures used in our experimental studies. This work is supported, in part, by NSF, ARO, DARPA, and AFOSR. Sandia is a multiprogram laboratory operated by Sandia Corporation, a Lockheed Martin Company, for the U.S. Department of Energy's National Nuclear Security Administration under contract DE-AC04-94AL85000.

References

1. Harris, S.E., Electromagnetically induced transparency, *Phys. Today,* 50(7), 36, 1997.
2. Scully, M.O. and Zubairy, M.S., *Quantum Optics,* Cambridge University Press, 1997.
3. Arimondo, E, *Progress in Optics,* 35, 259, 1996.
4. Lukin, M.D., Trapping and manipulating photons in an atomic ensembles, *Rev. Mod. Phys.,* 75, 457, 2003.
5. Hau, L.V. et al., Light speed reduction to 17 meters per second in an ultracold atomic gas, *Nature,* 397, 594, 1999.
6. Harris, S.E., Field, J.E., and Kasapi, A., Dispersive properties of electromagnetically induced transparency, *Phys. Rev. A,* 46, R29, 1992.
7. Kash, M.K. et al., Ultraslow group velocity and enhanced nonlinear optical effects in a coherently driven hot atomic gas, *Phys. Rev. Lett.,* 82, 5229, 1999.
8. Budker, D. et al., Nonlinear magneto-optics and reduced group velocity of light in atomic vapor with slow ground state relaxation, *Phys. Rev. Lett.,* 83, 1767, 1999.
9. Liu, C. et al., Observation of coherent optical information storage in an atomic medium using halted light pulses, *Nature,* 409, 490, 2001.
10. Phillips, D.F. et al., Storage of light in atomic vapor, *Phys. Rev. Lett.,* 86, 783, 2001.
11. Turukhin, A. et al., Observation of ultraslow and stored light pulses in a solid, *Phys. Rev. Lett.,* 88, 023602, 2002.
12. van der Wal, C.H. et al., Atomic memory for correlated photon states, *Science,* 301, 196, 2003.
13. Kuzmich, A. et al., Generation of nonclassical photon pairs for scalable quantum communication with atomic ensembles, *Nature,* 423, 731, 2003.
14. Lee, D.S. and Malloy, K.J., Analysis of reduced interband absorption mechanisms in semiconductor quantum wells, *IEEE J. Quantum Electron.,* 30, 85, 1994.
15. Zhao, Y., Huang, D.H., and Wu, C.K., Electric-field-induced quantum coherence of the intersubband transition in semiconductor quantum-wells, *Opt. Lett.,* 19, 816, 1994.
16. Lindberg, M. and Binder, R., Dark states in coherent semiconductor spectroscopy, *Phys. Rev. Lett.,* 75, 1403, 1995.
17. Agarwal, G.S., Electromagnetic-field-induced transparency in high-density exciton systems, *Phys. Rev. A,* 51, R2711, 1995.
18. Serapiglia, G.B. et al., Laser-induced quantum coherence in a semiconductor quantum well, *Phys. Rev. Lett.,* 84, 1019, 2000.
19. Donovan, M.E. et al., Evidence for intervalence band coherences in semiconductor quantum wells via coherently coupled optical stark shifts, *Phys. Rev. Lett.,* 87, 237402, 2001.
20. Imamoglu A. and Ram, R.J., Semiconductor lasers without population inversion, *Opt. Lett.,* 19, 1744, 1994.

21. Belyanin, A.A. et al., Infrared generation in low-dimensional semiconductor hetero-structures via quantum coherence, *Phys. Rev. A,* 63, 053803, 2001.
22. Hu, X.D. and Poetz, W., Coherent control of optical gain from electronic intersubband transitions in semiconductors, *Appl. Phys. Lett.,* 73, 876, 1998.
23. Liu, A. and Ning, C.Z., Exciton absorption in semiconductor quantum wells driven by a strong intersubband pump field, *J. Opt. Soc. Am. B,* 17, 433, 2000.
24. Silvestri, L. et al., Electromagnetically induced transparency in asymmetric double quantum wells, *Eur. Phys. J. B,* 27, 89, 2002.
25. Chemla, D.S. and Shah, J., Many-body and correlation effects in semiconductors, *Nature,* 411, 549, 2001.
26. Haug, H. and Koch, S.W., Quantum theory of the optical and electrical properties of semiconductors, *Word Scientific,* Singapore, 1993.
27. Wang, H. et al., Transient nonlinear optical response from excitation induced dephasing in GaAs, *Phys. Rev. Lett.,* 71, 1261, 1993.
28. Kikkawa, J.M. et al., Room-temperature spin memory in two-dimensional electron gases, *Science,* 277, 1284, 1997.
29. Kikkawa, J.M. and Awschalom, D.D., Resonant spin amplification in n-type GaAs, *Phys. Rev. Lett.,* 80, 4313, 1998.
30. Shah, J., *Ultrafast Spectroscopy of Semiconductors and Semiconductor Nanostructures,* 2nd ed., Springer, 1999.
31. Phillips, M.C. and Wang, H., Exciton spin coherence and electromagnetically induced transparency in the transient optical response of GaAs quantum wells, *Phys. Rev. B,* 69, 115337, 2004.
32. Phillips, M. and Wang, H., Spin coherence and electromagnetically induced transparency via exciton correlations, *Phys. Rev. Lett.,* 89, 186401, 2002.
33. Phillips, M. and Wang, H., Electromagnetically induced transparency due to intervalence band coherence in semiconductors, *Optics Lett.,* 28, 831, 2003.
34. Phillips, M.C. et al., Electromagnetically induced transparency in semiconductors via biexciton coherence, *Phys. Rev. Lett.,* 91, 183602, 2003.
35. Meystre P. and Sargent III, M., *Elements of Quantum Optics,* 3rd ed., Springer, 1999.
36. Axt, V.M. and Stahl, A., A dynamics-controlled truncation scheme for the hierarchy of density matrices in semiconductor optics, *Z. Phys. B: Condens. Matter,* 93, 195, 1994.
37. Kner, P. et al., Coherence of four-particle correlations in semiconductors, *Phys. Rev. Lett.,* 81, 5386, 1998.
38. Sieh, C. et al., Coulomb memory signatures in the excitonic optical Stark effect, *Phys. Rev. Lett.,* 82, 3112, 1999.
39. Bolton, S.R. et al., Demonstration of sixth-order coulomb correlations in a semiconductor single quantum well, *Phys. Rev. Lett.,* 85, 2002, 2000.
40. Combescot M. and Combescot, R., Excitonic stark shift — a coupling to semivirtual biexcitons, *Phys. Rev. Lett.,* 61, 117, 1988.
41. Bartels, G. et al., Identification of higher-order electronic coherences in semiconductors by their signature in four-wave-mixing signals, *Phys. Rev. Lett.,* 81, 5880, 1998.
42. Östreich, T., Schönhammer, K., and Sham, L.J., Exciton-exciton correlation in the nonlinear-optical regime, *Phys. Rev. Lett.,* 74, 4698, 1995.
43. Axt, V.M. and Mukamel, S., Nonlinear optics of semiconductor and molecular nanostructures; a common perspective, *Rev. Mod. Phys.,* 70, 145, 1998.
44. Sieh, C. et al., Influence of carrier correlations on the excitonic optical response including disorder and microcavity effects, *Eur. Phys. J. B,* 11, 407, 1999.

45. Damen, T.C. et al., Subpicosecond spin relaxation dynamics of excitons and free carriers in GaAs quantum wells, *Phys. Rev. Lett.*, 67, 3432, 1991.
46. Peyghambarian, N. et al., Blue shift of the exciton resonance due to exciton-exciton interactions in a multiple-quantum-well structure, *Phys. Rev. Lett.*, 53, 2433, 1984.
47. Saba, M. et al., Direct observation of the excitonic ac Stark splitting in a quantum well, *Phys. Rev. B,* 62, R16322, 2000.
48. Wang, H. et al., Polarization dependent coherent nonlinear optical response in GaAs quantum wells: dominant effects of two-photon coherence between the ground and biexciton states, *Solid State Commun.,* 91, 869, 1994.
49. Chen, G. et al., Biexciton quantum coherence in a single quantum dot, *Phys. Rev. Lett.,* 88, 117901, 2002.
50. Hulin, D. and Joffre, M., Excitonic optical stark redshift — the biexciton signature, *Phys. Rev. Lett.*, 65, 3425, 1990.
51. Shimano R. and Kuwata-Gonokami, M., Observation of Autler-Townes splitting of biexcitons in CuCl, *Phys. Rev. Lett.*, 72, 530, 1994.
52. Borri, P. et al., Ultralong Dephasing Time in InGaAs quantum Dots, *Phys. Rev. Lett.,* 87, 157401, 2001.
53. Palinginis, P. et al., Spectral hole burning and zero-phonon linewidth in semiconductor nanocrystals, *Phys. Rev. B,* 67, Rapid Comm., 201307, 2003.
54. Stievater, H. et al., Rabi oscillations of excitons in single quantum dots, *Phys. Rev. Lett.,* 87, 133603, 2001.
55. Kamada, H. et al., Exciton Rabi oscillation in a single quantum dot, *Phys. Rev. Lett.,* 87, 246401, 2001.
56. Htoon, H. et al., Interplay of Rabi oscillations and quantum interference in semiconductor quantum dots, *Phys. Rev. Lett.,* 88, 087401, 2002.
57. Chow, W.W., Schneider, H.C., and Phillips, M.C., Theory of quantum-coherence phenomena in semiconductor quantum dots, *Phys. Rev. A,* 68, 053802, 2003.
58. Palinginis, P. and Wang, H., Coherent Raman resonance from electron spin coherence in GaAs quantum wells, *Phys. Rev. B,* 70, 153307, 2004.
59. Li, T. et al., Electromagnetically induced transparency from electron spin coherence in a quantum well waveguide, *Opt. Express,* 11, 3298, 2003.

Index

A

Acoustic echoes, 103
AiAs/GaAs systems, 145–146
AiGaN systems, 171
AIN/GaN superlattices, 171
AIN systems, 171
Anisimov studies, 103
Anti-Stokes line, 185, 187, 202
Arbouet, A., 137
   Artificial dielectrics, nonlinear optical
      properties
   basics, 49–53, *51*, 67
   discussion, 66–67
   experiments and results, 61–64, *62–66*
   theoretical considerations, 53–61
Auger process, 111

B

BAP, *see* Bir, Aronov, and Pickus (BAP)
   mechanisms
Bartels studies, 145
Belotskii and Tomchuk studies, 133
Bessel function, 57
BIA, *see* Bulk Inversion Asymmetry (BIA) term
Biexcitonic coherence, 239–244, *240–246*
Binder, Rolf, 247
Bipolar opto-electronic devices, 89–90
Bir, Aronov, and Pickus (BAP) mechanisms, 2
Bloch angles and electrons, 45, 109
Bolometers, *153*, 153–155, 168
Boltzmann equation
   carrier dynamics and non-equilbrium phonons,
      193
   ultrafast non-equilibrium electron dynamics,
      116, 130, 135
Boundary conditions, 148–150
Bound biexciton states, *232–236*, 233–234
Bragg reflection, 148
Bragg scattering, 50, 54
Brillouin scattering
   basics, 16, 18, 34–35, 43–44
   diffraction, *31*, 39–43, *40*, *42*
   setup, *34*, 35–36
   shock waves/soliton trains transition, *30–31*,
      36–37, 36–39, *39*
   soliton formation, *31*, 39–43, *40*, *42*

Brillouin zone
   high-amplitude, ultrashort strain solitons, 23
   monochromatic acoustic phonons, 159–160
   ultrafast non-equilibrium electron dynamics,
      119
Broyer, M., 138
Bulk Inversion Asymmetry (BIA) term, 3
Bulk transport properties, 202
Burgers equation, 27

C

Carrier dynamics and non-equilbrium phonons
   basics, 180, *181*, 210
   carriers, Raman spectroscopy, 181–185, *182*
   drift velocities, large electrons, 206–209,
      207–208
   dynamics, carrier, *192–194*, 192–195
   experimental methods, 181–192
   InGaAsN analysis, *192–194*, 192–195
   InGaN/GaN analysis, 195, *195–196*, 197, *198*
   InN analysis, 202, *203–205*, 204–209, *207–208*
   InN/GaN analysis, 197–198, *199*, 200–202,
      201
   large electron drift velocities, 206–207,
      207–208
   lattice vibrations, *185*, 185–189
   light sources, 189
   non-equilibrium longitudinal phonons, 202,
      203–205, 204–206
   photon-counting devices, 190–191
   quantum-mechanical approaches, 182–185,
      185, 187–189
   Raman spectroscopy, 181–189
   results, 192–209
   spectrometer, 189–190
   transport, carrier, 195, *195–196*, 197–198,
      198–199, 200–202, *201*
Carriers, Raman spectroscopy, 181–185, *182*
Cathodoluminescence (CL) mapping technique,
   82–84
CBE, *see* Chemical beam epitaxy (CBE)
CDF, *see* Charge-density function (CDF)
Challis studies, 143–144
Charge-density function (CDF), 183
Chemical beam epitaxy (CBE), 180
Chemical vapor deposition (CVD) technique, 87

251

Chiral tubes, 52
Cho studies, 69–92
Christofilos, D., 137
Clausius-Mossotti relation
    artificial dielectrics, nonlinear optical
        properties, 53
    ultrafast non-equilibrium electron dynamics,
        104
Coherent electron-light coupling, *106,* 111–115,
    112, 114
Cole-Hopf transformation, 27
Collision-dominated regime, 3
Collision-free regime, 6, *7*
Compton scattering, 181
Cottancin, E., 138
Coulomb interactions
    carrier dynamics and non-equilbrium phonons,
        183
    electromagnetically induced transparency, 216,
        230, 232, 234, 239, 244–245
    self-assembled quantum dots, 72
    ultrafast non-equilibrium electron dynamics,
        117, 127–128
Cunningham, J.E., 247
Current-injected p-n junction LEDs, 88
CVD, *see* Chemical vapor deposition (CVD)
    technique

D

Daly studies, 18
Damilano studies, 91
Dang studies, 69–92
Daudin studies, 73
Debye properties, 118, 131
DECP, *see* Displacive excitation coherent phonons
    (DECP)
Deformation potential, 187
De Kadomtsev-Petviashvili equation, 31
De Kok, C.R., 45
Density matrix, 218–219
Density of states (DOS), 70
Determination, monochromatic acoustic phonons,
    154, 159–162, *160–162*
Dielectric function, *106–107,* 106–109
Diffraction, *31,* 39–43, *40, 42*
Dijkhuis, Muskens and, studies, 37–38, 41
Dijkhuis studies, 15–45
Discrete models, 28–29
Displacive excitation coherent phonons (DECP),
    165
DPK, *see* D'yakonov, Perel' and Karchorovskii
    (DPK) mechanisms
Dresselhaus term, 3
Drift velocities, large electrons, 206–209, *207–208*

Driscoll studies, 29
Drude expression, 107, 109, 111
Duquesne and Perrin studies, 18
D'yakonov, Perel' and Karchorovskii (DPK)
    mechanisms, 2–4
Dynamics, carrier, *192–194, 192–195*

E

EDF, *see* Energy-density fluctuations (EDF)
E1-HH1 transition, 148
Einstein summation convention, 20
Electromagnetically induced transparency (EIT),
    quantum wells
    basics, 215–217, 245–246
    biexcitonic coherence, 239–244, *240–246*
    bound biexciton states, *232–236,* 233–234
    density matrix, 218–219
    exciton spin coherence, 230, *232,*
        232–239
    experimental methods, 225–227
    intervalence band coherence, 227–230,
        228–231
    spin coherence, 230, *232,* 232–239
    steady-state solutions, *220–221,* 220–222
    Λ-system, 218–219
    theory, *217,* 217–225
    transient solutions, 222–225, *223–227*
    unbound biexciton states, *232,* 234–237,
        237–240, 239
Electron creation, 111–116
Electron dynamics, metal nanoparticles, *see*
    Ultrafast non-equilibrium electron
    dynamics, metal nanoparticles
Electron-electron energy exchanges, 125–130,
    126, 128–129
Electron gas, *see* Spin evolution
Electronic properties control, 89
Electron kinetics, 116–125
Electron-lattice energy exchanges, 130–135,
    132–134
Electron-phonon interactions, 183
Elliott-Yeffet (EY) mechanism, 2
El-Sayed, Link and, studies, 127, 133
Energy-density fluctuations (EDF), 183
Energy relaxation kinetics, *107,* 116–118, *119*
Ensemble Monte Carlo (EMC) simulations, 180,
    197, 200, 202, 208–210, *see also* Monte
    Carlo procedures; Quantum-mechanical
    Monte Carlo simulation
EOS, *see* Equation of state (EOS)
Equation of state (EOS), 19
Esaki studies, 147
Euler-Lagrange equations, 21
Exciton spin coherence, 230, *232,* 232–239

Experimental investigation, *5, 6–12, 7–9, 11–12*
EY mechanism, *see* Elliott-Yeffet (EY) mechanism

F

Fabry-Pérot interferometer, 36
Fabry-Pérot modes, 55–56
Faraday studies, 102
Fatti, N. Del, 137
Femtosecond lasers, 102, 152
Fermi, Pasta and Ulam studies, 28
Fermi-Dirac distribution, 115–117, 193
Fermi golden rule, 188
Fermi levels and properties
    spin dynamics, 6, 8–9, 11
    ultrafast non-equilibrium electron dynamics,
        107–108, 112–113, 120
Ferry, D.K., 210
FIR lasers, 148
Flytzanis, C., 137
Fourier transformations
    electromagnetically induced transparency, 222,
        226
    high-amplitude, ultrashort strain solitons, 37,
        42
Fraunhofer formula, 32, 42
Fröhlich interactions, 187
Full-width-at-half-maximum (FWHM) pulse, 193,
    195
Future outlook, monochromatic acoustic phonons,
    173–174
FWHM, *see* Full-width-at-half-maximum
    (FWHM) pulse

G

Gallium nitride and alloys, *155,* 170–173, *171–172*
Garnett theory, 53
Gas-source molecular beam epitaxy (GSMBE),
    180
Gaudry, M., 138
Gaussian properties
    artificial dielectrics, nonlinear optical
        properties, 60
    electromagnetically induced transparency, 222
    high-amplitude, ultrashort strain solitons, 25,
        31–32, 34
Glazov and Ivchenko studies, 4–5
Grebel studies, 49–67
Gridnev studies, 6
Grill and Weis studies, 16
Growth, self-assembled quantum dots
    basics, 73–75, *75*
    nucleation control, 87–89, *88*

GSMBE, *see* Gas-source molecular beam epitaxy
    (GSMBE)
Guillon, C., 137

H

Hall measurements, 7–8, 12
Hamanaka, Y., 138
Hamiltonian
    carrier dynamics and non-equilbrium phonons,
        183, 187
    electromagnetically induced transparency, 218
    spin dynamics, 2
    ultrafast non-equilibrium electron dynamics,
        118
Hao and Maris studies, 18–19, 33
Harley studies, 1–13
Hartland studies, 133
Hawker studies, 156, 174
Heaviside function, 127
Heisenburg, Kramers and, studies, 185
HH1-E1 transition, 151, 157–159
High-amplitude, ultrashort strain solitons
    basics, 15–16, 18–20, 43–45, *44*
    Brillouin-scattering experiments, 34–43
    diffraction, *31,* 39–43, *40, 42*
    discrete models, 28–29
    historical perspectives, *16,* 16–17
    multidimensional models, 28–29
    nano-ultrasonics, *17,* 17–18
    nonlinear elasticity, 20–22
    one-dimensional propagation, 23–24
    prospects, 43–45, *44*
    sapphire, 29–31, *30–31*
    shock waves, 18–20, *26,* 26–27, *30–31, 36–37,*
        36–39, *39*
    simulations, 29–34
    soliton formation, *31,* 39–43, *40, 42*
    soliton trains, 24–26, *26,* 29–32, *30–32, 36–37,*
        36–39, *39*
    stability, individual solitons, 32–34, *33*
    theories, 20–29
High-mobility, two-dimensional electron gas, *see*
    Spin evolution
Historical perspectives, *16,* 16–17
Hooke's law, 21

I

III-V semiconductors, mechanisms, 2–4
Impulsive stimulated Raman scattering (ISRS),
    145, 165
Index of refraction, 58–60, 221–222
Infrared unipolar opto-electronic devices, 90–91
InGaAsN analysis, *192–194,* 192–195

InGaN/GaN analysis
  carrier dynamics and non-equilbrium phonons,
    195, *195–196*, 197, *198*
  monochromatic acoustic phonons, 145, 170
InN analysis, 202, *203–205*, 204–209, *207–208*
InN/GaN analysis, 197–198, *199*, 200–202, *201*
Interband transitions, 89–90
Intersubband transitions, 90–91
Intervalence band coherence, 227–230, *228–231*
Inushima studies, 200
IR unipolar opto-electronic devices, 90–91
ISRS, *see* Impulsive stimulated Raman scattering
    (ISRS)
Ivchenko, Glazov and, studies, 4–5

J

Jacobsen and Stevens studies, 17
Johnson noise, 10
Jurrius, P., 45

K

Kalliakos studies, 79
Kane's notations and theory, 184, 209
Kapon studies, 88
Kasic studies, 200
Kawabata and Kubo studies, 112
KdV, *see* Korteweg-de Vries (KdV) equation
Kent studies, 143–174
Kerr effect, 52, 122
Kim and Yu studies, 193
Korteweg-de Vries-Burgers (KdV-Burgers)
    equation, 27, 29, 37–38
Korteweg-de Vries (KdV) equation, 16, 20, 24–25,
    28–29, 32, 34, 43
Kramers and Heisenburg studies, 185
Kramers-Kronig transformation, 120
Kronig-Penney model, 148
Kruskal, Zabusky and, studies, 29
Kubo, Kawabata and, studies, 112

L

LA, *see* Longitudinal acoustic (LA) phonons
Lagrangian density, 22
Lamb mode vibration, 103
Landau properties, 111, 113, 144
Langot, P., 137
Lansberg and Mandel'shtam studies, 185
Laplacian operator, 28
Large electron drift velocities, 206–207, *207–208*
Larmor vector, 2
Laser diodes (LDs), 70, 89
Lattice vibrations, *185*, 185–189

LEDs, *see* Light-emitting diodes (LEDs)
Lermé, M., 138
Light-emitting diodes (LEDs), 89, 91
Light sources, 189
Link and El-Sayed studies, 127, 133
Liouville equation, 218
Longitudinal acoustic (LA) phonons, *see also*
    Carrier dynamics and non-equilbrium
    phonons
  high-amplitude, ultrashort strain solitons, 23
  monochromatic acoustic phonons, 160–167,
    169–170
  non-equilibrium, 202, *203–205*, 204–206
Lorentz approach, 53–54
Lorentz-Lorentz approach, 104

M

Maillard, M., 138
Mair studies, 180
Mandel'shtam, Lansberg and, studies, 185
Maris, Hao and, studies, 18–19, 33
Material acoustic vibration, 103
Maxwell-Boltzmann distribution function, 182
Maxwell Garnett expression, 105
MBE, *see* Molecular beam epitaxy (MBE)
Mean free path measurements, 167–170,
    168–170
Meijer studies, 28
Metal nanoparticles, ultrafast non-equilibrium
    electron dynamics
  basics, 101–104, *103*, 136–137
  coherent electron-light coupling, *106*,
    111–115, *112, 114*
  dielectric function, *106–107*, 106–109
  electron creation, 111–116
  electron-electron energy exchanges, 125–130,
    126, 128–129
  electron kinetics, 116–125
  electron-lattice energy exchanges, 130–135,
    132–134
  energy relaxation kinetics, *107*, 116–118, *119*
  non-equilibrium electron gas energy, *107*,
    115–116, *116*
  non-equilibrium electron-lattice energy
    exchanges, 135–136, *136*
  nonlinear optical response, 110–111
  optical probing, *114*, 118–125, *121–122, 124*
  optical properties, 104–111
  response, optical, 104–106, *106*
  time-dependent optical response, 125, *126*
Metal-organic chemical vapor deposition
    (MOCVD)
  carrier dynamics and non-equilibrium phonons,
    180

monochromatic acoustic phonons, 146
self-assembled quantum dots, 74, 88, 91
Micro-PL techniques, 74, 85
Mid-Brillouin transitions, 52
Mie scattering, 50, 54, 67
Mie studies, 102, 104
Mini-Brillouin zones, 145
Mizoguchi studies, 146
Mizuno and Tamura studies, 150
MOCVD, *see* Metal-organic chemical vapor
    deposition (MOCVD)
Models, *see also* Simulations
    high-amplitude, ultrashort strain solitons,
        28–29
    Kronig-Penney model, 148
    Rosei models, 120
Molecular beam epitaxy (MBE)
    carrier dynamics and non-equilbrium phonons,
        191, 194
    monochromatic acoustic phonons, 146, 152,
        171
    self-assembled quantum dots, 73–74, *74*, 76,
        91
    spin dynamics, 6
Monochromatic acoustic phonons
    basics, 143–146, 173–174
    determination, monochromatic nature, *154,*
        159–162, 160–162
    experimental details, 151–155, *153–155*
    future outlook, 173–174
    gallium nitride and alloys, *155,* 170–173,
        171–172
    mean free path measurements, 167–170,
        168–170
    phonon optics demonstration, 167–170,
        168–170
    resolving phonon modes, *160, 162,* 163–166,
        164–165
    results, 156–157
    substrates, 166–167, *167*
    superlattice properties, 146–151, *147–149,*
        151–152
    terahertz phonons, 167–170, *168–170*
    time of flight techniques, *155,* 156–159,
        157–159
    transverse contribution, 166–167, *167*
    transverse mode enhancement, *160, 162,*
        163–166, *164–165*
Monte Carlo procedures, 11, *see also* Ensemble
    Monte Carlo (EMC) simulations;
    Quantum-mechanical Monte Carlo
    simulation
Motional-narrowing regime, 3
MQWs, *see* Multiple quantum wells (MQWs)
Multidimensional models, 28–29

Multiple quantum wells (MQWs)
    artificial dielectrics, nonlinear optical
        properties, 66
    monochromatic acoustic phonons, 170
    self-assembled quantum dots, 90
Muskens and Dijkhuis studies, 37–38, 41
Muskens studies, 15–45

N

Nakamura, A., 138
Nano-explosions, 45
Nano-ultrasonics, *17,* 17–18
"Nano within nano" concept, 50
Narayanamurti studies, 151
Natural Interface Asymmetry (NIA) term, 3
Newton's law, 22
NIA, *see* Natural Interface Asymmetry (NIA) term
Non-equilibrium electron dynamics, *see* Ultrafast
    non-equilibrium electron dynamics, metal
    nanoparticles
Non-equilibrium electron gas energy, *107,*
    115–116, *116*
Non-equilibrium electron-lattice energy
    exchanges, 135–136, *136*
Non-equilibrium longitudinal phonons, 202,
    203–205, 204–206, *see also* Carrier
    dynamics and non-equilbrium phonons
Nonlinear elasticity, 20–22
Nonlinear optical properties, artificial dielectrics
    basics, 49–53, *51,* 67
    discussion, 66–67
    experiments and results, 61–64, *62–66*
    theoretical considerations, 53–61
Nonlinear optical response, 110–111

O

Omi, S., 138
One-dimensional propagation, 23–24
Opals, 62
Optical Kerr effect, 122
Optical probing, *114,* 118–125, *121–122, 124*
Optical properties
    self-assembled quantum dots, 75–80, *77–79*
    ultrafast non-equilibrium electron dynamics,
        metal nanoparticles, 104–111
Optical properties, self-assembled quantum dots
    basics, 70–71, 91–92
    electronic properties control, 89
    growth, 73–75, *75,* 87–89, *88*
    infrared unipolar opto-electronic devices,
        90–91
    interband transitions, 89–90
    intersubband transitions, 90–91

MBE growth, 73–74, *74*
MOCVD growth, 74
optical properties, 75–80, *77–79*
physical properties, 71–73, *72*
prospects, 87–91
QD nucleation control, 87–89, *88*
rare earth doping, 79–80, *80*
single QD spectroscopy, *81–82,* 84–87, *86–87*
space-resolved optical studies, *77,* 82–84, *84–85*
time-resolved optical studies, *78–79,* 80–87, *81–83*
UV bipolar opto-electronic devices, 89–90
white light-emitting diodes, 91
Optical studies, carrier dynamics and non-equilibrium phonons
basics, 180, *181,* 210
carriers, Raman spectroscopy, 181–185, *182*
drift velocities, large electrons, 206–209, *207–208*
dynamics, carrier, *192–194,* 192–195
experimental methods, 181–192
InGaAsN analysis, *192–194,* 192–195
InGaN/GaN analysis, 195, *195–196,* 197, *198*
InN analysis, 202, *203–205,* 204–209, *207–208*
InN/GaN analysis, 197–198, *199,* 200–202, *201*
large electron drift velocities, 206–207, *207–208*
lattice vibrations, *185,* 185–189
light sources, 189
non-equilibrium longitudinal phonons, 202, *203–205,* 204–206
photon-counting devices, 190–191
quantum-mechanical approaches, 182–185, *185,* 187–189
Raman spectroscopy, 181–189
results, 192–209
spectrometer, 189–190
transport, carrier, 195, *195–196,* 197–198, *198–199,* 200–202, *201*

P

Pasta, Ulam, Fermi and, studies, 28
Pauli exclusion principle
spin dynamics, 4, 10
ultrafast non-equilibrium electron dynamics, 107, 117, 127
Pellarin, M., 138
Perkin-Elmer Eclipse, 152
Perrin, Duquesne and, studies, 18
Phillips studies, 215–246
Phonon echoes, 45
Phonon optics demonstration, 167–170, *168–170*

Photographic plates, 190–191
Photoluminescence (PL) spectra
monochromatic acoustic phonons, 148, 172–173
self-assembled quantum dots, 71–72, 74, 76, 78–80, 82, 84, 91–92
Photon-counting devices, 190–191
Physical properties, 71–73, *72*
Pico-ultrasonic techniques, 144–145
Pileni, M.P., 138
Planckian frequency spectrum, 144
P-n junction LEDs, current-injected, 88
Point defect (Rayleigh type) phonon scattering, 167
Potapov studies, 28
Prével, B., 138
Prospects
high-amplitude, ultrashort strain solitons, 43–45, *44*
self-assembled quantum dots, 87–91

Q

QCSE, *see* Quantum confined Stark effect (QCSE)
QD-based QUBITs, 174
QDIPs, *see* Quantum dot infrared photodetectors (QDIPS)
Quantum confined Stark effect (QCSE)
monochromatic acoustic phonons, 171
self-assembled quantum dots, 70–72, 74, 79, 81, 86
Quantum dot infrared photodetectors (QDIPS), 90
Quantum dots, 87–89, *88, see also* Optical properties, self-assembled quantum dots
Quantum-mechanical approaches, 182–185, *185,* 187–189
Quantum-mechanical Monte Carlo simulation, 193, *see also* Ensemble Monte Carlo (EMC) simulations; Monte Carlo procedures
Quantum wells, 78, 89–90
Quantum wells, electromagnetically induced transparency (EIT)
basics, 215–217, 245–246
biexcitonic coherence, 239–244, *240–246*
bound biexciton states, *232–236,* 233–234
density matrix, 218–219
exciton spin coherence, 230, *232,* 232–239
experimental methods, 225–227
intervalence band coherence, 227–230, *228–231*
spin coherence, 230, *232,* 232–239
steady-state solutions, *220–221,* 220–222
Λ-system, 218–219
theory, *217,* 217–225

transient solutions, 222–225, *223–227*
unbound two-exciton states, *232,* 234–237,
    237–240, 239
Quasi-Lorentzian shape, 123
QUBITs, 174

R

Rabi frequencies and oscillations, 216, 218, 222,
    225, 228–229, 245
Raman adiabatic passage, stimulated, 216
Raman scattering
    carrier dynamics and non-equilbrium phonons,
        182, 184–185, 187–190, 192, 198, 200,
        202, 204–206, 208–209
    monochromatic acoustic phonons, 145, 151
Raman spectrometer, 190
Raman spectroscopy
    artificial dielectrics, nonlinear optical
        properties, 64
    carrier dynamics and non-equilbrium phonons,
        181–189, 210
    self-assembled quantum dots, 74
Raman studies, 185, 190
Rare earth doping, 79–80, *80*
Rashba term, 3
Rayleigh scattering, 187
Rayleigh-type phonon scattering, 167
Refractive index, 58–60, 221–222
Resolving phonon modes, *160, 162,* 163–166,
    164–165
Response, optical, 104–106, *106*
Reynolds number, 27
Rosei models, 120
Rosei studies, 119
Rotating wave approximation (RWA), 218
Ruf studies, 145
Runge-Kutta scheme, 29
RWA, *see* Rotating wave approximation (RWA)
Rydberg length, 50

S

Sapphire, 29–31, *30–31*
Scanning electron microscope (SEM), 83
Schaff, William J., 210
Schottky gate, transparent, 7
Schrödinger equation, 24
SDF, *see* Spin-density fluctuations (SDF)
Self-assembled quantum dots, *see* Optical
    properties, self-assembled quantum dots
Self-induced transparency, 45
SEM, *see* Scanning electron microscope (SEM)
Semiconductor quantum wells,
    electromagnetically induced transparency

basics, 215–217, 245–246
biexcitonic coherence, 239–244, *240–246*
bound biexciton states, *232–236,* 233–234
density matrix, 218–219
exciton spin coherence, 230, *232,* 232–239
experimental methods, 225–227
intervalence band coherence, 227–230,
    228–231
spin coherence, 230, *232,* 232–239
steady-state solutions, *220–221,* 220–222
Λ-system, 218–219
theory, *217,* 217–225
transient solutions, 222–225, *223–227*
unbound two-exciton states, *232,* 234–237,
    237–240, 239
SERS, *see* Surface-enhanced Raman spectroscopy
    (SERS)
Shiren studies, 17
Shock waves
    development, *26,* 26–27
    soliton train transition, *30–31, 36–37,* 36–39,
        39
    strain solitons, 18–20
SIA, *see* Structural Inversion Asymmetry (SIA)
    term
SIFE, *see* Specular inverse Faraday effect (SIFE)
Simulations, *see also* Models
    diffraction, 31–32, *32*
    soliton trains, 29–31, *30–31*
    stability, individual solitons, 32–34, *33*
Single-particle scattering (SPS), 183–184, 189,
    192, 195, 198, 200, 208
Single QD spectroscopy, *81–82,* 84–87, *86–87*
Single-wall carbon nanotubes (SWCNTs), 50,
    52–54, 62, 64, 66–67
SK, *see* Stranski-Krastanow (SK) growth mode
SL, *see* Superlattice (SL) properties
Smekal studies, 185
Solids, *see* High-amplitude, ultrashort strain
    solitons
Soliton formation, *31,* 39–43, *40, 42*
Soliton phonon laser, 45
Soliton trains
    basics, 24–26, *26*
    sapphire, 29–31, *30–31*
    shock wave transition, *36–37,* 36–39, *39*
Space-resolved optical studies, *77,* 82–84, *84–85*
Spectrometer, 189–190
Specular inverse Faraday effect (SIFE), 9
Spin coherence, 230, *232,* 232–239
Spin-density fluctuations (SDF), 183–184
Spin evolution
    basics, 1–2, 12–13
    collision-free regime, 6, *7*

experimental investigation, *5*, 6–12, *7–9*,
11–12
III-V semiconductors, mechanisms, 2–4
strong scattering regime, 4–5, *5*
theories, 2–6
SPR, *see* Surface plasmon resonance (SPR)
SPS, *see* Single-particle scattering (SPS)
Stability, individual solitons, 32–34, *33*
Stanton studies, 143–174
Stark shifts, 228, 244
Steady-state solutions, *220–221*, 220–222
Steel, D.G., 247
Stevens, Jacobsen and, studies, 17
Stimulated Raman adiabatic passage, 216
Stokes-like shift, 78
Stokes line, 185
Stokes Raman process, 187–188, 202
Strain solitons, *see also* High-amplitude, ultrashort
    strain solitons
    basics, 20
    discrete models, 28–29
    multidimensional models, 28–29
    nonlinear elasticity, 20–22
    one-dimensional propagation, 23–24
    shock waves, 18–20, *26*, 26–27
    soliton trains, 24–26, *26*
Stranski-Krastanow (SK) growth mode, 73–76, 88,
    91–92
Strong scattering regime, 4–5, *5*
Structural Inversion Asymmetry (SIA) term, 3
Subpicosecond Raman spectroscopy, 210
Substrates, monochromatic acoustic phonons,
    166–167, *167*
Sun studies, 145
Superlattice (SL) properties, 145–151, *147–149*,
    151–152, 157–172
Surface charge waves, 60–61
Surface-enhanced Raman spectroscopy (SERS),
    50, 52–53, 67
Surface plasmon resonance (SPR), 105–106, *106*,
    109–110, 121, 123, 128
SWCNTs, *see* Single-wall carbon nanotubes
    (SWCNTs)
Λ-System, 218–219

T

Tamura, Mizuno and, studies, 150
Tamura studies, 148
TEM, *see* Transmission electron microscopy
    (TEM)
Terahertz (THz) phonons, 143, 167–170, *168–170*
Theories
    electromagnetically induced transparency, *217*,
    217–225

spin evolution, 2–6
strain solitons, 20–29
THz, *see* Terahertz (THz) phonons
Time-dependent optical response, 125, *126*
Time of flight techniques, *155*, 156–159, *157–159*
Time-resolved optical studies, 78–79, 80–87,
    81–83
Tomchuk, Belotskii and, studies, 133
Transient solutions, 222–225, *223–227*
Transmission electron microscopy (TEM),
    74–75
Transparent Schottky gate, 7
Transport, carrier, 195, *195–196*, 197–198,
    198–199, 200–202, *201*
Transverse acoustic (TA) phonons, 144, 163–167,
    169–170, 173
Transverse Bragg waveguides and condition, 54,
    59
Transverse contribution, 166–167, *167*
Transverse mode enhancement, *160, 162*,
    163–166, *164–165*
Tsen, K.T. studies, 179–210
Tsen studies, 179–210
Tucker studies, 17
Two-dimensional electron gas, *see* Spin evolution

U

Ulam and Pasta, Fermi, studies, 28
Ultrafast non-equilibrium electron dynamics,
    metal nanoparticles
    basics, 101–104, *103*, 136–137
    coherent electron-light coupling, *106*,
    111–115, *112, 114*
    dielectric function, *106–107*, 106–109
    electron creation, 111–116
    electron-electron energy exchanges, 125–130,
    126, 128–129
    electron kinetics, 116–125
    electron-lattice energy exchanges, 130–135,
    132–134
    energy relaxation kinetics, *107*, 116–118, *119*
    non-equilibrium electron gas energy, *107*,
    115–116, *116*
    non-equilibrium electron-lattice energy
    exchanges, 135–136, *136*
    nonlinear optical response, 110–111
    optical probing, *114*, 118–125, *121–122, 124*
    optical properties, 104–111
    response, optical, 104–106, *106*
    time-dependent optical response, 125, *126*
Ultrashort strain solitons, *see* High-amplitude,
    ultrashort strain solitons
Unbound biexciton states, *232*, 234–237, *237–240*,
    239

Unipolar opto-electronic devices, 90–91
UV bipolar opto-electronic devices, 89–90

V

Vallée studies, 101–137
Vellée studies, 101–137
Voigt notation, 22
Voisin, C., 137

W

Wallace studies, 20
Wang studies, 215–246
Weis, Grill and, studies, 16

Wetting layer (WL), 76, 84–85
White light-emitting diodes, 91
Winterling studies, 166
WL, *see* Wetting layer (WL)

Y

Yamamoto studies, 145
Yu, Kim and, studies, 193

Z

Zabusky and Kruskal studies, 29
Zhang studies, 133